SpringerWienNewYork

Computational Microelectronics

Edited by

Siegfried Selberherr
Technical University Vienna
Vienna, Austria

For further volumes:
www.springer.com/series/1263

Viktor Sverdlov

Strain-Induced Effects in Advanced MOSFETs

SpringerWienNewYork

Viktor Sverdlov
Technical University Vienna
Institute for Microelectronics
Gusshausstrasse 27-29
1040 Vienna
Austria
sverdlov@iue.tuwien.ac.at

Printed in Germany

SpringerWienNewYork is a part of Springer Science+Business Media
springer.at

Cover: WMXDesign GmbH, Heidelberg, Germany

Typesetting: SPi, Chennai, India

Printed on acid-free and chlorine-free bleached paper
SPIN: 80018573

With 101 Figures

Library of Congress Control Number: 2010938373

ISBN 978-3-7091-0381-4 e-ISBN 978-3-7091-0382-1
DOI 10.1007/978-3-7091-0382-1
SpringerWienNewYork

To Alexandra, Karin, Ludmila, & Nikolai

Preface

Strain is the main tool to boost current and enhance performance of advanced silicon-based metal-oxide-semiconductor field-effect transistors (MOSFETs). Modeling and understanding of strain effects on band structure and mobility has become the important task of modern simulation tools used to design ultra-scaled MOSFETs. This book focuses on modern modeling approaches and methods describing strain in silicon. Contrary to the valence band, strain-induced conduction band modifications have received substantially less attention. Peculiarities of subband structures in thin semiconductor films under stress are investigated in detail using numerical pseudopotential calculations as well as a $\mathbf{k}\cdot\mathbf{p}$ theory, which includes the two lowest conduction bands. Implementation of strain in transport modeling for modern microelectronics design tools is overviewed. Application ranges from device modeling to applied mathematics and software development.

The book is based on my research and partly on my course of lectures given for the Master's and PhD students in electrical engineering, microelectronics, physics, and applied mathematics at the Institute for Microelectronics, Technische Universität Wien. This book would not have been written without the support of the Institute for Microelectronics and its Director Univ.Prof. Dipl.-Ing. Dr.techn. E. Langer. I would like to thank Univ.Prof. Dipl.-Ing. Dr.techn. T. Grasser, Univ.Prof. Dipl.-Ing. Dr.techn. H. Kosina, and Univ.Prof. Dipl.-Ing. Dr.techn. Dr.h.c. S. Selberherr for their overwhelming encouragement, support, and help in writing the book.

I would like to acknowledge the contributions to the book made by my colleagues and co-authors: O. Baumgartner, J. Cervenka, S. Dhar, T. Grasser, A. Gehring, G. Karlowatz, M. Karner, H. Kosina, K. Likharev, M. Nedjalkov, M. Pourfath, F. Schanovsky, S. Selberherr, Z. Stanojevic, M. Vasicek, E. Ungersboeck, and T. Windbacher. I also would like to thank H. Ceric, R. Entner, O. Ertl, W. Goes, P. Hehenberger, R. Heinzl, S. Holzer, A. Makarov, G. Milovanovich, N. Neophytou, R. Orio, V. Palankovski, K. Rupp, P. Schwaha, I. Starkov, F. Stimpfl, O. Triebl, S. Tyaginov, S. Vitanov, P. Wagner, S. Wagner, J. Weinbub, and W. Wessner for many fruitful and stimulating discussions and C. Haslinger, E. Haslinger, M. Katterbauer, and R. Winkler for technical support in preparing the manuscript. Special thanks go to O. Baumgartner, M. Nedjalkov, and K. Sitzwohl, who kindly agreed to take the heavy duty of proof-reading and improving the manuscript.

The new scientific results described in the last sections of the book would have been impossible to obtain without financial support from the Austrian Science Fund FWF through the projects P-17285-N02 and P-19997-N14, from the European Research Council through the grant 247056 MOSILSPIN, from the European Commission, project SINANO IST-50684, and from the European Science Foundation EUROCORES Program FoNE funded by the Austrian Science Fund FWF (project I79-N16), CNR, EPSRC and the EC Sixth Framework Program.

Vienna
July 2010

Viktor Sverdlov

Contents

1 **Introduction** ... 1

2 **Scaling, Power Consumption, and Mobility Enhancement Techniques** 5
2.1 Power Scaling ... 5
2.2 Strain Engineering ... 6
2.3 Global Strain Techniques and Substrate Engineering ... 8
2.4 Local Stress Techniques ... 10
2.5 Advanced Stress Techniques ... 12
2.6 Hybrid Orientation Technology and Alternative Channel Materials. 14
References ... 16

3 **Strain and Stress** ... 23
3.1 Strain Definition ... 23
3.2 Stress ... 25
3.3 Relation Between Strain and Stress Tensor in Silicon and Germanium ... 27
3.4 Strain and Stress Tensors: Examples ... 28
3.4.1 Uniform All-Around Compression ... 28
3.4.2 Biaxial Strain Resulting From Epitaxial Growth ... 29
3.4.3 Uniaxial Stress ... 32
References ... 34

4 **Basic Properties of the Silicon Lattice** ... 35
4.1 Crystal Structure of Silicon and Germanium ... 35
4.2 Reciprocal Lattice and First Brillouin Zone ... 39
4.3 Particle in a Periodic Potential ... 41
References ... 44

5 **Band Structure of Relaxed Silicon** ... 45
5.1 Conduction and Valence Bands ... 45
5.2 First-Principle Band Structure Calculations ... 46
5.3 Pseudopotential Band Structure Calculations ... 49
5.4 Semi-Empirical Tight Binding Method ... 56

5.5 Comparison Between Different Numerical Methods 58
References 61

6 Perturbative Methods for Band Structure Calculations in Silicon 63
6.1 The **k·p** Method for a Non-Degenerate Band 63
6.2 Effective Mass Theory for Non-Degenerate Bands 64
6.2.1 Electron Effective Mass in Relaxed Silicon 66
6.2.2 Approximations for the Conduction Band Dispersion at Higher Energies 67
6.3 Valence Band 70
6.3.1 Spin–Orbit Coupling in the Valence Band 72
6.3.2 Dispersion of the Valence Band in Silicon 75
6.3.3 Luttinger Parameters 76
References 80

7 Strain Effects on the Silicon Crystal Structure 83
7.1 Strain-Induced Symmetry Reduction of Silicon Crystal Lattice 83
7.1.1 O_h Symmetry 83
7.1.2 D_{4h} Symmetry 84
7.1.3 D_{3d} Symmetry 85
7.1.4 D_{2h} Symmetry 85
7.1.5 C_{2h} Symmetry 86
7.2 Internal Strain Parameter 86
7.3 Strain and Symmetry of the Brillouin Zone 88
References 90

8 Strain Effects on the Silicon Band Structure 91
8.1 Linear Deformation Potential Theory 91
8.1.1 Conduction Band 91
8.1.2 Valence Band 93
8.1.3 Stress-Induced Band Splitting of the Valence Bands 94
8.2 Inclusion of Strain into Perturbative Band Structure Calculations ... 97
8.3 Empirical Pseudopotential Method with Strain 102
References 103

9 Strain Effects on the Conduction Band of Silicon 105
9.1 Limitation of the Effective Mass Approximation for the Conduction Band of Silicon 105
9.2 The Two-Band **k·p** Model 107
9.2.1 Valley Shift Due to Shear Strain 108
9.2.2 Stress-Dependent Transversal Effective Masses 111
9.2.3 Dependence on Strain of the Longitudinal Effective Mass 112
9.2.4 Stress and Non-Parabolicity 115

9.2.5 Comparison of the Two-Band **k·p** Model with Strain to the Empirical Pseudo-Potential Calculations 118
References 120

10 Electron Subbands in Silicon in the Effective Mass Approximation 123
10.1 Arbitrary Substrate Orientation 123
10.2 Substrate Orientation (001) 126
10.3 Substrate Orientation (110) 127
10.4 Substrate Orientation (111) 128
References 129

11 Electron Subbands in Thin Silicon Films 131
11.1 Numerical Methods for Subband Structure Calculations 131
11.2 "Linear Combination of Bulk Bands" Method 132
11.3 Unprimed Subbands in (001) Films: Analytical Consideration 137
11.3.1 Dispersion Relations from an Auxiliary Tight-Binding Model 141
11.4 Strain-Induced Valley Splitting 144
11.4.1 Small Strain Values 144
11.4.2 High Values of Shear Strain 144
11.4.3 Numerical Solutions 145
11.5 Effective Mass of the Unprimed Subbands 147
11.6 Valley Splitting in Magnetic Field and Point Contacts 152
11.6.1 Valley Splitting in Magnetic Fields 154
11.6.2 Valley Splitting in a Point Contact 154
11.7 Primed Subbands in Ultra-Thin (001) Silicon Films 155
11.7.1 Effective Mass of Primed Subbands 156
11.8 Substrate Orientations Different from (001) 157
11.8.1 Rotation of the Hamiltonian 158
11.8.2 Thin (110) Oriented Silicon Films 159
11.9 Appendix 162
11.9.1 Re-Expressing X_1 as a Function of X_2 162
11.9.2 Expressing the Dispersion Equations in Terms of $X_1 \pm X_2$ 164
References 165

12 Demands of Transport Modeling in Advanced MOSFETs 169
12.1 TCAD Tools: Technological Motivation and General Outlook 169
12.1.1 Brief History of TCAD Transport Modeling 171
12.1.2 Transport Modeling: Formulation of the Problem 172
12.2 Semi-Classical Transport 173
12.2.1 From Drift-Diffusion to Higher Moments Equations 174
12.2.2 Model Verification 178
12.3 Mobility in Strained Silicon 182
12.3.1 Mobility and Piezoresistance 183

12.3.2 Compact Mobility Modeling 184
12.3.3 Monte Carlo Methods for Transport Calculations 187
12.4 Mixed Quantum-Semi-Classical Description and Quantum Corrections in Current Transport Models 192
12.4.1 Subband Monte Carlo and Degeneracy Effects 195
12.4.2 Simulation Results for Mobilities in Single- and Double-Gate FETs 200
12.4.3 Electron Mobility Enhancement in FETs with Ultra-Thin Silicon Body 206
12.4.4 Stress-Induced Mobility and Drive Current Enhancement 207
12.5 Quantum Transport Models..208
12.5.1 Ballistic Transport and Tunneling 209
12.5.2 Quantum Transport Models with Scattering 216
12.5.3 Non-Equilibrium Green's Function Method 222
12.5.4 Conclusion and Trends 226
References...228

Author Index...239

Subject Index 251

List of Symbols

Notation

x	Scalar
$\mathbf{x}$	Vector
$\hat{x}$	Tensor
$\mathbf{A}$	Matrix
A_{ij}	Elements of the matrix $\mathbf{A}$
$\mathbf{x} \cdot \mathbf{y}$	Scalar product
$[hk\ell]$	Miller indices to specify a crystal direction
$\langle hk\ell \rangle$	Miller indices to specify equivalent crystal directions
$(hk\ell)$	Miller indices to specify a crystal plane
$\{hk\ell\}$	Miller indices to specify equivalent crystal planes

Physical Quantities

Symbol	Unit	Description
$\mathscr{C}[f]$	s^{-1}	Collision operator
e_{ij}, γ_{ij}	1	Engineering strain component (i, j)
ε_{ij}	1	Component (*ij*) of the strain tensor
σ_{ij}	GPa	Component (*ij*) of the stress tensor
$\sigma_x, \sigma_y, \sigma_z$		Pauli matrices
C_{ijkl}	GPa	Component (*ijkl*) of the elastic stiffness tensor
c_{ij}	GPa	Component (*ij*) of the contracted stiffness tensor
S_{ijkl}	GPa^{-1}	Component (*ijkl*) of the elastic compliance tensor
s_{ij}	GPa^{-1}	Component (*ij*) of the contracted compliance tensor
D_n	m^2s^{-1}	Electron diffusion coefficient
D	eV	Shear deformation potential
E	eV	Energy
$E_n(\mathbf{k})$	eV	Energy dispersion
E_F	eV	Fermi energy
E_g	eV	Band gap energy
$\mathbf{E}$	Vm^{-1}	Electric field
$\mathbf{F}$	N	Force
$f(\mathbf{r}, \mathbf{k}, t)$	1	Distribution function
$f_W(\mathbf{r}, \mathbf{k}, t)$	1	Wigner distribution function

ϕ	V	Electrostatic potential
g	$m^{-3}eV^{-1}$	Density of states
k	m^{-1}	Wave number
$\mathbf{k}$	m^{-1}	Wave number vector
κ	$AsV^{-1}m^{-1}$	Dielectric permittivity
μ_n	$m^2V^{-1}s^{-1}$	Electron mobility
μ_p	$m^2V^{-1}s^{-1}$	Hole mobility
m	kg	Mass
n	m^{-3}	Electron concentration
N_D	m^{-3}	Concentration of donors
N_A	m^{-3}	Concentration of acceptors
Ψ	$m^{-1/2}$	Wave function
$\mathbf{r}$	m	Space vector
a_0	m	Lattice constant
t	m	Film thickness
α	ev^{-1}	Non-parabolicity parameter
T	K	Temperature
$\mathbf{v}$	ms^{-1}	Velocity vector

Constants

h	Planck's constant	$6.6260755 \times 10^{-34}$ Js
$\hbar$	Reduced Planck's constant	$h/(2\pi)$
k_B	Boltzmann's constant	1.380662×10^{-23} JK^{-1}
e	Elementary charge	$1.6021892 \times 10^{-19}$ C
m_0	Electron rest mass	$9.1093897 \times 10^{-31}$ kg
κ_0	Dielectric constant of vacuum	$8.8541878 \times 10^{-12}$ $AsV^{-1}m^{-1}$
i	$\sqrt{-1}$	

Chapter 1
Introduction

Introduced in mass production at the beginning of the 1970s, the Metal-Oxide-Semiconductor Field Effect Transistor (MOSFET) is the key element of modern integrated circuits. Although the transistor feature size has shrunk dramatically over the past three decades, its overall design stayed nearly the same until recently. Even the 90 nm technology node MOSFETs introduced in 2004–2005 and still found in nowadays computers are based on the same principle and consist of the same basic elements as three decades ago. The inversion channel, which connects the source and drain electrodes, is formed at the silicon interface by applying a certain voltage to the gate electrode. The gate electrode made of heavily doped poly-silicon is electrically separated from the inversion channel by an oxide layer. A high quality silicon dioxide is resilient against an electrical break-through even at high electric fields and possesses little defects at the Si/SiO_2 interface. The good quality of this interface guarantees high mobility of the carriers in the inversion channel. Due to their perfect compatibility, the pair Si/SiO2 has quickly become the main stream microelectronic element of Si-based MOSFETs. Low defect density, high yield, and a relatively simple and inexpensive fabrication process have put MOSFETs into the heart of all modern high density integrated circuits.

Although the basic design of the transistor did not change, the operation speed and performance have increased dramatically. This became possible thanks to the scalability of the MOSFETs. Gordon Moore, one of the founders of Intel, has postulated the rule known as the Moore's law, according to which the MOSFET size reduces exponentially. A new generation of transistors with improved performance is introduced every two to three years which allows to double the number of transistors on integrated circuits every two years, decrease costs per transistor and increase performance for the same costs. With the 32 nm technology node presented at the International Electron Devices Meeting in December 2008 by Intel, the Moore's law did not loose its actuality and MOSFET scaling is successfully continuing.

Nevertheless, although the scaling is keeping pace with Moore's law, new technological solutions for MOSFET design had to be introduced beginning from the 90 nm technology node. These crucial changes are addressing growing heat generation caused by rapidly increasing leakage currents in scaled devices.

For a high-speed operation it is indispensable to have high drive current in the open, or on state of the transistor. In scaled devices the reduced gate length however

V. Sverdlov, *Strain-Induced Effects in Advanced MOSFETs*, Computational Microelectronics, DOI 10.1007/978-3-7091-0382-1_1, © Springer-Verlag/Wien 2011

results in a gradual channel control worsening which leads to high source-to-drain current in the passive, or off-state, for similar gate voltages. One option to keep the ratio between the on- and off-currents sufficiently high for operation is by decreasing the off-current, which can be done by increasing the gate voltage swing between the on- and off-state of the transistor. However, this again leads to high power production and thus is unacceptable. In order to continue scaling under the constrain of reduced heat generation the transport properties of the channel in the on-state must be improved. Since scattering with defects and surface roughness are already optimized, future progress requires a profound modification of the electron band structure leading to the increase of the carrier velocity.

Application of strain allows to increase the on-current significantly without changing the transistor design and meeting the projected performance increase. Although it has been long known that the electrical properties of silicon strongly depend on applied stress, strain as a mobility booster was first introduced in the MOSFET fabrication process at the 90 nm technology node. Since then strain engineering has become an integral part of the MOSFET fabrication process.

FinFET and ultra-thin body MOSFET multigate non-conventional structures possess superior channel control and reduced leakage as compared to bulk planar MOSFETs and are therefore suitable candidates for providing successful scaling to the end of the ITRS roadmap. Stress can be easily incorporated in non-conventional MOSFETs and is thus completely compatible with the upcoming non-classical MOSFET structures. Therefore, strain engineering is expected to keep its pace and remain one of the key elements of Complimentary MOS (CMOS) technology at the 22 nm technology node and beyond.

Strain is not the only new element introduced recently into CMOS production process. In order to guarantee a proper control over the channel in the 65 nm node transistor the silicon dioxide layer has become so thin that the gate leakage current and related heat generation could no longer be ignored. This prevents future silicon dioxide size reduction, and a new paradigm of scaling under the constraint of heat generation must appear. The solution is to replace the native silicon dioxide by another oxide with higher dielectric permittivity. This replacement allows to further reduce the equivalent electrostatic dielectric thickness thus improving electrostatic channel control while keeping the physical oxide dimension thick enough to prevent tunneling. At the same time, to reduce the depletion layer in the gate and to partly recover the channel mobility the polysilicon gate is replaced by a metal gate.

Although this step looks natural and simple, the introduction of a new dielectric and metal gates represents the most revolutionary change in the history of semiconductor industry and MOSFET production process since the replacement of germanium by silicon. Intel first introduced the new hafnium-based dielectrics with metal gates for its 45 nm technology node, and high-k materials with improved properties are now used in the 32 nm transistor. Together with new dielectric and metal gates, an improved technique to induce more strain into the channel for obtaining enhanced performance are employed for the 45 nm and the next 32 nm technology node.

Manufacturing complexity and production yield increase development cycle time and costs. Statistical parameter fluctuations are becoming more pronounced with shrinking transistor dimensions causing broader variations in device and circuit performance. It is customary to have a tool which allows predicting transistor properties thus making design easier. Technology modeling and simulations help reducing R&D costs and shorten the design cycle. Therefore, Technology Computer Aided Design (TCAD) tools are indispensable for development and optimization of upcoming generations of devices and integrated circuits.

In order to be predictive, TCAD tools must be based on accurate physical models. Although piezo-resistive coefficients describe modifications of electrical properties of bulk silicon on stress for small strain values, it is not enough to model transport in inversion channels, where the corresponding coefficients depend on carrier concentration, doping, channel length, etc. More detailed transport models are therefore required to describe current enhancement in inversion layers as well as in FinFETs and ultra-thin body FETs. The transport model must include carrier quantization in the confined direction. It has to include all appropriate carrier scattering mechanisms. For strain engineering the models must include stress induced modification of the band structure. These modifications have a profound impact on subband quantization energies, effective masses, non-parabolicity parameters, wave functions, and thus on scattering matrix elements. Although strain engineering is a mature technology to increase CMOS performance, the maximum performance enhancement has not been yet analyzed. A careful analysis to determine optimal conditions that lead to enhanced transport properties and the current boost is therefore needed.

Chapter 2
Scaling, Power Consumption, and Mobility Enhancement Techniques

2.1 Power Scaling

The power dissipation of a CMOS circuit consists of the dynamic (due to switching) and the static contribution in the off-state and can be written as [68]

$$P = \sum_i \alpha_i C_i V_{DD}^2 f + I_{OFF} V_{DD}, \tag{2.1}$$

where $0 < \alpha_i < 1$ is the "switching activity factor" of the ith circuit block, C_i is the total effective capacitance including that of all the interconnects and input capacitance of transistors, f is the clock frequency, and I_{OFF} is the total current in the off-state of all the transistors biased by the power supply voltage V_{DD}. In contrast to I_{OFF}, the on-current $I_{ON} = \sum_i (I_{ON})_i$ participates in (2.1) indirectly, via the speed requirement

$$f = p/\tau, \tag{2.2}$$

where

$$\tau = C_i V_{DD}/(I_{ON})_i, \tag{2.3}$$

and $p \ll 1$ is the fraction of the fraction of the clock period $1/f$ taken by the capacitance recharging constant τ.

The model of the power consumption described by (2.1)–(2.3) is approximate, however it captures the basic balance between the static and dynamic components of power generation.

At the beginning of the CMOS era the power consumption was reduced by scaling the transistor dimensions and thus the supply voltage V_{DD} down. However, with approaching 100 nm channel size, the V_{DD} scaling has slowed down. One of the reasons was a gradual increase of the currents in the *off*-state. This increase was mostly due to parasitic leakages, the most important is due to carrier tunneling through a thinner oxide. Indeed, in order to maintain a proper electrostatic control over the channel the thickness of the gate dielectric separating the gate from the channel must be reduced together with scaling of the gate length, which leads to a sharp increase of tunneling through a thin dielectric. With an increase of the *off*-current one option to preserve the high ratio I_{ON}/I_{OFF} is to increase the supply voltage

V. Sverdlov, *Strain-Induced Effects in Advanced MOSFETs*, Computational Microelectronics, DOI 10.1007/978-3-7091-0382-1_2, © Springer-Verlag/Wien 2011

V_{DD}. This option is, however, unacceptable, since, according to (2.1), it leads to an increase of the power consumption.

The industry has faced the problem of increase of heat generation already at the 90 nm technology node. The engineering solution to continue scaling, increase performance, and keep the heat generation under control was the introduction of strain into the channel [20]. Strain modifies the transport properties of the transistor in the open state, while keeping them practically unchanged in the off-state. If the I_{ON} current is increased by applying stress, it leads, according to (2.2), to higher speed and performance. Therefore, if a higher I_{ON} is achieved for the same I_{OFF} and V_{DD}, the performance gain is accomplished at nearly no increase of the power generation. Alternatively, the performance similar to an unstrained device is achieved at lower V_{DD} and thus reduced power consumption.

Although a new technology of high-k dielectric/metal gate, which allows reducing I_{OFF} (and thus power consumption) while preserving the good control over the channel, was introduced at 45 nm technology node [43], stress technique remains one of the main boosters of performance enhancement with scaling. In the 32 nm technology node introduced by Intel at the end of 2008, the fourth generation of advanced channel stressors is employed [44] allowing to get tensions of 1.2–1.5 GPa in the channel. In 2009 nearly 2 GPa stress in the channel was achieved [50].

The on-current boost by stress is due to the strain-induced mobility enhancement in the channel. Depending on the stress conditions, up to the fourfold mobility enhancement for holes and nearly twofold for electrons was reported [71] and up to 50% increase in transistor drive current [22, 43, 62, 76, 80] was documented. The mobility enhancement is predicted up to the stress level of at least 3 GPa [71], which is higher than the level currently delivered into the channel. It makes stress a viable, competitive, and important technology which will certainly be used to boost the performance of future technology generations beyond the 32 nm technology node currently in production.

In this chapter we will briefly review the history of stress in silicon and the main techniques to introduce stress currently utilized in laboratories and industry. Stress is not the only option to enhance mobility in the channel. As shown in Fig. 2.1, substrates different from the commonly used (001) wafers may also be used to obtain higher mobility. This hybrid orientation technology becomes important with the introduction of Fin-FET devices with the [110] channel direction, where the two fin interfaces are $(1\bar{1}0)$ oriented. Devices with channel directions different from [110] can also be considered. Finally, alternative channel materials with mobilities higher than silicon mobility, e.g., germanium or III–V semiconductors can be used.

2.2 Strain Engineering

Strain engineering technologies are based on enhancing the transport properties by mechanically stressing the silicon channel of a MOSFET. The advantage of these techniques is that they allow to get higher performance without changing the

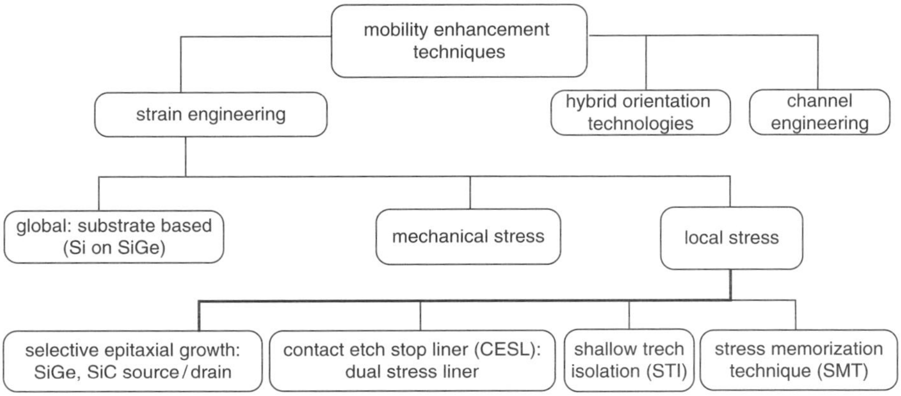

Fig. 2.1 Classification of stress techniques. Mechanical stress is used in laboratories

MOSFET size and architecture dramatically. Several techniques to deliver strain which require only little change in the process flow have been developed. This allows to integrate strained silicon into the manufacturing process at low additional production costs.

The influence of strain on transport in semiconductors has been a research topic for over half a century. Already in the beginning of the 1950s it was discovered that stress may influence the intrinsic silicon mobility [23,63]. To explain the effect, Herring and Vogt [26] have generalized the deformation potential theory initially proposed by Bardeen and Shockley [7] to describe the coupling between electrons and acoustic waves in solids and to express the relaxation times via the effective mass and deformation potentials. They have shown that the electron mobility change is due to repopulation between the valleys and reduced inter-valley scattering. Both effects are caused by stress-induced energy shifts which, depending on the stress condition, lead to the lifting of degeneracy of the six equivalent valleys. This interpretation of the mobility enhancement is often used to explain the mobility enhancement due to uniaxial stress as well, although, as we will show below, it is valid only for uniaxial stress in [001] direction, or, equivalently, for a biaxially, or inplane stressed sample. The effective mass change appears in [110] stressed samples, as was first demonstrated by Hensel [25] 1965 but since then well forgotten. Only recently [73] the Hensel-Hasegawa-Nakayma model of the conduction band was used to model the mobility enhancement in uniaxially stressed MOSFETs with technologically relevant [110] channel direction [72].

The stress-induced valence band shifts and warping are essential to understand the hole mobility modification. The $\mathbf{k}\cdot\mathbf{p}$-based model [40] with a Hamiltonian including strain [8] has been a reliable and inexpensive method to address the stress-induced valence band modification since 1963 [24], which is successfully used nowadays to describe the subband structure in inversion layers [66].

The transport properties of strained silicon can be reasonably well predicted by piezoresistance coefficients for small stress values. However, the value of piezoresistances depends on the parameters like doping level or temperature and should be measured for each sample. Another problem is that the bulk values of the piezoresistances may not be used to predict the behavior of MOSFETs with confined carriers in the surface layer where the piezoresistance depends on the effective field as well.

Until the beginning of 1990 stressed silicon was studied by the physics community, but remained relatively unexplored for engineering applications [18]. In the pioneering work by Welser in 1992 it was demonstrated that an n-MOSFET with a channel built out of biaxially stressed silicon possesses nearly a 70% higher mobility [77]. In 1993 an increase of hole mobility in a p-MOSFET was reported [45,46]. The biaxial stress in silicon was achieved by growing the silicon layer on SiGe substrate. The drive current enhancement in pMOSFETs as a function of germanium concentration was investigated in [56], while short-channel n-MOSFETs were studied in [55]. History and the current status of the technology based on biaxially strained silicon, SiGe, and germanium channel MOSFETs is discussed in detail in a recent review [38].

By now the industry has adopted several technologies to introduce strain in the Si channel of MOSFETs. The key challenge is to make the technology compatible with the CMOS manufacturing process flow. For uniaxial stress the integration was successfully achieved [9, 20, 35, 64]. This is why uniaxial stress first introduced in [19,33,61] is currently employed by the silicon industry. Uniaxial stress results in a smaller threshold voltage shift [69] and higher mobility enhancement [66]. Modern stress techniques are compatible with the multi-gate architectures [30–32, 67] and were recently integrated with high-k dielectrics and metal gates [15,79].

Although many strain technologies were developed and introduced up to now, they can be conveniently divided into two distinct categories: global techniques where stress is introduced into the whole wafer, and local techniques, where stress is delivered to each transistor separately and independently (Fig. 2.1). Local stress is usually introduced during the process of MOSFET fabrication and is sometimes called process-induced stress.

Stress must be beneficial for the transport boost in both n- and p-type channels. It turns out that to get the performance improvement n-MOSFETs should be stretched, while p-MOSFETs must be compressed. Obviously, the global stress technique cannot provide the current improvement for both n- and p-MOSFETs. Therefore, industry uses local stress techniques, although biaxially stressed Si can also be used to increase mobility of n-type transistors [16]. We begin with biaxially stressed Si on SiGe technology.

2.3 Global Strain Techniques and Substrate Engineering

High quality silicon wafers are the primary elements used in chip manufacturing. Due to growing needs for channels with improved transport properties and rapidly increasing expertise in synthesizing new materials with enhanced electrical, mechanical, or chemical characteristics several ways to engineer silicon wafers were recently explored. This results in a substrate with unique properties which cannot be achieved by using silicon alone. At the beginning of 1990 the system of a silicon layer grown on a thick SiGe virtual substrate attracted attention to enhanced mobility in strained silicon [77, 78]. The lattice constant of relaxed SiGe is slightly larger than the one in relaxed silicon. Thus, a thin silicon film grown epitaxially on top of a SiGe substrate becomes tensely strained due to the lattice mismatch between silicon and SiGe. Because of the lattice symmetry of silicon a (001) silicon film is equally elongated along [100] and [010] axes which results in biaxial strain. This type of strain is introduced globally through the whole wafer. Biaxial strain results in the conduction band modification which finally leads to improved electron transport. The drive current is increased by up to 25% in sub-100 nm strained silicon MOSFETs [54]. Global stress techniques are not restricted to standard bulk CMOS technology. Thanks to layer transfer and wafer bonding global stress is successfully integrated into SOI wafers. Recently, the performance enhancement in a 60 nm gate length n-MOSFET with an ultra-thin strained silicon layer grown on a SiGe substrate on insulator was demonstrated [21, 58].

Current enhancement alone is not sufficient for a technology to go into mass production. The new technology must be economically competitive [57] and deliver benefits exceeding production costs. Regardless of the proven electron current enhancement in biaxially strained silicon, the presence of the SiGe layer in a substrate introduces several challenges for process integration. One problem is that the SiGe layer induces a high density of defects in strained silicon [18]. The diffusion of Ge atoms into the strained silicon film reduces the thermal budget window. Due to the lower thermal conductivity of SiGe device self-heating may become a problem, especially in the SiGe on insulator structures. Finally, the diffusion rate of dopant atoms (boron, arsenic) is significantly different from that of silicon [74].

Several alternative approaches to introduce biaxial strain in silicon without SiGe layer were proposed. In the "strained silicon directly on insulator" technology the SiGe layer is eliminated before transistor fabrication. This technology delivers a 25% drive current enhancement while avoiding the difficulties of SiGe process integration.

Another back-end technique introduces strain into an already processed wafer. In this approach the wafer is mechanically stressed, after it was thinned down and put onto a polymer film. After that the wafer can safely be bonded to a final substrate. The advantage of the method is that it allows to introduce uniaxial as well as biaxial strain according to the mechanical deformation, and a 100% performance enhancement has been demonstrated [1, 53], however, yield and reliability issues have so far prevented the technique from being used in IC manufacturing.

As it was already pointed out, global stress techniques are able to provide only one type of strain through the whole wafer. However, n- and p-type channels are affected by strain alternatively: an in-plane biaxial tensile strain is beneficial for n-MOS but detrimental for a p-MOS, and vice versa. We briefly review local strain techniques delivering a particular stress to each MOSFET.

2.4 Local Stress Techniques

Already in the 1990s it was found that certain process steps and IC elements appearing during wafer processing result in channel stressing and thus performance increase. Shallow trench isolation [41, 59], silicidation at the source and drain region [65], and formation of nitride contact-etch-stop layer [33, 61] were among earlier local stress techniques investigated (Fig. 2.2). Although the process induced stresses were moderate and could not provide sufficient drive current boost at the earlier stage, local techniques have certain advantages over global ones. Process-induced stress can be independently delivered to p- and n-MOSFETs guaranteeing the performance enhancement in both types of transistors. Additionally, stress can be introduced along three coordinate axes. This allows to optimize performance enhancement and costs, reduce the threshold voltage shifts [39], and improve integration into the process flow [36]. Importantly, the interest in stress technology was supported and motivated by industry needs to optimize the ratio of the performance to heat generation for the upcoming 90 nm technology node. Several process-induced local stress techniques, such as stressed nitride contact etch stop liner, stress memorization technique, selective epitaxial growth for embedded SiGe in the source and drain contacts, and stress from shallow trench isolation were introduced in mass production of integrated circuits.

In modern sub-100 nm technologies the transistor dimensions are so small that the mechanical stress induced by shallow trench isolation becomes important [9,75]. Stress can be induced both parallel and orthogonal to the channel lateral directions.

Another way to introduce compressive uniaxial stress into a p-channel is by filling the source and drain regions with SiGe [6,17,27,49,70,85]. For this purpose, the source and drain regions are etched out and a recess area is created. This recess is later filled by SiGe grown epitaxially in the source and drain regions [6,49]. Alternatively, SiGe can also be grown on top of source and drain [12]. Depending on the

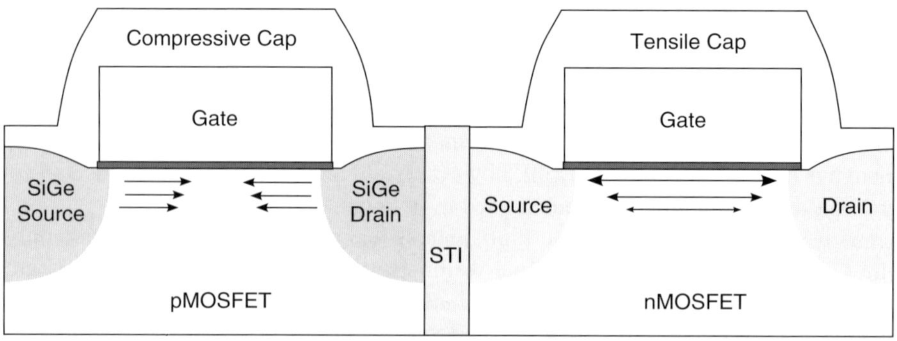

Fig. 2.2 Process-induced stressors employed by the semiconductor industry. Shallow-trench isolation, highly compressive and tensile capping layers, and compressive stress due to SiGe embedded in the source and drain regions are used in the CMOS process

thickness of the epitaxial $Si_{1-x}Ge_x$ and the Ge content x large uniaxial stress can be created using this method.

A part of the mechanical stress from a permanently stressed layer grown on top of a transistor can be transferred into the channel. The value of stress transferred depends on the thickness and the material properties of the liner [33]. To boost performance of an n-MOSFET a tensile cup layer is needed, while for a p-MOSFET the compressive layer is required. Thus, two different types of stress liners should be used to get performance enhancement in n-channel and p-channel MOSFETs simultaneously. Industry adopted a Dual Stress Liner (DSL) process, where a highly compressive nitride is deposited on top of the p-channel MOSFET, while a highly tensile nitride is deposited on top of the n-channel MOSFET. Silicon nitride (Si_3N_4) capping layers can produce both tensile and compressive strain depending on deposition conditions. In the fabrication process, a tensile silicon nitride layer is created by thermal chemical vapor deposition over the whole wafer. Parts of the layer are removed above p-MOSFETs by selective etching. After that a compressive Si_3N_4 layer is created by plasma-assisted chemical vapor deposition, followed by selective etching of the compressive layer above n-MOSFETs. Dual stress liners technology alone can improve the drive current by 11% in n-MOSFETs and by 20% in p-MOSFETs [60, 81].

Si_3N_4 layers with more than 2.0 GPa tensile and 2.5 GPa compressive stress which introduce approximately 1.0 GPa stress in the MOSFET channel are routinely used in 65 nm process [4]. This technique is successfully combined with selective epitaxial growth for embedded SiGe in the source and drain contacts [43]. Thus, strain engineering techniques may not only be combined for the same transistor, but can be superimposed to yield even larger performance boost [27].

Residual channel stress may by preserved after removal of the nitride layer. This fact is exploited in the stress memorization technique [11, 27, 36, 48]. In a process using this technique, the conventional dopant activation spike anneal is performed after the deposition of a tensile stressor capping layer. This layer is subsequently removed before an eventual salicidation process. Even though the stressor nitride layer is removed from the final structure, the stress has been transferred from the nitride film to the channel during annealing and memorized by the re-crystallization of source, drain and the poly gate amorphized layers. Stress from the capping layer can be memorized in the channel. Stress is preserved in the channel even after the stressed layer is removed from the final structure providing a 15% improvement of the drive current in n-channel MOSFETs [10].

Process-induced local stress techniques depend strongly on device geometry and must be adjusted and optimized to maximize beneficial effects from stressors [17]. However, regardless of the challenges of local stressors integration into the manufacturing flow, local stress techniques have proven useful for industrial applications and promising for future technology nodes.

2.5 Advanced Stress Techniques

Stress was introduced into the fabrication process flow at the 90 nm technology node. Since then stress is a compulsory technique to get the MOSFET performance enhanced included in all technology nodes. Stress techniques were constantly improved and perfected through the 65 nm and 45 nm technology nodes in order to transfer more strain into the channel. The germanium concentration in the source/drain regions of p-MOSFETs was constantly increased from 17% at the 90 nm technology node to 23% at the 65 nm which resulted in a 60% increase of the channel strain. At the same time an enhanced process flow adopted for the Si_3Ni_4 capping layers increase the channel strain in n-MOSFETs by 80% [6]. Strain techniques are compatible with high-k dielectrics/metal gate technology and were successfully integrated in the process flow at the 45 nm technology node, resulting in the third generation strained silicon [43].

At the International Electron Devices Meeting in 2008 Intel has reported its second generation of high-k dielectrics/metal gate 32 nm transistors. The fourth generation of stress technology allowed to get approximately 14% in performance improvement [44] as compared to the 45 nm transistors. The technique allowed to build the largest SRAM with more than 1.9 billions transistors. Multiple stressors are combined to produce even higher strain in the channel. The fourth generation stress technology includes improved stress liners for both n- and p-MOSFETs. Compared to the 45 nm technology node where the dual stress liners with 1.5 GPa tensile and 2.8 GPa of compressive stress were used [43], capping layers with more than 2 GPa tensile and 3.5 GPa compressive stress are introduced for the 32 nm node. In combination with SiGe source/drain regions with high (approximately 30%) germanium concentration uniaxial stress of approximately 1.5 GPa is produced in the channel. The replacement metal gate, or gate last, process when the poly-silicon gate of a transistor is removed and later substituted by a metal gate allows to produce even more uniaxial compressive stress [5, 44], as demonstrated in Fig. 2.3. This allows to obtain the best drive currents of 1.55 mA/μm for n-MOSFETs and 1.21 mA/μm for p-MOSFETs reported for 32 nm technology node at the end of 2008 [44].

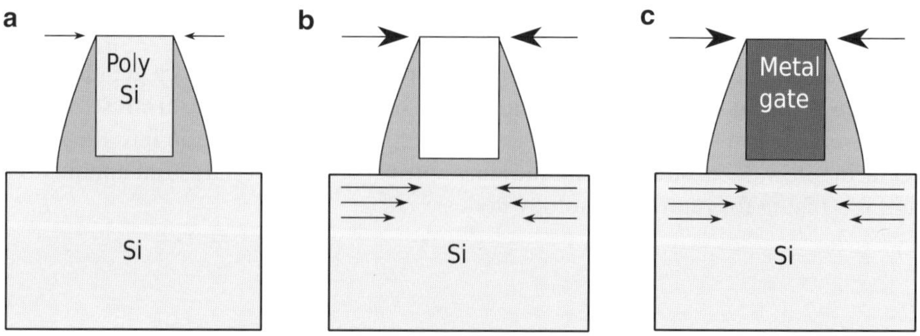

Fig. 2.3 Illustration of additional tensile strain introduced in the gate-last process [5, 44]

Tensile stress can also be generated in the n-channel MOSFET by using $Si_{1-x}C_x$ stressors with a small mole fraction x [2]. It was demonstrated that for n-channel MOSFETs the implementation of the SiC source/drain regions provides significant drive current enhancement of up to 50% at a gate length of 50 nm [14]. Thus, $Si_{0.99}C_{0.01}$ induces as much of tensile stress as $Si_{0.75}Ge_{0.25}$ - compressive stress [3]. This method was not incorporated yet into mass production and possesses a large potential to induce tensile stress for technology nodes beyond 32 nm.

For technology nodes beyond 32 nm the gate becomes less than 30 nm, and improved channel control by the gate is required. Although alternative channel materials with improved transport properties may be the key to extend the planar MOSFETs down to 22 nm and even to 16 nm technology nodes, multi-gate FinFETs and ultra-thin body SOI based technologies provide a better channel control and are thus considered as viable candidates for the next generation technology nodes. An integration of stress into an SOI CMOS fabrication process was demonstrated by [27]. Like in bulk devices, an embedded SiGe process and a compressively stressed liner film are used to introduce compressive strain in the p-MOSFET, whereas a stress memorization and a tensile stressed liner are inducing tensile strain in the n-channel MOSFET. An optimization of the process-induced stresses yields an improvement in saturation drive current of 53% for p-channel and 32% for n-channel MOSFETs, respectively.

Although the process of introducing global stress by growing silicon on relaxed SiGe substrate did not receive appreciation by industry, it has prompted the introduction of a technique called by IBM a reverse-embedded SiGe approach [16]. This technique employs a burried SiGe layer to induce tensile stress in n-channel SOI MOSFETs. In the reverse-embedded SiGe approach the n-FET is fabricated at SiGe source/drain areas of p-MOSFETs following silicon re-growth. SiGe, which is typically used to generate compressive stress in p-channels, is very efficient to impart tensile stress of approximately 400 MPa in the n-channel, leading to 15% increase of the drive current (Fig. 2.4).

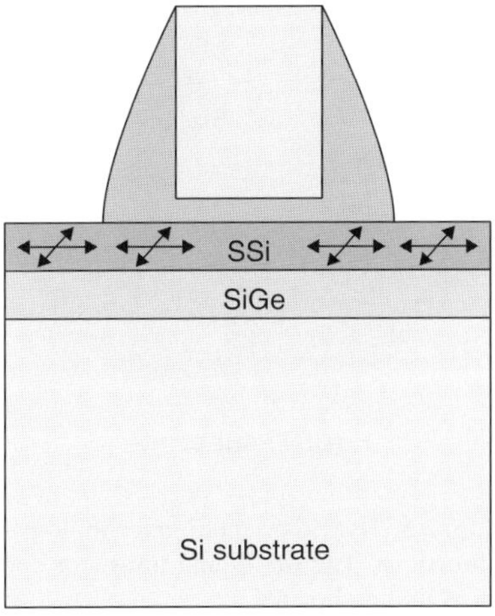

Fig. 2.4 Illustration of the reverse-embedded SiGe approach [16], when an n-FET with strained silicon (SSi) is fabricated at the SiGe source/drain areas of p-MOSFETs following silicon re-growth

With the device dimensions further reduced, the volume of SiGe source/drain regions shrinks as well, thus degrading its stressing capabilities. Although strain remains an efficient performance booster for the 22 nm and even for 16 nm technology nodes, it remains to be seen if stress is an efficient performance booster for short-channel devices beyond 16 nm. Thus, other techniques including hybrid orientation technology and alternative channel materials become important.

2.6 Hybrid Orientation Technology and Alternative Channel Materials

Carrier mobility in silicon surface layers depends strongly on the substrate crystal orientation and the direction of the current flow. For instance, hole mobility increases if one selects the [100] instead the [110] transport direction for a standard (001) wafer [37, 41, 84], a fact which was already employed in the 90 nm technology [34].

Electron mobility in the silicon channel is known to be the highest for the traditional substrate (001)/[110] channel direction configuration, while hole mobility is maximal in $(\bar{1}10)$ substrate for [110] channel direction, which is 2.5 times higher that the mobility for the traditional (001) substrate and [110] channel direction. Thus, by adjusting the crystal orientation, channel direction (Fig. 2.5), and strain one can optimize the transport properties for n- and p-MOSFETs [81, 82].

In case of a [110] oriented FinFET the two faces are (001) oriented, while the other two are of $(\bar{1}10)$ crystal orientation. Therefore, the relative contribution of (001) and $(\bar{1}10)$ surfaces into transport depends on the aspect ratio of the fin width

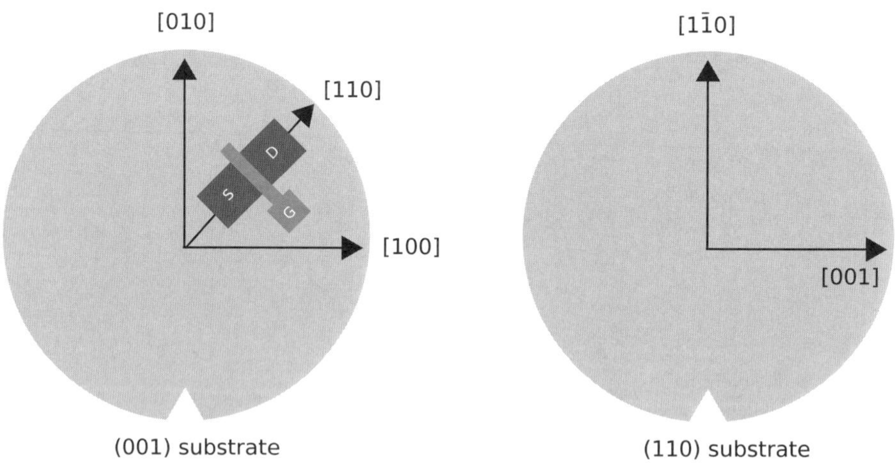

Fig. 2.5 Hybrid orientation technology uses different substrate crystal orientations and/or channel directions. The traditional transistor configuration on the (001) substrate with the [110] channel direction is shown

to the height. This additional option to choose the transport interface and thus to control the FinFET by performance makes them attractive for the technology nodes beyond 22 nm.

Benefits of applying stress to (110) p-MOSFETs are under intensive investigation. Recent computational studies on strain-induced low field mobility enhancement demonstrated that, although $(\bar{1}10)/[110]$ hole mobility is superior over the (001)/[110] mobility in relaxed silicon, they become approximately equal in a highly stressed p-channel [66]. These results are supported by a comparative study between Si (110) and (100) substrates on mobility and velocity enhancements for short-channel highly-strained p-MOSFETs [42], where it is demonstrated that mobility and velocity enhancements under high channel stress for (100) substrate are larger than those for (110). Thus, saturation current on (100) is similar or slightly higher than that on (110) for p-MOSFETs with higher channel stress, and benefits of the hybrid orientation technique combined with stress are becoming less obvious.

In order to extend the planar technology to the 16 nm node and beyond one has to use alternative materials for the channel. The use of III–V semiconductors with high intrinsic mobility integrated on silicon substrate is one possible solution explored recently. Intel has presented InGaAs quantum well device structure on silicon [29]. No mobility degradation in quantum well grown on silicon as compared to the well grown on III–V substrates was observed. Compared to n-MOSFETs, the proposed quantum well FET on silicon operates at low supply voltage of 0.5 V, exhibits more than ten times DC power reduction for the same speed performance, and experiences twofold performance gain for the same power. Recently a high mobility III–V-on-insulator FET on silicon substrate with metal source/drain contacts using direct wafer bonding was demonstrated [83]. Lateral tensile strain introduced in $In_{0.53}Ga_{0.47}As$ n-MOSFETs with in-situ doped lattice-mismatched source and drain stressors and interface engineering is shown to boost the performance of III–V n-FETs [83]. Compatibility of stress techniques with the III–V FETs makes them very attractive for future ultra-scaled high-speed applications.

For high-speed p-MOSFETs germanium channels are considered. The ultimate advantage of germanium as compared to silicon are two times higher electron bulk mobility and, most importantly, the fourfold increased hole mobility. However, despite the intrinsic speed advantages, germanium does not grow a stable oxide. Thus a manufacturing method similar to the traditional one employed for silicon MOSFETs is not applicable. Although high-k dielectrics have been implemented with germanium channel transistors [13], satisfactory quality of the gate stacks has not yet been achieved. This results in a mobility enhancement smaller than anticipated. Finally, germanium application to n-channel devices is prohibited by the low drive current of Ge-based n-MOSFET. The poor electron transport property in Ge-based n-channels is primarily due to the intrinsically low density of state and high conductivity effective masses [47].

At the International Electron Devices Meeting 2008 Intel has presented the first high-speed low power III–V p-channel quantum well FET [51, 52]. The compressively strained InSb quantum well device possesses hole mobility as high as 1230 cm^2/Vs. The highest cut-off frequency of 140 GHz for III–V p-channel FETs was demonstrated at a supply voltage of 0.5 eV. As compared to a p-channel silicon MOSFET, the quantum well FET produces ten times less heat at the same speed and develops two-times higher speed for the same power. Therefore, both n- and p-type III–V quantum well FETs are perfect candidates for future high speed and low power logic applications. However, before they become applicable to the silicon industry, many technical challenges must be overcome for III–V compound semiconductors to be integrated onto large silicon wafers [28].

References

1. Andrieu, F., Ernst, T., Faynot, O., Rozeau, O., Bogumilowicz, Y., Hartmann, J.M., Brevard, L., Toffoli, A., Lafond, D., Ghyselen, B.: Performance and physics of sub-50 nm strained Si on $Si_{1-x}Ge_x$ on insulator (SGOI) nMOSFETs. Solid State Electron. **50**, 566–572 (2006)
2. Ang, K.W., Chui, K.J., Bliznetsov, V., Du, A., Balasubramanian, N., Li, M.F., Samudra, G., Yeo, Y.C.: Enhanced performance in 50 nm n-MOSFETs with silicon-carbon source/drain regions. In: Int. Electron Devices Meeting, pp. 1069–1071 (2004)
3. Ang, K.W., Chui, K.J., Bliznetsov, V., Tung, C.H., Du, A., Balasubramanian, N., Samudra, G., Li, M.F., Yeo, Y.C.: Lattice strain analysis of transistor structures with silicon–germanium and silicon–carbon source/drain stressors. Appl. Phys. Lett. **86**(9), 093102 (2005)
4. Arghavani, R., Xia, L., Saad, H., Balseanu, M., Karunasiri, G., Mascarenhas, A., Thompson, S.E.: A reliable and manufacturable method to induce a stress of >1 Gpa on a p-channel MOSFET in high volume manufacturing. IEEE Electron Device Lett. **27**(2), 114–116 (2006)
5. Auth, C., Cappellani, A., Chun, J.S., Dalis, A., Davis, A., Ghani, T., Glass, G., Glassman, T., Harper, M., Hattendorf, M., Hentges, P., Jaloviar, S., Joshi, S., Klaus, J., Kuhn, K., Lavric, D., Lu, M., Mariappan, H., Mistry, K., Norris, B., Rahhal-orabi, N., Ranade, P., Sandford, J., Shifren, L., Souw, V., Tone, K., Tambwe, F., Thompson, A., Towner, D., Troeger, T., Vandervoorn, P., Wallace, C., Wiedemer, J., Wiegand, C.: 45 nm high-k + metal gate strain-enhanced transistors. In: VLSI Technology, 2008 Symposium on, pp. 128–129 (2008)
6. Bai, P., Auth, C., Balakrishnan, S., Bost, M., Brain, R., Chikarmane, V., Heussner, R., Hussein, M., Hwang, J., Ingerly, D., James, R., Jeong, J., Kenyon, C., Lee, E., Lee, S.H., Lindert, N., Liu, M., Ma, Z., Marieb, T., Murthy, A., Nagisetty, R., Natarajan, S., Neirynck, J., Ott, A., Parker, C., Sebastian, J., Shaheed, R., Sivakumar, S., Steigerwald, J., Tyagi, S., Weber, C., Woolery, B., Yeoh, A., Zhang, K., Bohr, M.: A 65 nm logic technology featuring 35 nm gate lengths, enhanced channel strain, 8 Cu interconnect layers, low-k ILD and $0.57\mu m^2$ SRAM cell. In: Intl. Electron Devices Meeting, pp. 657–660 (2004)
7. Bardeen, J., Shockley, W.: Deformation potentials and mobilities in non-polar crystals. Phys. Rev. **80**(1), 72–80 (1950)
8. Bir, G.L., Pikus, G.E.: Symmetry and strain-induced effects in semiconductors. Willey, New York - Toronto (1974)
9. Chan, V., Rengarajan, R., Rovedo, N., Jin, W., Hook, T., Nguyen, P., Chen, J., Nowak, E., Chen, X.D., Lea, D., Chakravarti, A., Ku, V., Yang, S., Steegen, A., Baiocco, C., Shafer, P., Ng, H., Huang, S.F., Wann, C.: High speed 45 nm gate length CMOSFETs integrated into a 90 nm bulk technology incorporating strain engineering. In: Intl. Electron Devices Meeting, pp. 77–80 (2003)
10. Chan, V., Rim, K., Ieong, M., Yang, S., Malik, R., Teh, Y.W., Yang, M., Ouyang, Q.C.: Strain for CMOS performance improvement. pp. 667 – 674 (2005)
11. Chen, C.H., Lee, T., Hou, T., Chen, C., Chen, C., Hsu, J., Cheng, K., Chiu, Y., Tao, H., Jin, Y., Diaz, C., Chen, S., Liang, M.S.: Stress memorization technique (smt) by selectively strained-nitride capping for sub-65 nm high-performance strained-si device application. In:

VLSI Technology, 2004. Digest of Technical Papers. 2004 Symposium on, pp. 56–57 (2004). DOI 10.1109/VLSIT.2004.1345390
12. Chidambaram, P.R., Bowen, C., Chakravarthi, S., Machala, C., Wise, R.: Fundamentals of silicon material properties for successful exploitation of strain engineering in modern CMOS manufacturing. IEEE Trans. Electron Devices **53**(5), 944–964 (2006)
13. Chui, C.O., Ramanathan, S., Triplett, B.B., McIntyre, P.C., Saraswat, K.C.: Ultrathin high-k gate dielectric technology for germanium MOS applications. In: Proc. Device Research Conf., pp. 191–192 (2002)
14. Chui, K.J., Ang, K.W., Balasubramanian, N., Li, M.F., Samudra, G.S., Yeo, Y.C.: n-MOSFET with silicon–carbon source/drain for enhancement of carrier transport. IEEE Trans. Electron Devices **54**, 249–256 (2007)
15. Datta, S., Dewey, G., Doczy, M., Doyle, B., Jin, B., Kavalieros, J., Kotlyar, R., Metz, M., Zelick, N., Chau, R.: High mobility Si/SiGe strained channel MOS transistors with HfO_2/TiN gate stack. In: Intl. Electron Devices Meeting, pp. 28.1.1–28.1.4 (2003)
16. Donaton, R.A., Chidambarrao, D., Johnson, J., Chang, P., Liu, Y., Henson, W.K., Holt, J., Li, X., Li, J., Domenicucci, A., Madan, A., Rim, K., Wann, C.: Design and fabrication of MOSFETs with a reverse embedded sige (rev. e-SiGe) structure. In: Intl. Electron Devices Meeting, pp. 1–4 (2006)
17. Eneman, S., Verheyen, P., Rooyackers, R., Nouri, F., Washington, L., Degraeve, R., Kaczer, B., Moroz, V., De Keersgieter, A., Schreutelkamp, R., Kawaguchi, M., Kim, Y., Samoilov, A., Smith, L., Absil, P.P., De Meyer, K., Jurczak, M., Biesemans, S.: Layout impact on the performance of a locally strained PMOSFET. In: Proc. Symposium on VLSI Technology, pp. 22–23 (2005)
18. Fitzgerald, E., Xie, Y., Green, M., Brasen, D., Kortan, A., Michel, J., Mii, Y., Weir, B.: Totally relaxed Ge_xSi_{1-x} layers with low threading dislocation densities grown on Si substrates. Appl. Phys. Lett. **59**(7), 811–813 (1991)
19. Gannavaram, S., Pesovic, N., Ozturk, C.: Low temperature (800C) recessed junction selective silicon-germanium source/drain technology for sub-70 nm CMOS. In: Intl. Electron Devices Meeting, pp. 437–440 (2000). DOI 10.1109/IEDM.2000.904350
20. Ghani, T., Armstrong, M., Auth, C., Bost, M., Charvat, P., Glass, G., Hoffmann, T., Johnson, K., Kenyon, C., Klaus, J., McIntyre, B., Mistry, K., Murthy, A., Sandford, J., Silberstein, M., Sivakumar, S., Smith, P., Zawadzki, K., Thompson, S., Bohr, M.: A 90 nm high volume manufacturing logic technology featuring novel 45 nm gate length strained silicon CMOS transistors. In: Intl. Electron Devices Meeting, pp. 11.6.1–11.6.3 (2003)
21. Ghyselen, B., Hartmann, J.M., Ernst, T., Aulnette, C., Osternaud, B., Bogumilowicz, Y., Abbadie, A., Besson, P., Rayssac, O., Tiberj, A.: Engineering strained silicon on insulator wafers with the smart cut technology. Solid State Electron. **48**, 1285–1296 (2004)
22. Gusev, E.P., Narayanan, V., Frank, M.M.: Advanced high-k dielectric stacks with polysi and metal gates: Recent progress and current challenges. IBM J. Res. Dev. **50**(4–5), 387–410 (2006)
23. Hall, H.H., Bardeen, J., Pearson, G.L.: The effects of pressure and temperature on the resistance of $p-n$ junctions in germanium. Phys. Rev. **84**(1), 129–132 (1951). DOI 10.1103/PhysRev.84.129
24. Hensel, J.C., Feher, G.: Cyclotron resonance experiments in uniaxially stressed silicon: Valence band inverse mass parameters and deformation potentials. Phys. Rev. **129**(3), 1041–1062 (1963). DOI 10.1103/PhysRev.129.1041
25. Hensel, J.C., Hasegawa, H., Nakayama, M.: Cyclotron resonance in uniaxially stressed silicon. II. Nature of the covalent bond. Phys. Rev. **138**(1A), A225–A238 (1965)
26. Herring, C., Vogt, E.: Transport and deformation-potential theory for many-valley semiconductors with anisotropic scattering. Phys. Rev. **101**(3), 944–961 (1956)
27. Horstmann, M., Wei, A., Kammler, T., Höntschel, J., Bierstedt, H., Feudel, T., Frohberg, K., Gerhardt, M., Hellmich, A., Hempel, K., Hohage, J., Javorka, P., Klais, J., Koerner, G., Lenski, M., Neu, A., Otterbach, R., Press, P., Reichel, C., Trentsch, M., Trui, B., Salz, H., Schaller, M., Engelmann, H.J., Herzog, O., Ruelke, H., Hübler, P., Stephan, R., Greenlaw, D.,

Raab, M., Kepler, N.: Integration and optimization of embedded-SiGe, compressive and tensile stressed liner films, and stress memorization in advanced SOI CMOS technologies. In: Intl. Electron Devices Meeting, pp. 233–236 (2005)
28. Hudait, M., Chau, R.: Integrating III-V on silicon for future nanoelectronics. In: Compound Semiconductor Integrated Circuits Symposium, 2008. CSICS '08. IEEE, pp. 1–2 (2008)
29. Hudait, M., Dewey, G., Datta, S., Fastenau, J., Kavalieros, J., Liu, W., Lubyshev, D., Pillarisetty, R., Rachmady, W., Radosavljevic, M., Rakshit, T., Chau, R.: Heterogeneous integration of enhancement mode $In_{0.7}Ga_{0.3}As$ quantum well transistor on silicon substrate using thin (less than 2 μm) composite buffer architecture for high-speed and low-voltage (0.5V) logic applications. In: Intl. Electron Devices Meeting, pp. 625–628 (2007)
30. Irisawa, T., Numata, T., Tezuka, T., Usuda, K., Nakaharai, S., Hirashita, N., Sugiyama, N., Toyoda, E., Takagi, S.: High performance multi-gate pMOSFET using uniaxially-strained SGOI channels. In: Intl. Electron Devices Meeting, pp. 709–712 (2005)
31. Irisawa, T., Numata, T., Tezuka, T., Usuda, K., Sugiyama, N., Takagi, S.I.: Device design and electron transport properties of uniaxially strained-SOI tri-gate nMOSFETs. IEEE Trans. Electron Devices **55**(2), 649–654 (2008)
32. Irisawa, T., Okano, K., Horiuchi, T., Itokawa, H., Mizushima, I., Usuda, K., Tezuka, T., Sugiyama, N., Takagi, S.I.: Electron mobility and short-channel device characteristics of SOI FinFETs with uniaxially strained (110) channels. IEEE Trans. Electron Devices **56**(8), 1651–1658 (2009)
33. Ito, S., Namba, H., Yamaguchi, K., Hirata, T., Ando, K., Koyama, S., Kuroki, S., Ikezawa, N., Suzuki, T., Saitoh, T., Horiuchi, T.: Mechanical stress effect of etch-stop nitride and its impact on deep submicron transistor design. In: Intl. Electron Devices Meeting, pp. 247–251 (2000)
34. James, D.: 2004 - the year of 90 nm: a review of 90 nm devices. In: Advanced Semiconductor Manufacturing Conference and Workshop, 2005 IEEE/SEMI, pp. 72–76 (2005)
35. Jan, C.H., Bai, P., Choi, J., Curello, G., Jacobs, S., Jeong, J., Johnson, K., Jones, D., Klopcic, S., Lin, J., Lindert, N., Lio, A., Natarajan, S., Neirynck, J., Packan, P., Park, J., Post, I., Patel, M., Ramey, S., Reese, P., Rockford, L., Roskowski, A., Sacks, G., Turkot, B., Wang, Y., Wei, L., Yip, J., Young, I., Zhang, K., Zhang, Y., Bohr, M., Holt, B.: A 65 nm ultra low power logic platform technology using uni-axial strained silicon transistors. In: Intl. Electron Devices Meeting, pp. 60–63 (2005)
36. Khamankar, R., Bu, H., Bowen, C., Chakravarthi, S., Chidambaram, P.R., Bevan, M., Krishnan, A., Niimi, H., Smith, B., Blatchford, J., Hornung, B., Lu, J.P., Nicollian, P., Kirkpatrick, B., Miles, D., Hewson, M., Farber, D., Hall, L., Alshareef, H., Varghese, A., Gurba, A., Ukraintsev, V., Rathsack, B., DeLoach, J., Tran, J., Kaneshige, C., Somervell, M., Aur, S., Machala, C., Grider, T.: An enhanced 90 nm high performance technology with strong performance improvements from stress and mobility increase through simple process changes. In: Proc. Symposium on VLSI Technology, pp. 162–163 (2004)
37. Komoda, T., Oishi, A., Sanuki, T., Kasai, K., Yoshimura, H., Ohno, K., Iwai, A., Saito, M., Matsuoka, F., Nagashima, N., Noguchi, T.: Mobility improvement for 45 nm node by combination of optimized stress and channel orientation design. In: Intl. Electron Devices Meeting, pp. 217–220 (2004)
38. Lee, M.L., Fitzgerald, E.A., Bulsara, M.T., Currie, M.T., Lochtefeld, A.: Strained si, sige, and ge channels for high-mobility metal-oxide-semiconductor field-effect transistors. J. Appl. Phys. **97**(1), 011101 (2005)
39. Lim, J.S., Thompson, S.E., Fossum, J.G.: Comparison of threshold-voltage shifts for uniaxial and biaxial tensile-stressed n-MOSFETs. IEEE Electron Device Lett. **25**, 731–733 (2004)
40. Luttinger, J.M., Kohn, W.: Motion of electrons and holes in perturbed periodic fields. Phys. Rev. **97**(4), 869–883 (1955)
41. Matsumoto, T., Maeda, S., Dang, H., Uchida, T., Ota, K., Hirano, Y., Sayama, H., Iwamatsu, T., Ipposhi, T., Oda, H., Maegawa, S., Inoue, Y., Nishmura, T.: Novel SOI wafer engineering using low stress and high mobility CMOSFET with $\langle 100 \rangle$ channel for embedded RF/analog applications. In: Intl. Electron Devices Meeting, pp. 663–666 (2002)

42. Mayuzumi, S., Yamakawa, S., Kosemura, D., Takei, M., Nagata, K., Akamatsu, H., Aamari, K., Tateshita, Y., Wakabayashi, H., Tsukamoto, M., Ohno, T., Saitoh, M., Ogura, A., Nagashima, N.: Comparative study between si (110) and (100) substrates on mobility and velocity enhancements for short-channel highly-strained pfets. In: VLSI Technology, 2009 Symposium on, pp. 14–15 (2006)
43. Mistry, K., Allen, C., Auth, C., Beattie, B., Bergstrom, D., Bost, M., Brazier, M., Buehler, M., Cappellani, A., Chau, R., Choi, C.H., Ding, G., Fischer, K., Ghani, T., Grover, R., Han, W., Hanken, D., Hattendorf, M., He, J., Hicks, J., Huessner, R., Ingerly, D., Jain, P., James, R., Jong, L., Joshi, S., Kenyon, C., Kuhn, K., Lee, K., Liu, H., Maiz, J., McIntyre, B., Moon, P., Neirynck, J., Pae, S., Parker, C., Parsons, D., Prasad, C., Pipes, L., Prince, M., Ranade, P., Reynolds, T., Sandford, J., Shifren, L., Sebastian, J., Seiple, J., Simon, D., Sivakumar, S., Smith, P., Thomas, C., Troeger, T., Vandervoorn, P., Williams, S., Zawadzki, K.: A 45 nm logic technology with high-k+metal gate transistors, strained silicon, 9 Cu interconnect layers, 193 nm dry patterning, and 100% Pb-free packaging. In: Intl. Electron Devices Meeting, pp. 247–250 (2007)
44. Natarajan, S., Armstrong, K., Bost, M., Brain, R., Brazier, M., Chang, C.H., Chikarmane, V., Childs, M., Deshpande, H., Dev, K., Ding, G., Ghani, T., Golonzka, O., Han, W., He, J., Heussner, R., James, R., Jin, I., Kenyon, C., Klopcic, S., Lee, S.H., Liu, M., Lodha, S., McFadden, B., Murthy, A., Neiberg, L., Neirynck, J., Packan, P., Pae, S., Parker, C., Pelto, C., Pipes, L., Sebastian, J., Seiple, J., Sell, B., Sivakumar, S., Song, B., Tone, K., Troeger, T., Weber, C., Yang, M., Yeoh, A., Zhang, K.: A 32 nm logic technology featuring 2nd-generation high-k + metal-gate transistors, enhanced channel strain and $0.171\,\mu m^2$ SRAM cell size in a 291Mb array. In: Intl. Electron Devices Meeting, pp. 941–943 (2008)
45. Nayak, D., Goto, K., Yutani, A., Murota, J., Shiraki, Y.: High-mobility strained-Si PMOSFET's. IEEE Trans. Electron Devices **43**(10), 1709–1716 (1996)
46. Nayak, D., Woo, J., Park, J., Wang, K., MacWilliams, K.: High-mobility p-channel metal-oxide-semiconductor field-effect-transistor on strained Si. Jpn. J. Appl. Phys. **33**, 2412–2414 (1994)
47. Oh, J., Ok, I., Kang, C.Y., Jamil, M., Lee, S.H., Loh, W.Y., Huang, J., Sassman, B., Smith, L., Parthasarathy, S., Coss, B., Choi, W.H., Lee, H.D., Cho, M., Banerjee, S., Majhi, P., Kirsch, P., Tseng, H.H., Jammy, R.: Mechanisms for low on-state current of Ge (SiGe) nMOSFETs: A comparative study on gate stack, resistance, and orientation-dependent effective masses. In: VLSI Technology, 2009 Symposium on, pp. 238–239 (2006)
48. Ota, K., Sugihara, K., Sayama, H., Uchida, T., Oda, H., Eimori, T., Morimoto, H., Inoue, Y.: Novel locally strained channel technique for high performance 55 nm CMOS. In: Intl. Electron Devices Meeting, pp. 27–30 (2002)
49. Ouyang, Q., Yang, M., Holt, J., Panda, S., Chen, H., Utomo, H., Fischetti, M., Rovedo, N., Li, J., Klymko, N., Wildman, H., Kanarsky, T., Costrini, G., Fried, D., Bryant, A., Ott, J., Ieong, M., Sung, C.: Investigation of CMOS devices with embedded SiGe source/drain on hybrid orientation substrates. In: Proc. Symposium on VLSI Technology, pp. 28–29 (2005)
50. Packan, P., Akbar, S., Armstrong, M., Bergstrom, D., Brazier, M., Deshpande, H., Dev, K., Ding, G., Ghani, T., Golonzka, O., Han, W., He, J., Heussner, R., James, R., Jopling, J., Kenyon, C., Lee, S.H., Liu, M., Lodha, S., Mattis, B., Murthy, A., Neiberg, L., Neirynck, J., Pae, S., Parker, C., Pipes, L., Sebastian, J., Seiple, J., Sell, B., Sharma, A., Sivakumar, S., Song, B., St. Amour, A., Tone, K., Troeger, T., Weber, C., Zhang, K., Luo, Y., Natarajan, S.: High performance 32 nm logic technology featuring 2nd generation high-k + metal gate transistors. pp. 1 –4 (2009)
51. Radosavljevic, M., Ashley, T., Andreev, A., Coomber, S., Dewey, G., Emeny, M., Fearn, M., Hayes, D., Hilton, K., Hudait, M., Jefferies, R., Martin, T., Pillarisetty, R., Rachmady, W., Rakshit, T., Smith, S., Uren, M., Wallis, D., Wilding, P., Chau, R.: High-performance 40 nm gate length insb p-channel compressively strained quantum well field effect transistors for low-power ($v_{CC} = 0.5V$) logic applications. In: Intl. Electron Devices Meeting, pp. 1–4 (2008)
52. Radosavljevic, M., Chu-Kung, B., Corcoran, S., Dewey, G., Hudait, M., Fastenau, J., Kavalieros, J., Liu, W., Lubyshev, D., Metz, M., Millard, K., Mukherjee, N., Rachmady, W.,

Shah, U., Chau, R.: Advanced high-k gate dielectric for high-performance short-channel in0.7ga0.3as quantum well field effect transistors on silicon substrate for low power logic applications. In: Intl. Electron Devices Meeting, pp. 1 –4 (2009)
53. Rim, K., Chan, K., Shi, L., Boyd, D., Ott, J., Klymko, N., Cardone, F., Tai, L., Koester, S., Cobb, M., Canaperi, D., To, B., Duch, E., Babich, I., Carruthers, R., Saunders, P., Walker, G., Zhang, Y., Steen, M., Ieong, M.: Fabrication and mobility characteristics of ultra-thin strained Si directly on insulator (SSDOI) MOSFETs. In: Intl. Electron Devices Meeting, pp. 49–52 (2003)
54. Rim, K., Chu, J., Chen, H., Jenkins, K., Kanarsky, T., Lee, K., Mocuta, A., Zhu, H., Roy, R., Newbury, J., Ott, J., Petrarca, K., Mooney, P., Lacey, D., Koester, S., Chan, K., Boyd, D., Ieong, M., Wong, H.: Characteristics and device design of sub-100 nm strained Si n- and p-MOSFETs. In: Proc. Symposium on VLSI Technology, pp. 98–99 (2002)
55. Rim, K., Hoyt, J., Gibbons, J.: Transconductance enhancement in deep submicron strained Si n-MOSFETs. In: Intl. Electron Devices Meeting, pp. 707–710 (1998)
56. Rim, K., Welser, J., Hoyt, J., Gibbons, J.: Enhanced hole mobilities in surface-channel strained-Si p-MOSFETs. In: Intl. Electron Devices Meeting, pp. 517–520 (1995)
57. Rupp, K., Selberherr, S.: The economic limit to Moore's law. Proc. of the IEEE **98**(3), 351–353 (2010)
58. Sadaka, M., Thean, A., Barr, A., Tekleab, D., Kalpat, S., White, T.: Fabrication and operation of sub-50 nm strained-Si on $Si_{1-x}Ge_x$ on insulator (SGOI) CMOSFETs. In: Proc. IEEE International SOI Conference, pp. 209–211 (2004)
59. Scott, G., Lutze, J., Rubin, M., Nouri, F., Manley, M.: NMOS drive current reduction caused by transistor layout and trench isolation induced stress. In: Intl. Electron Devices Meeting, pp. 827–830 (1999)
60. Sheraw, C., Yang, M., Fried, D., Costrini, G., Kanarsky, T., Lee, W., Chan, V., Fischetti, M., Holt, J., et al.: Dual stress liner enhancement in hybrid orientation technology. In: Proc. Symposium on VLSI Technology, pp. 12–13 (2005)
61. Shimizu, A., Hachimine, K., Ohki, N., Ohta, H., Koguchi, M., Nonaka, Y., Sato, H., Ootsuka, F.: Local mechanical-stress control (LMC): A new technique for CMOS-performance enhancement. In: Intl. Electron Devices Meeting, pp. 433–436 (2001)
62. Sleight, J., Lauer, I., Dokumaci, O., Fried, D., Guo, D., Haran, B., Narasimha, S., Sheraw, C., Singh, D., Steigerwalt, M., Wang, X., Oldiges, P., Sadana, D., Sung, C., Haensch, W., Khare, M.: Challenges and opportunities for high performance 32 nm CMOS technology. In: Intl. Electron Devices Meeting, pp. 697–700 (2006)
63. Smith, C.S.: Piezoresistance effect in germanium and silicon. Phys. Rev. **94**(1), 42–49 (1954)
64. Steegen, A., Mo, R., Mann, R., Sun, M.C., Eller, M., Leake, G., Vietzke, D., Tilke, A., Guarin, F., Fischer, A., Pompl, T., Massey, G., Vayshenker, A., Tan, W., Ebert, A., Lin, W., Gao, W., Lian, J., Kim, J.P., Wrschka, P., Yang, J.H., Ajmera, A., Knoefler, R., Teh, Y.W., Jamin, F., Park, J., Hooper, K., Griffin, C., Nguyen, P., Klee, V., Ku, V., Baiocco, C., Johnson, G., Tai, L., Benedict, J., Scheer, S., Zhuang, H., Ramanchandran, V., Matusiewicz, G., Lin, Y.H., Siew, Y., Zhang, F., Leong, L., Liew, S., Park, K., Lee, K.W., Hong, D., Choi, S.M., Kaltalioglu, E., Kim, S., Naujok, M., Sherony, M., Cowley, A., Thomas, A., Sudijohno, J., Schiml, T., Ku, J.H., Yang, I.: 65 nm CMOS technology for low power applications. In: Intl. Electron Devices Meeting, pp. 64–67 (2005)
65. Steegen, A., Stucchi, M., Lauwers, A., Maex, K.: Silicide induced pattern density and orientation dependent transconductance in MOS transistors. In: Intl. Electron Devices Meeting, pp. 497–500 (1999)
66. Sun, G., Sun, Y., Nishida, T., Thompson, S.E.: Hole mobility in silicon inversion layers: Stress and surface orientation. J. Appl. Phys. **102**(8), 084501 (2007)
67. Suthram, S., Hussain, M.M., Harris, H.R., Smith, C., Tseng, H.H., Jammy, R., Thompson, S.E.: Comparison of Uniaxial Wafer Bending and Contact-Etch-Stop-Liner Stress Induced Performance Enhancement on Double-Gate FinFETs. IEEE Electron Device Lett. **29**, 480–482 (2008)
68. Sverdlov, V.A., Walls, T.J., Likharev, K.K.: Nanoscale silicon MOSFETs: A theoretical study. IEEE Trans. Electron Devices **50**(9), 1926–1933 (2003)

69. Thompson, S., Sun, G., Wu, K., Lim, J., Nishida, T.: Key differences for process-induced uniaxial vs. substrate-induced biaxial stressed Si and Ge channel MOSFETs. In: Intl. Electron Devices Meeting, pp. 221–224 (2004)
70. Thompson, S.E., Armstrong, M., Auth, C., Cea, S., Chau, R., Glass, G., Hoffmann, T., Klaus, J., Ma, Z., McIntyre, B., Murthy, A., Obradovic, B., Shifren, L., Sivakumar, S., Tyagi, S., Ghani, T., Mistry, K., Bohr, M., El-Mansy, Y.: A logic nanotechnology featuring strained-silicon. IEEE Electron Device Lett. **25**(4), 191–193 (2004)
71. Thompson, S.E., Suthram, S., Sun, Y., Sun, G., Parthasarathy, S., Chu, M., Nishida, T.: Future of strained Si/semiconductors in nanoscale MOSFETs. In: Intl. Electron Devices Meeting, pp. 681–684 (2006)
72. Uchida, K., Krishnamohan, T., Saraswat, K.C., Nishi, Y.: Physical mechanisms of electron mobility enhancement in uniaxial stressed MOSFETs and impact of uniaxial stress engineering in ballistic regime. In: Intl. Electron Devices Meeting, pp. 129–132 (2005)
73. Ungersboeck, E., Dhar, S., Karlowatz, G., Sverdlov, V., Kosina, H., Selberherr, S.: The effect of general strain on band structure and electron mobility of silicon. IEEE Trans. Electron Devices **54**(9), 2183–2190 (2007)
74. Uppal, S., Bollani, M., Willoughby, A., Bonar, J., Morris, R., Dowsett, M.: Diffusion of ion-implanted boron in high Ge content SiGe alloys. In: Electrocemical Society Proc., vol. 07, pp. 159–165 (2004)
75. Wang, J., Lundstrom, M.: Ballistic transport in high electron mobility transistors. IEEE Trans. Electron Devices **50**(7), 1604–1609 (2003)
76. Washington, L., Nouri, F., Thirupapuliyur, S., Eneman, G., Verheyen, P., Moroz, V., Smith, L., Xiaopeng, X., Kawaguchi, M., Huang, T., Ahmed, K., Balseanu, M., Li-Qun, X., Shen, M., Kim, Y., Rooyackers, R., Meyer, K.D., Schreutelkamp, R.: pMOSFET with 200% mobility enhancement induced by multiple stressors. IEEE Electron Device Lett. **27**(6), 511–513 (2006)
77. Welser, J., Hoyt, J., Gibbons, J.: NMOS and PMOS transistors fabricated in strained silicon/relaxed silicon-germanium structures. In: Intl. Electron Devices Meeting, pp. 1000–1002 (1992)
78. Welser, J., Hoyt, J., Gibbons, J.: Electron mobility enhancement in strained-Si n-type metal-oxide-semiconductor field-effect transistors. IEEE Electron Device Lett. **15**(3), 100–102 (1994)
79. Xiang, Q., Goo, J.S., Pan, J., Yu, B., Ahmed, S., Zhang, J., Lin, M.R.: Strained silicon nmos with nickel-silicide metal gate. In: VLSI Technology, 2003. Digest of Technical Papers. 2003 Symposium on, pp. 101–102 (2003)
80. Yang, H.S., Malik, R., Narasimha, S., Li, Y., Divakaruni, R., Agnello, P., Allen, S., Antreasyan, A., Arnold, J.C., Bandy, K., Belyansky, M., Bonnoit, A., Bronner, G., Chan, V., Chen, X., Chen, Z., Chidambarrao, D., Chou, A., Clark, W., Crowder, S.W., Engel, B., Harifuchi, H., Huang, S.F., Jagannathan, R., Jamin, F.F., Kohyama, Y., Kuroda, H., Lai, C.W., Lee, H.K., Lee, W.H., Lim, E.H., Lai, W., Mallikarjunan, A., Matsumoto, K., McKnight, A., Nayak, J., Ng, H.Y., Panda, S., Rengarajan, R., Steigerwalt, M., Subbanna, S., Subramanian, K., Sudijono, J., Sudo, G., Sun, S.P., Tessier, B., Toyoshima, Y., Tran, P., Wise, R., Wong, R., Yang, I.Y., Wann, C.H., Su, L.T., Horstmann, M., Feudel, T., Wei, A., Frohberg, K., Burbach, G., Gerhardt, M., Lenski, M., Stephan, R., Wieczorek, K., Schaller, M., Salz, H., Hohage, J., Ruelke, H., Klais, J., Huebler, P., Luning, S., van Bentum, R., Grasshoff, G., Schwan, C., Ehrichs, E., Goad, S., Buller, J., Krishnan, S., Greenlaw, D., Raab, M., Kepler, N.: Dual stress liner for high performance sub-45 nm gate length SOI CMOS manufacturing. In: Intl. Electron Devices Meeting, pp. 1075–1077 (2004)
81. Yang, J.W., Fossum, J.G., Workman, G.O., Huang, C.L.: A physical model for gate-to-body tunneling current and its effects on floating-body PD/SOI CMOS devices and circuits. Solid State Electron. **48**(2), 259–270 (2004)
82. Yang, M., Chan, V., Chan, K., Shi, L., Fried, D., Stathis, J., et al.: Hybrid-orientation technology (HOT): Opportunities and challenges. IEEE Trans. Electron Devices **53**, 965–978 (2006)
83. Yokoyama, M., Yasuda, T., Takagi, H., Yamada, H., Fukuhara, N., Hata, M., Sugiyama, M., Nakano, Y., Takenaka, M., Takagi, S.: High mobility metal S/D IIIV-On-Insulator MOSFETs

on a Si substrate using direct wafer bonding. In: VLSI Technology, 2009 Symposium on, pp. 242–243 (2006)
84. Yu, B., Wang, H., Milic, O., Xiang, Q., Wang, W., An, J., Lin, M.R.: 50 nm gate-length CMOS transistor with super-halo: design, process, and reliability. In: Intl. Electron Devices Meeting, pp. 653–656 (1999)
85. Zhang, D., Nguyen, B., White, T., Goolsby, B., et al.: Embedded SiGe S/D PMOS on thin body SOI substrate with drive current enhancement. In: Proc. Symposium on VLSI Technology, pp. 26–27 (2005)

Chapter 3
Strain and Stress

3.1 Strain Definition

All parts of a relaxed solid are in mechanical equilibrium with each other. Let us characterize a point within the solid by a radius vector $\mathbf{r}$ with the coordinates (x, y, z) in a Cartesian coordinate system. Under application of external forces the solid gets deformed: it changes its form and the volume. Due to deformation the point $\mathbf{r}$ moves to another point $\mathbf{r}' = (x', y', z')$. The difference $\mathbf{r} - \mathbf{r}'$ is called displacement $\mathbf{u}$. The displacement

$$\mathbf{u}(\mathbf{r}) = \mathbf{r}'(\mathbf{r}) - \mathbf{r} \tag{3.1}$$

defined as the function of a coordinate $\mathbf{r}$ characterizes the deformation of a solid quantitatively.

The dependence of the displacement $\mathbf{u}(\mathbf{r})$ on the position $\mathbf{r}$ is the reason of modification of the relative distances between the two points in a solid. The distance between the two points $\Delta L' = \sqrt{\Delta x'^2 + \Delta y'^2 + \Delta z'^2}$ in a deformed solid can be expressed via the distance $\Delta L = \sqrt{(\Delta x)^2 + (\Delta y)^2 + (\Delta z)^2}$ in the relaxed solid and the displacement $\mathbf{u}$ as

$$\Delta L' = \sqrt{\Delta L^2 + \left(\frac{\partial u_i}{\partial x_j} + \frac{\partial u_j}{\partial x_i} + \frac{\partial u_k}{\partial x_i} \frac{\partial u_k}{\partial x_j} \right) \Delta x_i \Delta x_j}, \tag{3.2}$$

where summation over repeating indices is assumed $(i, j = x, y, z)$. The expression (3.2) can be rewritten as:

$$\Delta L' = \sqrt{\Delta L'^2 + 2\varepsilon_{ik} \Delta x_i \Delta x_j}, \tag{3.3}$$

where the strain tensor

$$\varepsilon_{ij} = \frac{1}{2} \left(\frac{\partial u_i}{\partial x_j} + \frac{\partial u_j}{\partial x_i} + \frac{\partial u_k}{\partial x_i} \frac{\partial u_k}{\partial x_j} \right) \tag{3.4}$$

V. Sverdlov, *Strain-Induced Effects in Advanced MOSFETs*, Computational Microelectronics, DOI 10.1007/978-3-7091-0382-1_3, © Springer-Verlag/Wien 2011

is introduced. By the definition (3.4) the strain tensor is symmetric:

$$\varepsilon_{ij} = \varepsilon_{ji}. \tag{3.5}$$

Therefore, the strain tensor (3.4) can be diagonalized by an appropriate transformation of the coordinate system. The coordinate system in which the strain tensor is diagonal depends on the position **r**. In the system where the strain tensor is diagonal the modification of the distance between the two points $\Delta L'$ is expressed as:

$$\Delta L' = \sqrt{(1 + 2\varepsilon_{\xi\xi})\Delta L_\xi^2 + (1 + 2\varepsilon_{\nu\nu})\Delta L_\nu^2 + (1 + 2\varepsilon_{\zeta\zeta})\Delta L_\zeta^2}. \tag{3.6}$$

Thus, an arbitrary deformation of a small volume around the point **r** can be represented as a combination of three simple independent deformations along the three orthogonal axes diagonalyzing the strain tensor at this point. A relative change of the length along an axis $\alpha = \xi, \nu, \zeta$ is

$$\frac{\Delta L_\alpha - \Delta L'_\alpha}{\Delta L_\alpha} = \sqrt{1 + 2\varepsilon_{\alpha\alpha}} - 1. \tag{3.7}$$

In case of small displacements, which is a typical situation of up to a few percents deformation, the relative change in the length L_α is proportional to $\varepsilon_{\alpha\alpha}$:

$$\frac{\Delta L_\alpha - \Delta L'_\alpha}{\Delta L_\alpha} = \varepsilon_{\alpha\alpha}. \tag{3.8}$$

The strain tensor in this case simplifies to

$$\varepsilon_{ij} = \frac{1}{2}\left(\frac{\partial u_i}{\partial x_j} + \frac{\partial u_j}{\partial x_i}\right). \tag{3.9}$$

According to (3.8), the volume change $(\Delta V - \Delta V')/\Delta V$ is proportional to the sum of $\varepsilon_{\alpha\alpha}$. Since the trace of a tensor is invariant under a coordinate transformation, the relative volume change can be written as:

$$\frac{\Delta V - \Delta V'}{\Delta V} = \varepsilon_{xx} + \varepsilon_{yy} + \varepsilon_{zz}. \tag{3.10}$$

We mention other quantities used in literature to describe strain. Frequently, the engineering strain tensor e_{ij} is used. It is defined as:

$$\begin{pmatrix} e_{xx} & e_{xy} & e_{xz} \\ e_{xy} & e_{yy} & e_{yz} \\ e_{xz} & e_{yz} & e_{zz} \end{pmatrix} = \begin{pmatrix} \varepsilon_{xx} & 2\varepsilon_{xy} & 2\varepsilon_{xz} \\ 2\varepsilon_{xy} & \varepsilon_{yy} & 2\varepsilon_{yz} \\ 2\varepsilon_{xz} & 2\varepsilon_{yz} & \varepsilon_{zz} \end{pmatrix}. \tag{3.11}$$

Here, sometimes the off-diagonal terms of the engineering strains are denoted as $\gamma_{ij} = e_{ij} = 2\varepsilon_{ij}$.

Due to the symmetry $\varepsilon_{ij} = \varepsilon_{ji}$ the strain tensor is characterized by six independent components, which may be arranged into a vector:

$$(\varepsilon_{xx}, \varepsilon_{yy}, \varepsilon_{zz}, 2\varepsilon_{yz}, 2\varepsilon_{xz}, 2\varepsilon_{xy}) = (e_1, e_2, e_3, e_4, e_5, e_6)\,. \tag{3.12}$$

The six components notation is convenient, when writing the relation between strain and stress in cubic semiconductors.

3.2 Stress

In a deformed solid the internal structure is changed. Due to deformation the atoms are shifted from their equilibrium positions. This leads to the appearance of the internal forces trying to bring the atoms back to their equilibrium. In the absence of piezo-electric effects the forces of interatomic interaction are short-ranged. In a situation when we are interested in deformations of a volume which includes many atoms, the interaction radius can be set to zero. Thus, the internal forces due to deformation are local. This means that these forces can act on a certain part of a deformed solid only through the surface surrounding this part of a solid.

The total force **F** acting on the part of a solid can be written as an integral:

$$\mathbf{F} = \int \mathbf{f} dV, \tag{3.13}$$

where **f** is a force acting on an elementary volume of the selected part. Since the interactions between the elementary volumes inside the part of a solid can not produce the force **F** acting on the part, the total force **F** is due to the action of the external surrounding environment on the selected part. Due to the short-range nature of interaction forces, this action is local and is applied to the surface surrounding the selected part. Thus, the component $i = x, y, z$ of the total force **F** must be equal to an integral over the surface of a vector $\boldsymbol{\sigma}_i$:

$$F_i = \int \boldsymbol{\sigma}_i d\mathbf{S} = \sum_j \int \sigma_{ij} dS_j, \quad i, j = x, y, z, \tag{3.14}$$

where $d\mathbf{S}$ is a vector orthogonal to the surface element and pointed in the outer direction with an absolute value equal to the area of the surface element. The second-rank tensor σ_{ij} is called the stress tensor.

The definition (3.14) of the stress tensor allows a simple interpretation. Let us choose a small cube as the selected volume, with its faces orthogonal to the axes of the coordinate system. Let us consider a single face of the cube, say dS_x oriented along the positive direction of the axis x (the outer normal vector is in the positive

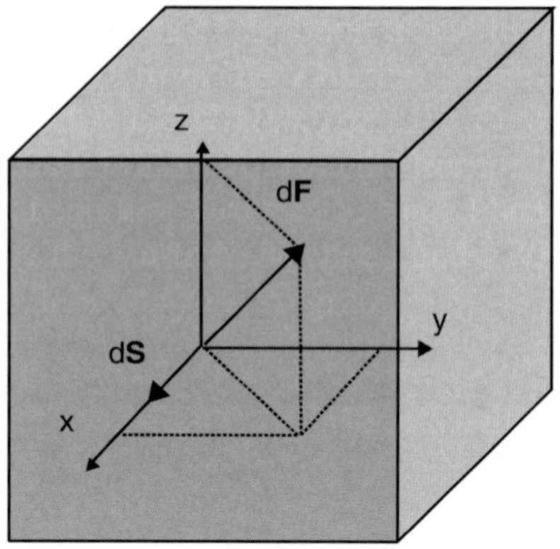

Fig. 3.1 Illustration of the force $d\mathbf{F}$ acting on a cube face $d\mathbf{S}$ used in defining the stress tensor (3.15)

direction of the x axis). The force $d\mathbf{F}$ acting on the face dS_x is not necessarily parallel to the x axis as shown in Fig. 3.1. This force is characterized by the component parallel to the face plane, the shear components, and the component orthogonal to the plane, the normal component. Then the components of the stress tensor σ_{ix} at the point surrounded by the selected cube are defined as the limit

$$\sigma_{ix} = \lim_{dS \to 0} \frac{dF_i}{dS}. \tag{3.15}$$

Other components are defined by analogy. The off-diagonal components are due to the projection of the force $d\mathbf{F}$ on the face and lead to a shear distortion of the selected cube. The off-diagonal stress components are also called shear.

From the Gauß theorem, (3.13) and (3.14) it follows that

$$F_i = \int f_i dV = \int \mathrm{div}\boldsymbol{\sigma}_i dV, \tag{3.16}$$

and

$$f_i = \mathrm{div}\boldsymbol{\sigma}_i.$$

The definition of the shear tensor σ_{ij} is not unique [4]. This uncertainty is removed by making the stress tensor symmetric:

$$\sigma_{ij} = \sigma_{ji}. \tag{3.17}$$

This symmetry is required by the condition that the total moment with respect to an arbitrary point in equilibrium is zero.

3.3 Relation Between Strain and Stress Tensor in Silicon and Germanium

External forces applied to a solid result in deformation and internal stress. Depending on the amplitude of the force applied to a relaxed solid it can either return into an initial relaxed state or, alternatively, it can preserve a certain level of deformation even when the external force that caused the deformation is removed. In the last case of large deformation the atoms are moved so far away from their equilibrium position that the solid finds another local energy minimum. The potential barrier separating the local minimum from the global one preserves the solid to return into relaxed state. In case of relatively small deformation the structure of the solid is not changed dramatically, atoms are shifted only little from their equilibrium positions. In this case of elastic deformation there is a unique relation between deformation and internal strain. The well-known elementary text-book example is a deformed spring where the relation between the applied force F and deformation u is described by Hooke's law $F = \kappa u$, where κ is the spring constant.

A generalization of Hooke's law describing the relation between the stress and strain tensors in a three-dimensional elastically deformed solid was suggested by Cauchy

$$\sigma_{ij} = C_{ijkl}\varepsilon_{kl}. \tag{3.18}$$

Here C_{ijkl} is the elastic stiffness tensor of the fourth order. Number of independent entries depends on the symmetry of the solid [3]. Silicon and germanium belong to the cubic semiconductors. In this case the stiffness tensor is characterized by only three independent components conveniently written as c_{11}, c_{12}, and c_{44}. The values of the stiffness constants of silicon and germanium [5] are summarized in Table 3.1.

The stiffness tensor allows to find the stress tensor if strain is known:

$$\begin{pmatrix} \sigma_{xx} \\ \sigma_{yy} \\ \sigma_{zz} \\ \sigma_{yz} \\ \sigma_{xz} \\ \sigma_{xy} \end{pmatrix} = \begin{pmatrix} c_{11} & c_{12} & c_{12} & 0 & 0 & 0 \\ c_{12} & c_{11} & c_{12} & 0 & 0 & 0 \\ c_{12} & c_{12} & c_{11} & 0 & 0 & 0 \\ 0 & 0 & 0 & c_{44} & 0 & 0 \\ 0 & 0 & 0 & 0 & c_{44} & 0 \\ 0 & 0 & 0 & 0 & 0 & c_{44} \end{pmatrix} \cdot \begin{pmatrix} \varepsilon_{xx} \\ \varepsilon_{yy} \\ \varepsilon_{zz} \\ 2\varepsilon_{yz} \\ 2\varepsilon_{xz} \\ 2\varepsilon_{xy} \end{pmatrix}. \tag{3.19}$$

Alternatively, the relation between strain and stress is given by the compliance tensor S_{ijkl}

$$\varepsilon_{ij} = S_{ijkl}\sigma_{kl}, \tag{3.20}$$

Table 3.1 Elastic stiffness constants of Si and Ge [5]

	Silicon	Germanium	Units
c_{11}	166.0	126.0	GPa
c_{12}	64.0	44.0	GPa
c_{44}	79.6	67.7	GPa

or in matrix form

$$\begin{pmatrix} \varepsilon_{xx} \\ \varepsilon_{yy} \\ \varepsilon_{zz} \\ 2\varepsilon_{yz} \\ 2\varepsilon_{xz} \\ 2\varepsilon_{xy} \end{pmatrix} = \begin{pmatrix} s_{11} & s_{12} & s_{12} & 0 & 0 & 0 \\ s_{12} & s_{11} & s_{12} & 0 & 0 & 0 \\ s_{12} & s_{12} & s_{11} & 0 & 0 & 0 \\ 0 & 0 & 0 & s_{44} & 0 & 0 \\ 0 & 0 & 0 & 0 & s_{44} & 0 \\ 0 & 0 & 0 & 0 & 0 & s_{44} \end{pmatrix} \cdot \begin{pmatrix} \sigma_{xx} \\ \sigma_{yy} \\ \sigma_{zz} \\ \sigma_{yz} \\ \sigma_{xz} \\ \sigma_{xy} \end{pmatrix}. \tag{3.21}$$

Taking into account that the compliance tensor is an inverse for the stiffness tensor, the compliance constants s_{ij} are expressed via the stiffness constants c_{ij} as

$$\begin{aligned} s_{11} &= \frac{c_{11} + c_{12}}{c_{11}^2 + c_{11}c_{12} - 2c_{12}^2}, \\ s_{12} &= \frac{-c_{12}}{c_{11}^2 + c_{11}c_{12} - 2c_{12}^2}, \quad \text{and} \\ s_{44} &= \frac{1}{c_{44}}. \end{aligned}$$

3.4 Strain and Stress Tensors: Examples

3.4.1 Uniform All-Around Compression

The components of the stress tensor can be easily determined in the case of uniform compression. Forces acting on any face of a selected small cubical volume are equal. The force $d\mathbf{F}$ acting on a face dS_x is equal to $d\mathbf{F} = (-pdS_x, 0, 0)$. According to (3.15), the stress tensor possesses only diagonal elements and can be written as:

$$\boldsymbol{\sigma} = \begin{pmatrix} -p & 0 & 0 \\ 0 & -p & 0 \\ 0 & 0 & -p \end{pmatrix}, \tag{3.22}$$

Strain and stress in a solid are completely determined by the elements of the corresponding tensors in a given coordinate system. It is convenient to choose the coordinate system in cubic semiconductors so that the axes are orthogonal to the cube faces. This system is called crystallographic coordinate system. Stress and strain tensors will be always given this system.

In order to characterize directions and plane orientations in cubic crystals the notation of Miller indices is used [1, 3]. A triplet of integers [*nmk*] defines the coordinates of the radius vector along the direction in the crystallographic coordinate system, while ⟨*nmk*⟩ is used to describe all the directions equivalent by symmetry to [*nmk*]. Similarly, to determine orientation of a plane, the triplet of integers (*nmk*) is used to define the coordinates of the vector normal to the plane, while {*nmk*}

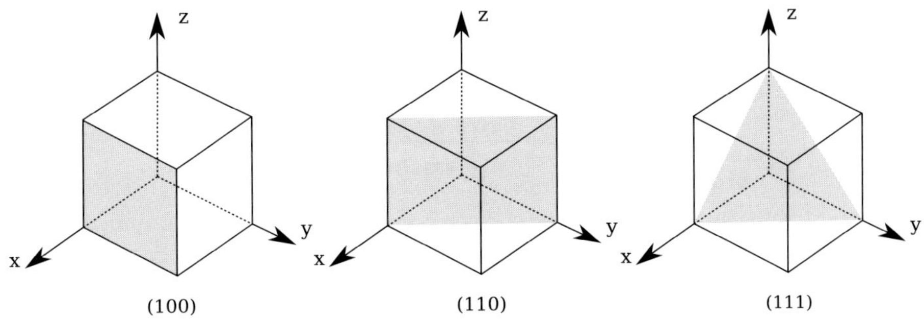

Fig. 3.2 Three most frequently used surface orientations characterized by the Miller indices (100), (110), and (111)

describes all the planes equivalent by symmetry to the (*nmk*) one. Negative indices are indicated with a bar above the number. Orientations of the most commonly used surfaces (100), (110), and (111) are shown in Fig. 3.2.

3.4.2 Biaxial Strain Resulting From Epitaxial Growth

In a global stress technique a thin silicon layer is grown epitaxially on a top of relaxed virtual SiGe substrate. Due to the lattice constant mismatch between silicon and SiGe the silicon layer is biaxially stretched. The strain tensor in the crystallographic coordinate system for an arbitrary substrate orientation can be evaluated [2] by first determining the strain tensor σ'_{nm} in the interface coordinate system and then transforming it by an appropriate coordinate transformation. The in-plane strain is described by the difference between the lattice constants a_r and a_s in relaxed [5] and strained silicon, respectively:

$$\varepsilon_{||} = \frac{a_s - a_r}{a_r}. \tag{3.23}$$

Thus, the in-plane components of σ'_{nm} are equal:

$$\varepsilon'_{11} = \varepsilon'_{22} = \varepsilon_{||}. \tag{3.24}$$

During the epitaxial growth shear distortions usually do not appear, which allows to set all the off-diagonal components to zero:

$$\varepsilon'_{ij} = 0, \quad i \neq j. \tag{3.25}$$

The only undetermined component of the strain tensor ε'_{33} is found from the condition that the stress component in the direction of growth is set to zero: $\sigma'_{33} = 0$. By using the generalized Hooke's law $\sigma'_{ij} = C'_{ijkm}\varepsilon'_{km}$ in the interface coordinate system, the unknown ε'_{33} is written in the form:

$$\varepsilon'_{33} = -\frac{c'_{3311} + c'_{3322}}{c'_{3333}}\varepsilon_{||}. \tag{3.26}$$

As soon as the strain tensor in the surface coordinate system is determined, it can be transformed to the crystallographic coordinate system by a unitary transformation U

$$\hat{\varepsilon} = U\hat{\varepsilon}'U^T, \tag{3.27}$$

where U is the rotation matrix. (3.27) can be written in its unfolded form:

$$\varepsilon_{kl} = U_{ik}\varepsilon'_{ij}U_{lj}, \tag{3.28}$$

where the summation over repeating indices is assumed.

The direction of the z' axis of the surface system is described by a polar angle ϕ and azimuthal angle θ in the crystallographic coordinate system. Thus, the surface coordinate system can be obtained from the crystallographic system by rotating it first around the z axis by a polar angle ϕ followed by a rotation around the new y' axis by an azimuthal angle θ shown in Fig. 3.3. The transformation matrix U can be written as:

$$U = \left(R_z(\phi)R_{y'}(\theta)\right)^T, \tag{3.29}$$

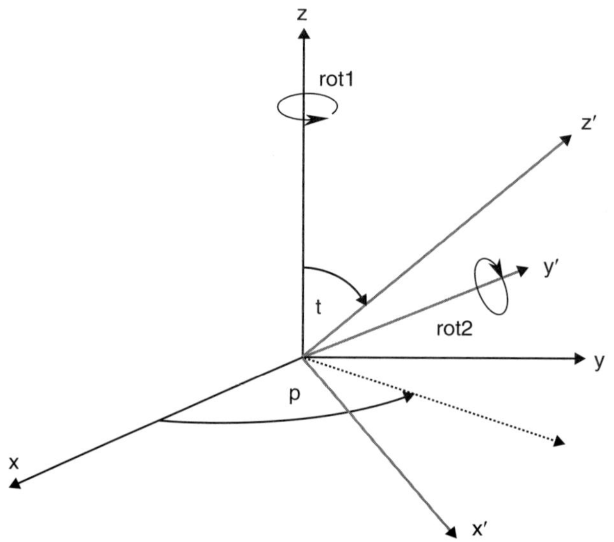

Fig. 3.3 Transformation of the crystallographic coordinate system to the interface system is described by a counterclockwise rotation around the z axis followed by a rotation around y' axis

where

$$R_z(\phi) = \begin{pmatrix} \cos(\phi) & \sin(\phi) & 0 \\ -\sin(\phi) & \cos(\phi) & 0 \\ 0 & 0 & 1 \end{pmatrix}, \tag{3.30}$$

$$R_{y'}(\theta) = \begin{pmatrix} \cos(\theta) & 0 & -\sin(\theta) \\ 0 & 1 & 0 \\ \sin(\theta) & 0 & \cos(\theta) \end{pmatrix}. \tag{3.31}$$

Therefore,

$$U = \begin{pmatrix} \cos(\theta)\cos(\phi) & -\sin(\phi) & \sin(\theta)\cos(\phi) \\ \cos(\theta)\sin(\phi) & \cos(\theta) & \sin(\theta)\sin(\phi) \\ -\sin(\theta) & 0 & \cos(\theta) \end{pmatrix}. \tag{3.32}$$

Using the transformation U, the stiffness tensor in the interface coordinate system needed in (3.26) can be expressed via the one in the crystallographic system as:

$$c_{ijkl} = U_{im} U_{kn} c'_{mpns} U_{jp} U_{ls}, \tag{3.33}$$

Below are the strain tensor examples for silicon grown epitaxially on (001), (110), and (111) SiGe virtual substrates:

$$\hat{\varepsilon}_{(001)} = \varepsilon_{||} \begin{pmatrix} 1 & 0 & 0 \\ 0 & 1 & 0 \\ 0 & 0 & -\dfrac{2c_{12}}{c_{11}} \end{pmatrix}$$

$$\hat{\varepsilon}_{(110)} = \varepsilon_{||} \begin{pmatrix} \dfrac{2c_{44}-c_{12}}{c_{11}+c_{12}+2c_{44}} & -\dfrac{c_{11}+2c_{12}}{c_{11}+c_{12}+2c_{44}} & 0 \\ -\dfrac{c_{11}+2c_{12}}{c_{11}+c_{12}+2c_{44}} & \dfrac{2c_{44}-c_{12}}{c_{11}+c_{12}+2c_{44}} & 0 \\ 0 & 0 & 1 \end{pmatrix}$$

$$\hat{\varepsilon}_{(111)} = \varepsilon_{||} \begin{pmatrix} \dfrac{4c_{44}}{c_{11}+2c_{12}+4c_{44}} & -\dfrac{c_{11}+2c_{12}}{c_{11}+2c_{12}+4c_{44}} & -\dfrac{c_{11}+2c_{12}}{c_{11}+2c_{12}+4c_{44}} \\ -\dfrac{c_{11}+2c_{12}}{c_{11}+2c_{12}+4c_{44}} & \dfrac{4c_{44}}{c_{11}+2c_{12}+4c_{44}} & -\dfrac{c_{11}+2c_{12}}{c_{11}+2c_{12}+4c_{44}} \\ -\dfrac{c_{11}+2c_{12}}{c_{11}+2c_{12}+4c_{44}} & -\dfrac{c_{11}+2c_{12}}{c_{11}+2c_{12}+4c_{44}} & \dfrac{4c_{44}}{c_{11}+2c_{12}+4c_{44}} \end{pmatrix} \tag{3.34}$$

The strain tensor contains shear components if silicon layer is grown on a (110) and (111) substrate. These components can be large for higher germanium concentration, as shown in Fig. 3.4.

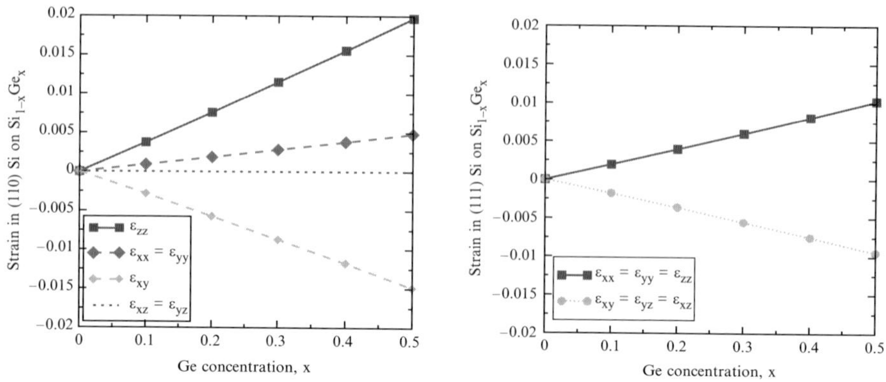

Fig. 3.4 Strain in a silicon film grown epitaxially on (110) (*left panel*) and (111) (*right panel*) $Si_{1-x}Ge_x$ as function of germanium concentration x. All shear strain components are non-zero in the (111) case

3.4.3 Uniaxial Stress

As we have mentioned in the previous chapter, uniaxial process-induced stress is currently employed by the semiconductor industry. Several important examples of the strain and stress tensor for different stress directions are briefly reviewed below.

It is convenient to choose the coordinate system in which one of the axes, say z, is parallel to the stress direction. In this coordinate system the stress tensor has only one non-zero component $\sigma_{zz}{}' = P$, where P is the stress magnitude. By transforming the stress tensor back to the crystallographic coordinate system. we obtain the following expressions in case of stress in [001], [110], [111], and [120] directions:

$$\boldsymbol{\sigma}_{[001]} = \begin{pmatrix} 0 & 0 & 0 \\ 0 & 0 & 0 \\ 0 & 0 & P \end{pmatrix} \qquad \boldsymbol{\sigma}_{[110]} = \begin{pmatrix} P/2 & P/2 & 0 \\ P/2 & P/2 & 0 \\ 0 & 0 & 0 \end{pmatrix}$$

$$\boldsymbol{\sigma}_{[111]} = \begin{pmatrix} P/3 & P/3 & P/3 \\ P/3 & P/3 & P/3 \\ P/3 & P/3 & P/3 \end{pmatrix} \qquad \boldsymbol{\sigma}_{[120]} = \begin{pmatrix} P/5 & 2P/5 & 0 \\ 2P/5 & 4P/5 & 0 \\ 0 & 0 & 0 \end{pmatrix} \tag{3.35}$$

The strain tensor is found from (3.35) and (3.20):

$$\hat{\boldsymbol{\varepsilon}}_{[001]} = P \begin{pmatrix} s_{12} & 0 & 0 \\ 0 & s_{12} & 0 \\ 0 & 0 & s_{11} \end{pmatrix} \quad \hat{\boldsymbol{\varepsilon}}_{[110]} = \frac{P}{2} \begin{pmatrix} s_{11}+s_{12} & s_{44}/2 & 0 \\ s_{44}/2 & s_{11}+s_{12} & 0 \\ 0 & 0 & 2s_{12} \end{pmatrix}$$

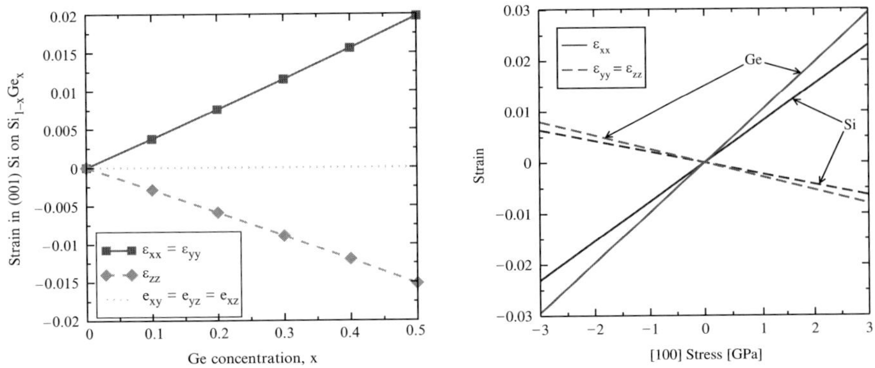

Fig. 3.5 *Left panel.* Strain in a silicon film grown on a (001) $Si_{1-x}Ge_x$ substrate as a function of germanium concentration x. *Right panel.* Strain as a function of [001] uniaxial stress. All shear strain components are zero

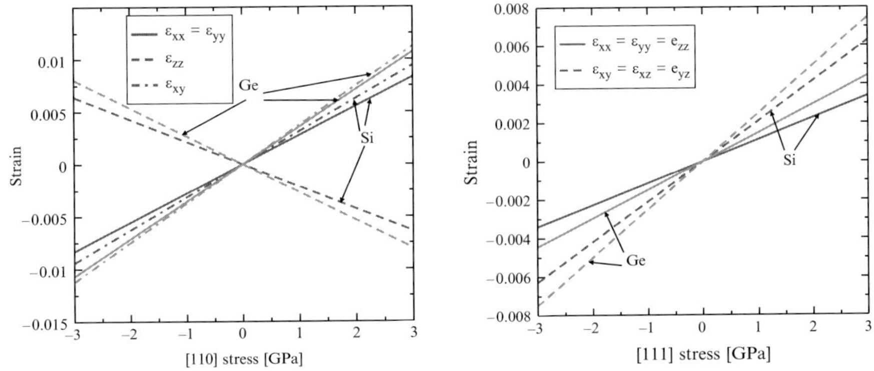

Fig. 3.6 Strain tensor components as a function of [110] and [111] uniaxial stress in silicon and germanium. Contrary to [100] uniaxial stress, shear strain components are non-zero

$$\hat{\boldsymbol{\varepsilon}}_{[111]} = \frac{P}{3}\begin{pmatrix} s_{11}+2s_{12} & s_{44}/2 & s_{44}/2 \\ s_{44}/2 & s_{11}+2s_{12} & s_{44}/2 \\ s_{44}/2 & s_{44}/2 & s_{11}+2s_{12} \end{pmatrix}$$

$$\hat{\boldsymbol{\varepsilon}}_{[120]} = \frac{P}{5}\begin{pmatrix} s_{11}+4s_{12} & s_{44} & 0 \\ s_{44} & s_{12}+4s_{11} & 0 \\ 0 & 0 & 5s_{12} \end{pmatrix} \tag{3.36}$$

Figure 3.5 demonstrates strain in silicon epitaxially grown on (001) $Si_{1-x}Ge_x$ substrate as a function of germanium concentration x and strain in silicon and germanium under [001] uniaxial stress. In both cases shear strain is absent. If, however, silicon or germanium are stressed along [110] or [111] axes, non-zero shear strain components appear, as displayed in Fig. 3.6. As will be shown below, this type of deformation plays an important role in mobility enhancement.

References

1. Ashcroft, N.W., Mermin, N.D.: Solid state physics. Harcourt College Publishers, Fort Worth (1976)
2. Hinckley, J., Singh, J.: Influence of substrate composition and crystallographic orientation on the band structure of pseudomorphic Si-Ge alloy films. Phys. Rev. B **42**, 3546–3566 (1990)
3. Kittel, C.: Introduction to solid state physics. 7'th edition. Wiley, New York (1996)
4. Landau, L., Pitaevskii, L., Lifshitz, E., Kosevich, A.: Theory of elasticity, Third edition, Theoretical Physics, Vol. 7. Butterworth-Heinemann, Oxford (1986)
5. Levinshtein, M., Rumyantsev, S., Shur, M. (eds.): Handbook series on semiconductor parameters, vol. 1,2. World Scientific, London (1999)

Chapter 4
Basic Properties of the Silicon Lattice

4.1 Crystal Structure of Silicon and Germanium

Atoms in crystalline silicon are arranged into a repeating three-dimensional pattern – a crystal. The crystallographic unit cell of the crystal structure of silicon is shown in Fig. 4.1. The crystal lattice can be represented as two face centered cubic lattices (fcc), with the cube side $a_0 = 0.543\mathring{A}$, where the second lattice made of atoms B is displaced relative to the first one made of atoms A by a translation vector of $\frac{a_0}{4}(1,1,1)$ along a diagonal. If the atoms A and B were different, the lattice structure is called zinc-blend lattice structure. The crystal lattice of GaAs, AlAs, InAs, InP, InSb, and many other III–V semiconductors is of the zinc-blend type.

When the atoms A and B are of the same type, like in diamond, the lattice type is called diamond. This is the case of the typical representatives of the group IV semiconductors such as silicon, germanium, and SiGe alloys in any proportions.

The translational symmetry of a three-dimensional crystal can be reproduced by taking a minimal number of atoms called a basis set which is repeatedly translated by three primitive lattice vectors $\mathbf{a}_i, i = 1,2,3$. In cases of silicon lattice structure the basis set consists of the two Si atoms A and B. The basis vectors of the Bravais lattice are shown in Fig. 4.2:

$$\mathbf{a}_1 = \frac{a_0}{2}\begin{pmatrix}0\\1\\1\end{pmatrix}, \quad \mathbf{a}_2 = \frac{a_0}{2}\begin{pmatrix}1\\0\\1\end{pmatrix}, \quad \text{and} \quad \mathbf{a}_3 = \frac{a_0}{2}\begin{pmatrix}1\\1\\0\end{pmatrix}, \tag{4.1}$$

where a_0 is the lattice constant of the relaxed lattice. The lattice is obtained by translation of the basis atoms set by multiples of the basis vectors and their linear combination:

$$\mathbf{a} = i\mathbf{a}_1 + j\mathbf{a}_2 + k\mathbf{a}_3, \tag{4.2}$$

where i, j and k are integers. In the relaxed lattice the distance from the basis atom at $\frac{a_0}{4}(1,1,1)$ to its four nearest neighbors is equal to $a \cdot \sqrt{3}/4$, and the atomic packaging factor is $\sqrt{3}\pi/16$, with eight atoms per unit cell.

V. Sverdlov, *Strain-Induced Effects in Advanced MOSFETs*, Computational Microelectronics, DOI 10.1007/978-3-7091-0382-1_4, © Springer-Verlag/Wien 2011

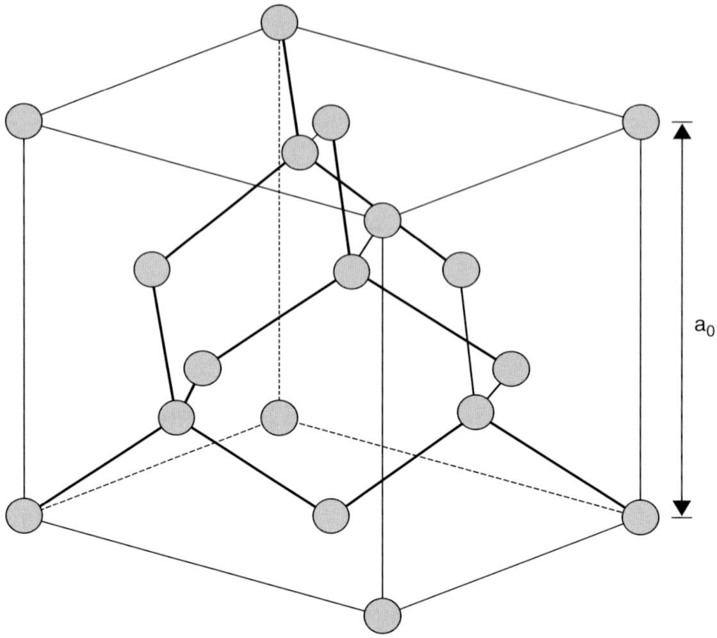

Fig. 4.1 Crystallographic unit cell (unit cube) of the diamond structure

Besides with the translations (4.2), the diamond crystal structure is invariant under point symmetry operations. These symmetry operations can be represented as a superposition of reflections, rotations, and inversion. Inversion over a point can also be represented as a rotation by π around an axis going through the inversion point followed by a reflection on a plane perpendicular to the rotation axis and crossing it at the inversion point. However, due to the importance of the inversion symmetry, it is usually considered as a separate symmetry operation.

Let us briefly review the symmetry classification of crystal lattices.

By definition a symmetry transformation applied to a lattice brings the lattice into coincidence with itself thus leaving it invariant. Any symmetry transformation may be decomposed into an elementary rotation about an axis, reflection in a plane, and translation. A rotation is defined by specifying the direction of the rotation axis and the angle of the rotation. A reflection in a plane is determined by specifying the plane. A translation by a vector displaces each point of a lattice by this translation vector. A combination of two symmetry transformations applied to a lattice in succession is again a symmetry transformation since it leaves the lattice invariant. An inverse to a symmetry transformation is again a symmetry transformation. A set of these symmetry transformations, or elements, form a group. Every group contains one and only one identity element.

A symmetry group possessing a fixed point common to all transformations of the group is called a point group. For a Bravais lattice, the totality of reflections and

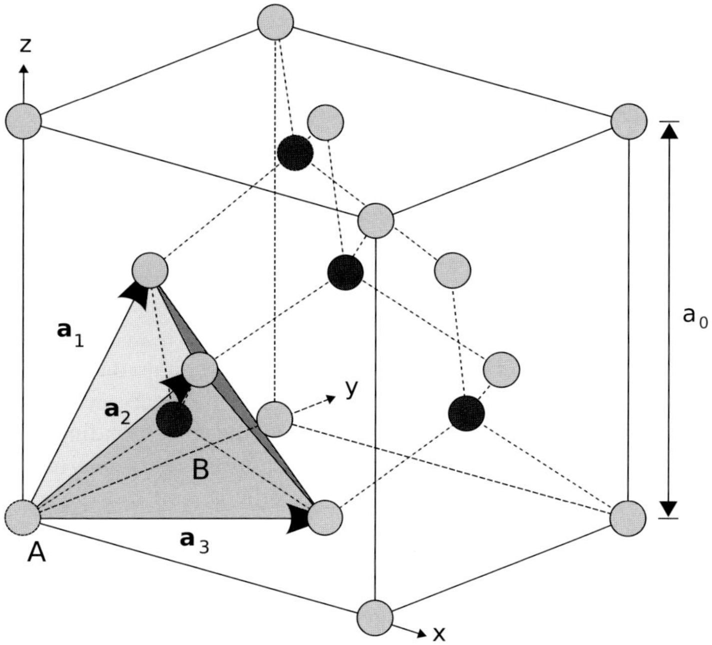

Fig. 4.2 The primitive basis vectors of the face centered cubic (fcc) lattice and the two basis atoms *A* and *B*

rotations which map the Bravais lattice onto itself have a fixed point and therefore form a point group. This point group of the Bravais lattice is also the symmetry group of the crystal symmetry group which characterizes the translational symmetry of the crystal [1]. There are seven point symmetry groups for all crystals, or seven systems of Bravais lattices: triclinic, monoclinic, orthorhombic, tetragonal, trigonal, hexagonal, and cubic. Each system may contain several types of Bravais lattices: simple, volume-centered, base-centered, and face-centered. For example, the cubic system includes simple, body-centered, and face-centered lattice types. The relaxed silicon diamond structure belongs to the face-centered cubic system. In total the seven systems subdivide into fourteen lattice types.

The system of the Bravais lattice and the lattice type, characterize the symmetry group of a crystal incompletely. In compound crystals, like III–V semiconductors, the primitive cell contains more than one unlike atom. The Bravais lattices built on the different atoms coincide, however, the equivalent points are not the same. Thus, inversion, which is an element of the symmetry group of the Bravais lattice, may not be a symmetry element of the symmetry group of a crystal. The elements of the symmetry group of the Bravais lattice, including translations, that transforms every direction in the crystal to an equivalent one form a subgroup of the initial group. The equivalent direction in the crystal is the direction along which all the crystal properties are identical. This subgroup is called the crystal class. In total there exist 32 different crystal classes [1].

Even the crystal class and lattice type, however, are not sufficient to characterize the space symmetry of a crystal completely. Within a point group all elements have a common point, while for a space group characterizing the symmetry of a crystal, the position of each rotation element within the primitive cell must be specified. Therefore, apart from elements of point group symmetry, the space group contains translations by a vector of the length less than a vector of the Bravais lattice. For example, a compound lattice containing at least two equivalent atoms per primitive cell, which is the case of the group IV semiconductors like silicon and germanium, may be viewed as a system of inter-penetrating identical Bravais lattices shifted by this translation vector. These additional translation operations map the positions of each component lattice onto the identical points of another lattice. Therefore, in this case a space group characterizing the symmetry of a crystal contains an invariant translation subgroup. By dividing all the elements of the space group into co-sets modulo the translation subgroup, one can construct the corresponding factor group, which is isomorphic to the crystallographic point group characterizing the crystal class. Altogether, there exist 230 different space groups.

The following symmetry operations are defined:

E Unity operation
n_k^+ Clockwise rotation of angle $2\pi/n$ around axis $\mathbf{e}_k$
n_k^- Counter-clockwise rotation of angle $2\pi/n$ around axis $\mathbf{e}_k$
I Inversion
$\bar{n}_k^+$ Clockwise rotation of angle $2\pi/n$ around axis $\mathbf{e}_k$ followed by inversion
$\bar{n}_k^-$ Counter-clockwise rotation of angle $2\pi/n$ around axis $\mathbf{e}_k$ followed by inversion

The rotation axes $\mathbf{e}_k$ are grouped into five classes and defined by the following Miller indices:

$\mathbf{e}_i$ $(1,0,0)$, $(0,1,0)$, $(0,0,1)$
$\mathbf{e}_{i''}$ $(0,1,0)$, $(\sqrt{3},-1,0)$, $(-\sqrt{3},-1,0)$
$\mathbf{e}_j$ $(1,1,1)$, $(-1,-1,1)$, $(1,-1,-1)$, $(-1,-1,-1)$
$\mathbf{e}_p$ $(1,1,0)$, $(-1,1,0)$, $(1,0,1)$, $(0,1,1)$, $(-1,0,1)$, $(0,-1,1)$
$\mathbf{e}_s$ $(1,1,0)$, $(-1,1,0)$

Thus, the symbol $n_i^{\pm}$ stands for rotations around three axes $\mathbf{e}_i$ and corresponds to three point symmetry operations.

The set of all point operations for a particular crystal lattice structure forms a group. Relaxed silicon crystal lattice belongs to the O_h symmetry group [5], in Schönflies notation. It is characterized by the symmetry elements listed in Table 4.1. A more detailed description of the O_h symmetry group is presented in Chapter 7.

The point group of the diamond structure which corresponds to relaxed silicon consists of 48 symmetry elements.

Table 4.1 Symmetry elements of the point symmetry group of relaxed silicon lattices

Point group	Symmetry elements	Number of elements	Stress direction
O_h	E 2_i 3_j^+ 3_j^- 2_p 4_i^+ 4_i^- I $\bar{2}_i$ $\bar{3}_j^+$ $\bar{3}_j^-$ $\bar{2}_p$ $\bar{4}_i^+$ $\bar{4}_i^-$	48	relaxed

4.2 Reciprocal Lattice and First Brillouin Zone

For a given Bravais lattice defined by the vectors **a** a reciprocal lattice can be constructed using the vectors **b**. The basis vectors of the reciprocal lattice $\mathbf{b}_i$ are defined by the relation:

$$\mathbf{b}_i = 2\pi \frac{\mathbf{a}_j \times \mathbf{a}_k}{(\mathbf{a}_1 \times \mathbf{a}_2) \cdot \mathbf{a}_3}, \tag{4.3}$$

where $\mathbf{a}_j \times \mathbf{a}_k$ denotes the vector product, $(\mathbf{a}_2 \times \mathbf{a}_2) \cdot \mathbf{a}_3$ is the volume of the primitive cell, and a cyclic rotation $(i, j, k) \to (k, i, j) \to (j, k, i)$ of the indices is assumed. Thus, the basis reciprocal lattice vectors are:

$$\mathbf{b}_1 = \frac{2\pi}{a_0}\begin{pmatrix} -1 \\ 1 \\ 1 \end{pmatrix}, \quad \mathbf{b}_2 = \frac{2\pi}{a_0}\begin{pmatrix} 1 \\ -1 \\ 1 \end{pmatrix}, \quad \text{and} \quad \mathbf{b}_3 = \frac{2\pi}{a_0}\begin{pmatrix} 1 \\ 1 \\ -1 \end{pmatrix}. \tag{4.4}$$

It follows from (4.4) that

$$\mathbf{b}_i \cdot \mathbf{a}_j = 2\pi \delta_{ij},$$

where δ_{ij} is the Kronecker symbol.

An arbitrary reciprocal lattice vector is written in the form

$$\mathbf{G} = l\mathbf{b}_1 + m\mathbf{b}_2 + n\mathbf{b}_3, \tag{4.5}$$

where l, m, and n are integers.

The symmetry point group of the reciprocal lattice coincides with the symmetry group of the Bravais lattice, however, the type of the reciprocal Bravais lattice in general is not the same as that of the Bravais lattice of the crystal structure.

The first Brillouin zone is the symmetrized Wigner-Seitz cell in the reciprocal lattice. To construct the Brillouin zone, one should connect a chosen point on the reciprocal lattice to all nearest points on the reciprocal lattice and define planes perpendicular to the vectors connecting the points. These planes lie equidistant between the chosen and the connected point. The polyhedron bounded by these planes represents the Brillouin zone. It follows from the construction of the Brillouin zone that there are no pairs of equivalent vectors inside the zone. There exist 24 different types of Brillouin zones.

The first Brillouin zone represents the primitive cell of the reciprocal lattice. The interface boundaries of the first Brillouin in silicon zone are determined by planes that are perpendicular to the reciprocal lattice vectors pointing from the center of the cell to the 14 lattice points nearest to the origin of the cell at their midpoints. These 14 faces are

$$|k_x|+|k_y|+|k_z| = \frac{3}{2}\frac{2\pi}{a_0}, \quad |k_x| = \frac{2\pi}{a_0}, \quad |k_y| = \frac{2\pi}{a_0}, \quad \text{and} \quad |k_z| = \frac{2\pi}{a_0}. \tag{4.6}$$

The first Brillouin zone of silicon is shown in Fig. 4.3.

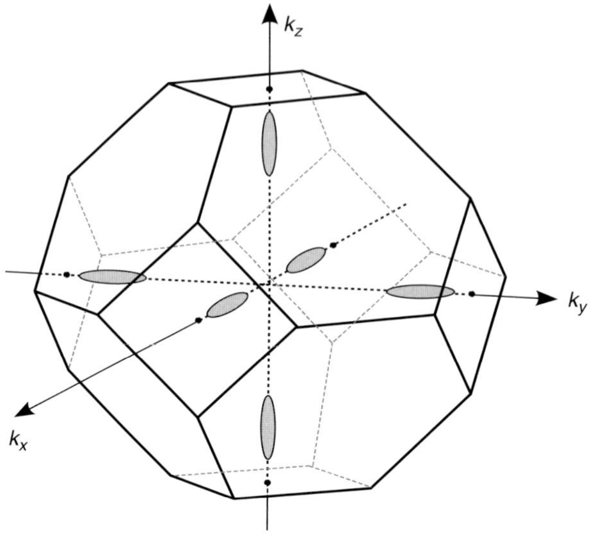

Fig. 4.3 First Brillouin zone of silicon crystal lattice. The six equivalent valleys of the conduction band of silicon along the three principal axes are also shown

Due to the translational invariance of the lattice the energy bands are periodic in the reciprocal space. It is sufficient to know the energy dispersions only in the first Brillouin zone for band structure calculations [5]. The notion of the first Brillouin zone is important to define all linearly independent solutions of the Schrödinger equation with the periodic crystal potential briefly reviewed in the next section.

Let us select a point **k** in the Brillouin zone. For each vector **k** there exists a set of symmetry operations from the space group that leaves **k** unchanged or maps it onto an equivalent vector. This set of elements is usually called the little group. Like the space group, the little group contains an invariant translation subgroup. By factorizing the translation subgroup one can construct the point symmetry group corresponding to the given point **k**. For $\mathbf{k} = 0$ this point symmetry group coincides with the crystallographic point group. In general, the point group for a given **k** is a subgroup of the crystallographic point group characterizing the crystal class. Points and lines of the Brillouin zone can therefore be classified according to their point symmetry groups.

The number of symmetry elements in the point subgroup of the little group depends on the wave vector **k**. Because of the unity operation the symmetry group contains at least one element. If the symmetry group of a given vector **k** contains more elements than the symmetry group of neighboring points, this specific vector **k** is referred to as symmetry point. Thus, points and lines of the Brillouin zone can be classified according to their symmetry.

In Fig. 4.4 the symmetry points (filled dots) and symmetry lines of the relaxed silicon lattice are shown. Strictly speaking, the points K and U are not symmetry points because they have the same symmetry as the points along the symmetry lines Σ and S, however, it is convenient to keep special symbols for these points due to

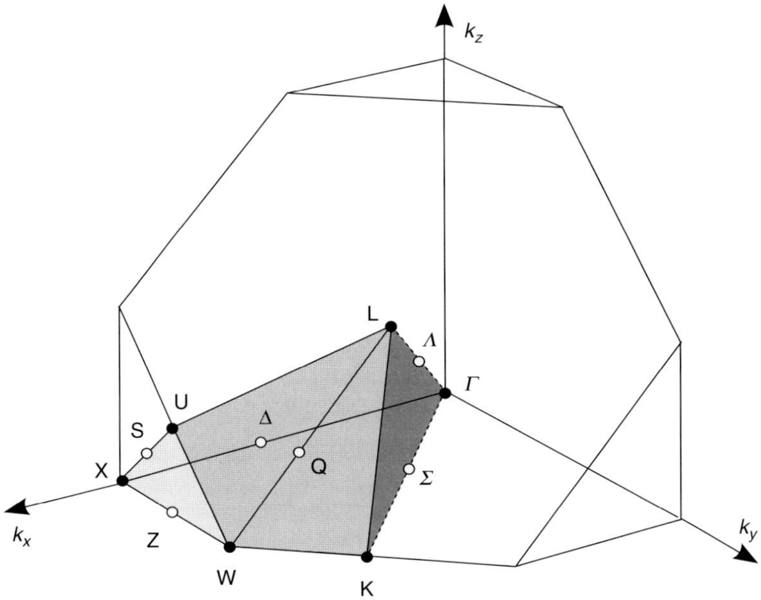

Fig. 4.4 The locations of certain symmetry points and symmetry lines in the first Brillouin zone of silicon crystal lattice

their symmetric positions in the first Brillouin zone. By convention, capital Greek letters are used for points and lines inside the Brillouin zone while capital Roman letters denote points and lines on the surface of the first Brillouin zone.

Since the center point Γ of the Brillouin zone is mapped onto itself at any point operation of the crystal lattice, all symmetry operations of the lattice are included in the point group $P(\Gamma)$. Because of the symmetry, it is not necessary to compute the dispersion relations in the whole Brillouin zone. The group $P(\Gamma)$ determines the volume of a part of the Brillouin zone sufficient to obtain the energy spectrum of electrons in a crystal. This part is called the irreducible wedge. The number of symmetry elements in the symmetry group determines the volume of the irreducible wedge as [5]

$$\Omega_{\mathrm{irred}} = \Omega_{\mathrm{BZ}}/|P(\Gamma)|, \tag{4.7}$$

where $|P(\Gamma)|$ is the number of elements of the symmetry group $P(\Gamma)$.

4.3 Particle in a Periodic Potential

In order to describe the motion of a carrier in a periodic crystal potential $U(\mathbf{r})$ one has to solve the corresponding Schrödinger equation for the wave function $\Psi(\mathbf{r})$:

$$H\Psi(\mathbf{r}) = \left(-\frac{\hbar^2\nabla^2}{2m_0} + U(\mathbf{r})\right)\Psi(\mathbf{r}) = E\Psi(\mathbf{r}), \tag{4.8}$$

where m_0 is the electron mass, $\hbar$ is the Plank constant, and E is the energy. The periodicity of the potential $U(\mathbf{r})$:

$$U(\mathbf{r}+\mathbf{a}) = U(\mathbf{r}), \tag{4.9}$$

where $\mathbf{a}$ is the periodicity vector (4.3), which makes the Schrödinger equation invariant under any translation $\mathbf{r} \rightarrow \mathbf{r}+\mathbf{a}$. It means that the wave function $\Psi(\mathbf{r}+\mathbf{a})$ describes the same state of a particle in a periodic potential as the wave function $\Psi(\mathbf{r})$. Therefore, the wave function $\Psi(\mathbf{r}+\mathbf{a})$ must be proportional to $\Psi(\mathbf{r})$. The proportionality constant can be complex, however, its absolute value must be equal to one, in order to prevent an infinite growth of the wave function at large values of the translation vector $\mathbf{a}$. It is convenient to choose the proportionality constant to be equal to $\exp(i\mathbf{ka})$, where $\mathbf{k}$ is an arbitrary constant vector. Then the wave function $\Psi(\mathbf{r})$ can be written as [3]

$$\Psi_{n\mathbf{k}}(\mathbf{r}) = e^{i\mathbf{kr}} u_{n\mathbf{k}}(\mathbf{r}), \tag{4.10}$$

where $u_{n\mathbf{k}}(\mathbf{r})$ is the periodic function:

$$u_{n\mathbf{k}}(\mathbf{r}+\mathbf{a}) = u_{n\mathbf{k}}(\mathbf{r}). \tag{4.11}$$

A possibility to write the solution of the Schrödinger equation with a periodic potential in the form of a product of a plane wave envelope and a periodic function (4.10) is the result of the Bloch theorem. The solution (4.10) is called the Bloch function, or the Bloch state, while the periodic part $u_{n\mathbf{k}}(\mathbf{r})$ is frequently named as the periodic amplitude of the Bloch function, or the periodic Bloch function.

At a fixed value of $\mathbf{k}$ the Schrödinger equation (4.1) has an infinite set of solutions numbered by an integer index n. The energy $E_n(\mathbf{k})$ corresponding to the solution $u_{n\mathbf{k}}(\mathbf{r})$ also depends on the vector $\mathbf{k}$ and index n. For an n fixed $E_n(\mathbf{k})$ takes the values within a certain interval called the energy band. Different bands may be separated by the energy gaps or may overlap. In the last case there is more than one state characterized by the same energy. If the bands intersection appears at the same value of the $\mathbf{k}$ vector, the bands become degenerate at these points, and their energy dispersions may develop peculiarities close to the intersection points.

Under the translation $\mathbf{r} \rightarrow \mathbf{r}+\mathbf{a}$ the Bloch function transforms as

$$\Psi(\mathbf{r}+\mathbf{a}) = e^{i\mathbf{ka}} \Psi(\mathbf{r}). \tag{4.12}$$

The phase factor can be multiplied by a factor $\exp i\mathbf{Ga} = 1$, where $\mathbf{G}$ is any reciprocal lattice vector, without altering the result. Thus, the $\mathbf{k}$ vector is defined up to a reciprocal lattice vector. However, since the wave function (4.12) is not changed and describes the same state, the Bloch function must be periodic in the reciprocal lattice:

$$\Psi_{n\mathbf{k}+\mathbf{G}}(\mathbf{r}) = \Psi_{n\mathbf{k}}(\mathbf{r}). \tag{4.13}$$

It then follows that the band energy is also a periodic function in the reciprocal lattice:

$$E_n(\mathbf{k} + \mathbf{G}) = E_n(\mathbf{k}). \tag{4.14}$$

Therefore, all non-equivalent values of **k** are lying within the first Brillouin zone.

Apart from the periodicity (4.14), the band energies $E_n(\mathbf{k})$ satisfy additional symmetry relations due to the point group symmetry of the crystal [4, 5]. Thanks to these additional symmetries the band structure calculations in relaxed silicon may only be performed on a $1/48$-th part of the first Brillouin zone. The energy dispersion in the whole Brillouin zone is recovered using the following eight reflections

$$E_n(k_x, k_y, k_z) = E_n(|k_x|, |k_y|, |k_z|) \tag{4.15}$$

and six permutations

$$\begin{aligned} E_n(k_x, k_y, k_z) &= E_n(k_x, k_z, k_y) = E_n(k_y, k_x, k_z) = \\ E_n(k_y, k_z, k_x) &= E_n(k_z, k_x, k_y) = E_n(k_z, k_y, k_x) \,. \end{aligned} \tag{4.16}$$

The irreducible wedge of the fcc lattice is shown in Fig. 4.4. It has six corners

$$\begin{aligned} &\Gamma = \frac{2\pi}{a_0}\begin{pmatrix}0\\0\\0\end{pmatrix}, \quad X = \frac{2\pi}{a_0}\begin{pmatrix}1\\0\\0\end{pmatrix}, \quad L = \frac{2\pi}{a_0}\begin{pmatrix}\frac{1}{2}\\\frac{1}{2}\\\frac{1}{2}\end{pmatrix}, \\ &W = \frac{2\pi}{a_0}\begin{pmatrix}1\\\frac{1}{2}\\0\end{pmatrix}, \quad K = \frac{2\pi}{a_0}\begin{pmatrix}\frac{3}{4}\\\frac{3}{4}\\0\end{pmatrix}, \quad U = \frac{2\pi}{a_0}\begin{pmatrix}1\\\frac{1}{4}\\\frac{1}{4}\end{pmatrix}, \end{aligned} \tag{4.17}$$

connected via symmetry lines

$$\begin{aligned} &\Lambda : \Gamma \to L, \qquad \Delta : \Gamma \to X, \qquad S : X \to U, K, \\ &\Sigma : \Gamma \to U, K, \qquad Q : L \to W, \qquad Z : X \to W. \end{aligned} \tag{4.18}$$

Here, the points U and K can be interchanged, since they are equivalent in the reciprocal space.

Due to the time-reversal symmetry [2]

$$E_{n\uparrow}(\mathbf{k}) = E_{n\downarrow}(-\mathbf{k}), \tag{4.19}$$

where $\uparrow$ and $\downarrow$ stand for the spin projections. In addition, in crystals with a center of inversion the band energies satisfy the relation [2]

$$E_{n\uparrow}(\mathbf{k}) = E_{n\downarrow}(\mathbf{k}). \tag{4.20}$$

The Bloch functions corresponding to different n and **k** within the first Brilloin zone are orthogonal [3]:

$$\int \Psi^*_{n'\mathbf{k}'}(\mathbf{r})\Psi_{n\mathbf{k}}(\mathbf{r})d\mathbf{r} = \int e^{i(\mathbf{k}-\mathbf{k}')\mathbf{r}} u^*_{n'\mathbf{k}'}(\mathbf{r})u_{n\mathbf{k}}(\mathbf{r})d\mathbf{r} = \delta_{nn'}\delta(\mathbf{k}-\mathbf{k}'), \quad (4.21)$$

where * stands for complex conjugate, $\delta(\mathbf{k}-\mathbf{k}')$ is the delta-function. It follows from (4.10) that the periodic Bloch amplitudes $u_{n\mathbf{k}}(\mathbf{r})$ with the same $\mathbf{k}$ are also orthogonal:

$$\int u^*_{n'\mathbf{k}}(\mathbf{r})u_{n\mathbf{k}}(\mathbf{r})dv = \frac{\Omega}{(2\pi)^3}\delta_{nn'}, \quad (4.22)$$

where, because of the periodicity of $u_{n\mathbf{k}}(\mathbf{r})$, the integration is restricted to a unit cell Ω [3].

The equation for the periodic Bloch function amplitude is in the form:

$$\left[-\frac{\hbar^2}{2m_0}(\nabla + i\mathbf{k})^2 + V(\mathbf{r})\right]u_{n\mathbf{k}}(\mathbf{r}) = E_n(\mathbf{k})u_{n\mathbf{k}}(\mathbf{r}). \quad (4.23)$$

References

1. Bir, G.L., Pikus, G.E.: Symmetry and Strain-Induced Effects in Semiconductors. Willey, New York - Toronto (1974)
2. Landau, L., Lifshitz, E.: Statistical Physics: Part 2, Second edition: Volume 9 (Theoretical Physics, Vol. 9). Butterworth-Heinemann, London (1986)
3. Luttinger, J.M., Kohn, W.: Motion of electrons and holes in perturbed periodic fields. Phys. Rev. **97**(4), 869–883 (1955)
4. Nowotny, H.: Theoretische festkörperphysik II (1998). Vorlesungsskript Technische Universität Wien
5. Yu, P., Cardona, M.: Fundamentals of semiconductors. Springer, Berlin (2003)

Chapter 5
Band Structure of Relaxed Silicon

5.1 Conduction and Valence Bands

Intrinsic silicon is a semiconductor. The band gap $E_g = 1.12\,\text{eV}$ at 300 K separates the maximum of the last filled band called valence band from the minimum of the first empty band called conduction band. The minimum of the conduction band is shifted with respect to the maximum of the valence band situated at the Γ point of the first Brillouin zone by the wave vector $K_0 = 0.85\,(2\pi)/a_0$, where $a_0 = 0.54\,\text{nm}$ is the lattice constant of the relaxed lattice of silicon. Because the conduction band minimum is shifted with respect to the Γ-point of location of the valence band maximum, silicon is a semiconductor with an undirect gap.

The conduction band consists of six equivalent minima located symmetrically along the axes [100], [010], and [001] at a distance K_0 from the Γ symmetry point, or at a distance $k_0 = 0.15\,(2\pi)/a_0$ from the corresponding X-point. Close to the minimum of the conduction band the dispersion $E(\mathbf{k})$ is usually described within a parabolic approximation

$$E(\mathbf{k}) = \frac{\hbar^2(k_z - k_0)^2}{2m_l} + \frac{\hbar^2(k_x^2 + k_y^2)}{2m_t}, \tag{5.1}$$

where the masses $m_l = 0.916m_0$ and $m_t = 0.19m_0$ are called longitudinal and transverse effective masses of silicon [1]. To account for a deviation of the measured density of states from a purely parabolic dispersion behavior at higher energies an isotropic non-parabolicity correction is introduced via the expression [1]:

$$E(\mathbf{k})(1 + \alpha E\mathbf{k}) = \frac{\hbar^2(k_z - k_0)^2}{2m_l} + \frac{\hbar^2(k_x^2 + k_y^2)}{2m_t}, \tag{5.2}$$

where the value of the non-parabolicity parameter $\alpha = 0.5\,\text{eV}^{-1}$ is determined phenomenologically. We will demonstrate, however, that the dispersion relations (5.1) and even (5.2) do not describe the modification of the conduction band of silicon due to uniaxial [110] stress, and a more complex description of the conduction band is needed.

V. Sverdlov, *Strain-Induced Effects in Advanced MOSFETs*, Computational Microelectronics, DOI 10.1007/978-3-7091-0382-1_5, © Springer-Verlag/Wien 2011

The three highest valence bands in silicon are the so-called heavy hole band, the light hole band, and the band split due to the spin-orbit interaction. The heavy and light hole bands are degenerate at the maximum situated at the Γ symmetry point of the first Brillouin zone. The maximum of the spin-orbit split band is also at the Γ point and lies $\Delta_{so} = 44\,\mathrm{meV}$ below the maximum of the light and heavy hole bands. Due to the degeneracy between the light and heavy hole bands their dispersions may not be described within a parabolic band approximation due to their non-analytical behavior near the maximum. A six-band k·p theory [19] usually employed to describe the bulk dispersions in the valence band includes the spin-orbit interaction between the three valence bands and is thus a good approximation.

The band structure of a crystal material can be qualitatively evaluated by using symmetry of a crystal [31]. For example, due to the point symmetry of the face-centered lattice the six directions [100], $[\bar{1}00]$, [010], $[0\bar{1}0]$, [001], and $[00\bar{1}]$ are equivalent. Consequently, the six conduction band valleys shown in Fig. 4.3 are equivalent. However, in order to find the correct dispersions, the solution of the Schrödinger equation (4.1) with the periodic crystal potential is required. This is the subject of electronic band structure calculations. The periodic potential consists of the core potential contribution from the ions forming the crystal and the electronic contribution. Both parts represent a challenge in describing the band structure. The core Coulomb potential is singular at the position of an ion. This singularity makes the wave function strongly oscillating close to the core thus requesting for a special stable numerical methods capable to capture such oscillations. The electronic part is the result of action of all the electrons on a particular carrier and must be found self-consistently. It turns out that the so-called correlation corrections to the self-consistent Hartree field are very important in determining the band structure correctly. This is why accurate band structure calculations represent a serious computational challenge even nowadays.

Electronic band structure calculations can be subdivided into the three distinct groups: first-principle, empirical numerical, and perturbative analytical methods. Methods based on the first-principle calculations, or ab-initio, are of great importance because they do not require any phenomenologically adjustable fitting parameters. These methods are based on the density-functional theory.

5.2 First-Principle Band Structure Calculations

The Hohenberg-Kohn theorems establish the ground of the density-functional methods. The first theorem states that the ground state density of a system of interacting particles in an external potential is uniquely determined by the external potential. Alternatively, the profile of the external potential is uniquely determined by the density of the system of interacting particles in the ground state. It means that there is one to one correspondence between the ground state density and the external potential. Therefore, the Hamiltonian describing the many-body system is completely determined by the ground state density. Thus, all the properties of the system in the ground and excited states can in principle be determined provided that the density in the ground state is known.

The second theorem states that the energy of the ground state of a system of interacting particles can be calculated as a minimum of a universal functional for the energy $E[n]$ which depends on the density $n(\mathbf{r})$ [13]. However, the theorem does not provide a recipy on how to build such a functional for a system of interacting particles.

In 1965, Kohn and Sham [16] proposed an approach which has enabled practical electronic structure calculations. The idea of the Kohn-Sham approach is to replace a complicated system of interacting particles with a simpler system that can be solved exactly. The assumption that Kohn and Sham employed is that an auxiliary system consists of an ensemble of non-interacting particles. It is assumed that the ground-state density of the ensemble of non-interacting particles is equal to the density in the original interacting system. By solving the equations for non-interacting particles one finds the density and the ground-state energy of the original system:

$$E[n] = T[n] + V_H[n] + \int V_{ext} n \, d\mathbf{r} + E_{xc}[n], \tag{5.3}$$

where V_{ext} is the external potential,

$$n(\mathbf{r}) = \sum_i^N \psi_i^*(\mathbf{r}) \psi_i(\mathbf{r}) \tag{5.4}$$

is the density on an N-particle system, $\psi_i(\mathbf{r})$ are the Kohn-Sham orbitals, $T[n]$ is the kinetic energy:

$$T[n] = \sum_i^N \int d\mathbf{r} \psi_i^*(\mathbf{r}) \left(-\frac{\nabla^2}{2m_0} \right) \psi_i(\mathbf{r}), \tag{5.5}$$

V_H is the Hartree contribution:

$$V_H[n] = \frac{e^2}{2} \int d\mathbf{r} \int d\mathbf{r}' \frac{n(\mathbf{r}) n(\mathbf{r}')}{|\mathbf{r} - \mathbf{r}'|}, \tag{5.6}$$

e is the electron charge, and $E_{xc}[n]$ is the unknown exchange correlation energy functional.

The Kohn-Sham single particle equations are obtained by varying the energy functional (5.3) with respect to a set of orbitals ψ_i:

$$\left\{ -\frac{\hbar^2}{2m_0} \nabla^2 + V_{\text{eff}}(\mathbf{r}) \right\} \psi_i(\mathbf{r}) = E_i \psi_i(\mathbf{r}), \tag{5.7}$$

where

$$V_{\text{eff}}(\mathbf{r}) = V_{ext}(\mathbf{r}) + V_H(\mathbf{r}) + \frac{\delta E_{xc}[n]}{\delta n(\mathbf{r})}. \tag{5.8}$$

After the equations for the single-particle orbitals are resolved, the wave function of the many-body system is build as a Slater determinant.

In the Kohn and Sham approach, all the difficult many-body terms beyond the external potential and the self-consistent Hartree field are put into the exchange-correlation potential. Therefore, the accuracy to find the exact ground state density and the energy are only limited by the approximations in the exchange-correlation functional. The exchange correlation functional is unknown and can be constructed approximately. The local density approximation or generalized-gradient approximations are quite accurate in describing the ground-state properties. However, the challenge is to develop density-functional-based methods to describe accurately excited state properties. The applicability of local density approximation to excited properties is limited due to several reasons. First, the approximate exchange-correlation potentials are jellium-based, thus, they incorporate an artificial self-interaction. The discontinuity of the exchange-correlation potential with respect to the number of carriers is also neglected. The incomplete cancellation of the self-interaction is the reason why the gap as well as the properties of localized states are incorrectly predicted.

The exact-exchange based density-functional theory in the optimized effective potential approach is free from the self-interaction error and greatly improves the Kohn-Sham energies with respect to the quasi-particle excitations [25]. In spirit of the Kohn and Sham theory, the total energy of the system is divided into known and unknown parts. In addition to the Kohn-Sham approach, the Fock exact-exchange energy contribution is isolated in the exchange-correlation functional:

$$E_X[n] = -\frac{e^2}{2}\sum_{ij}\int d\mathbf{r}\int d\mathbf{r}'\frac{\psi_i^*(\mathbf{r})\psi_j(\mathbf{r})\psi_j^*(\mathbf{r}')\psi_i(\mathbf{r}')}{|\mathbf{r}-\mathbf{r}'|}. \tag{5.9}$$

For the occupied $i = j$ states the exact-exchange term (5.9) cancels exactly the self-interaction contribution from the Hartree energy functional (5.6). Varying the energy functional $E[n] = E_{ext}[n] + V_H[n] + E_X[n]$ with respect an orbital one obtains the set of the Hartree-Fock equations for $\psi_i(\mathbf{r})$:

$$\left\{-\frac{\hbar^2}{2m_0}\nabla^2 + V_{ext}(\mathbf{r}) + V_H(\mathbf{r})\right\}\psi_i(\mathbf{r}) + \int d\mathbf{r}\frac{\Sigma_F(\mathbf{r},\mathbf{r}')}{|\mathbf{r}-\mathbf{r}'|} = E_i\psi_i(\mathbf{r}), \tag{5.10}$$

where $\Sigma_F(\mathbf{r},\mathbf{r}')$ is a non-local operator:

$$\Sigma_F(\mathbf{r},\mathbf{r}') = -\frac{e^2}{2}\sum_i\frac{\psi_i(\mathbf{r})\psi_i^*(\mathbf{r}')}{|\mathbf{r}-\mathbf{r}'|}. \tag{5.11}$$

An effective local exact-exchange potential can be derived by making the variation of (5.9) with respect to the density $n(\mathbf{r})$ and applying an optimized effective potential approach [6, 20]. The local correlations can be easily added into the exact-exchange potential-based density-functional theory by including the local density correlation energy [22] which improves results for the band gap.

To obtain accurate gaps for a large set of solid materials a more advanced many particle method called the GW approximation is currently employed. This method allows to evaluate the self-energy Σ of an interacting system in terms of the single-particle Green's function G and the dynamically screened Coulomb interaction W [10]. Using the self-energy Σ the Green's function can then be evaluated, which allows to address the band structure and the spectral properties of the system. Because in the GW approximation the spectrum of elementary excitations, or quasi-particles, is evaluated the method gives much better results for the gaps of semiconductors and insulators, when it is applied after DFT calculations [25]. The GW approximation is implemented in several packages for ab-initio calculations. We have used the Vienna Ab-nitio Simulation Package (VASP) [29] to obtain the results for the silicon band structure.

5.3 Pseudopotential Band Structure Calculations

The first-principle band structure calculations, especially with the GW approximation included became available only recently and are typically applied to study the band structure of new or unstudied materials with not very well known properties. The band structure of silicon is a subject of research for a long time and is well known experimentally. Several approximate numerical methods to evaluate the band structure have also been developed. The method of empirical pseudopotentials allows to reproduce all the characteristics of the band structure known experimentally including the gap, the spin-orbit split-off energy, the effective masses, the non-parabolicity parameter. The method employs only a relatively small set of parameters which can be calibrated to reproduce the properties of several materials. Below we follow the method of non-local empirical pseudopotentials of Rieger and Vogl [24], which allows to describe the band structure of silicon, germanium, and SiGe.

Pseudopotentials were first introduced by Fermi [9] to investigate high-lying atomic states. The idea of the pseudopotential method of band structure calculations is quite simple. Electrons on a silicon atom can be divided into a group of electrons localized on the ion core and a group of valence electrons involved in binding to nearest silicon atoms. The valence electrons are from the outer partly filled shells of silicon atoms and are nearly free. While moving in the crystal these electrons do not feel the pure Coulomb potential of the ion core. Instead, because the core is dressed by the localized electrons, the valence electrons are subject to the screened core potential. In order to describe the properties of silicon due to these nearly free valence electrons it is enough to approximate in the vicinity of the core the real singular Coulomb potential by a soft effective potential. The soft effective potential changes the behavior of the wave function close to the ion core removing the strong oscillations in the singular Coulomb potential. This allows to resolve the Schrödinger equation numerically. At the same time, the properties of

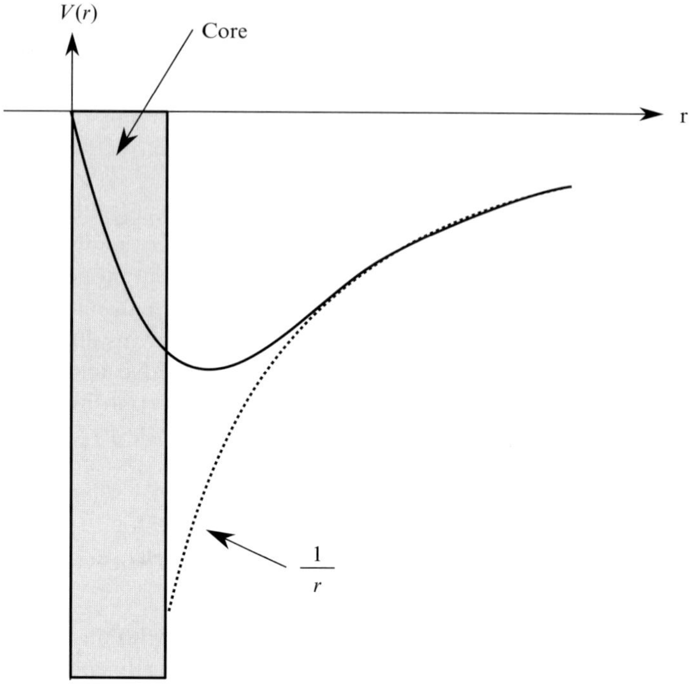

Fig. 5.1 The idea of the pseudopotential method is to replace the Coulomb potential by a smooth pseudopotential within the core region [31]

the nearly free valence and free conduction electrons are determined by the behavior of the wave function outside the ion cores and are therefore not affected by the substitution of the real core potential by a smooth effective potential. The effective potential called the pseudopotential is schematically shown in Fig. 5.1.

In order to construct the pseudopotential, Herring [12] has suggested to approximate the crystal wave function $\psi_{\mathbf{k}}(\mathbf{r})$ of the valence or conducting electrons as a linear combination of a smooth function $\phi_{\mathbf{k}}(\mathbf{r})$ augmented by the core states $\Psi_{j\mathbf{k}}$ localized at the ions:

$$\psi_{\mathbf{k}}(\mathbf{r}) = \phi_{\mathbf{k}}(\mathbf{r}) - \sum_j \langle \Psi_{j\mathbf{k}} | \phi_{\mathbf{k}} \rangle \, \Psi_{j\mathbf{k}}(\mathbf{r}), \tag{5.12}$$

where the coefficients are chosen from the condition that the crystal wave function $\psi_{\mathbf{k}}(\mathbf{r})$ is orthogonal to the wave functions of the localized electrons: $\langle \Psi_{j\mathbf{k}} | \psi_{\mathbf{k}} \rangle = 0$. The smooth wave function $\phi_{\mathbf{k}}(\mathbf{r})$ satisfies the following equation:

$$\left\{ -\frac{\hbar^2}{2m_0}\nabla^2 + V_C(\mathbf{r}) + \sum_j \frac{(E - E_j)\langle \Psi_{j\mathbf{k}} | \phi_{\mathbf{k}} \rangle \, \Psi_{j\mathbf{k}}(\mathbf{r})}{\phi_{\mathbf{k}}} \right\} \phi_{\mathbf{k}}(\mathbf{r}) = E(\mathbf{k})\phi_{\mathbf{k}}(\mathbf{r}), \tag{5.13}$$

where it is taken into account that for the localized states

$$\left\{-\frac{\hbar^2}{2m_0}\nabla^2 + V_C(\mathbf{r})\right\}\Psi_{j\mathbf{k}}(\mathbf{r}) = E_j\Psi_{j\mathbf{k}}. \tag{5.14}$$

According to the cancellation theorem [23], the value of the pseudopotential

$$V(\mathbf{r}) = V_C(\mathbf{r}) + \sum_j \frac{(E - E_j)\langle\Psi_{j\mathbf{k}}|\phi_{\mathbf{k}}\rangle\,\Psi_{j\mathbf{k}}(\mathbf{r})}{\phi_{\mathbf{k}}} \tag{5.15}$$

is small in the core region, while it has the correct Coulomb behavior e^2/r far from an ion. Therefore, (5.13) determines a pseudo-wave function $\phi_{\mathbf{k}}(\mathbf{r})$ which has a smooth behavior in the core region. Thus, the pseudo-wave function is a good approximation to the true wave function outside the core region and can be used, for example, to evaluate transport characteristics which depend on the properties of the valence and conduction electrons. Regardless the fact that the pseudopotential Schrödinger equation is for the pseudo-wave function and not the real crystal function, the eigenenergies in (5.13) correspond to the true energies of the crystal wave function.

The pseudopotential depends on the solutions of (5.13–5.14) and is therefore not known. It can, however, be approximated by a function with a few parameters. The parameters can be adjusted later to reproduce the known characteristics of a semiconductor, like the energy gap, effective masses of electron and holes. A simple but efficient approximation due to Heine and Abarenkov for a pseudopotential of a single atom is to assume it coincides with the correct Coulomb potential outside the core, while the behavior of the pseudopotential is modeled by a constant within the core region. This constant may depend on energy E. The so called empty core model is obtained, when this constant is set to zero, while the constant effective potential model is recovered, when the core potential is equal to the Coulomb potential at the core distance R from the ion.

By its definition (5.15) the crystal pseudopotential depends on angular momenta present in the core states and the energy E [7]. Inspite of the fact that the pseudopotential is non-local, many properties can be found by assuming V to be a simple function of position $\mathbf{r}$. The pseudo-potential can be constructed from the single-ion pseudopotentials $V_0(\mathbf{r})$:

$$V(\mathbf{r}) = \sum_i \left(V_0(\mathbf{r} - \mathbf{R_i} + \boldsymbol{\tau}) + V_0(\mathbf{r} - \mathbf{R_i} - \boldsymbol{\tau})\right), \tag{5.16}$$

where the summation is over the lattice vectors $\mathbf{R_i}$, and $\boldsymbol{\tau} = a/8(1,1,1)$ is the vector determining the positions of the two silicon basis atoms in the unit cell. The pseudopotential is a periodic function and can be expanded into a Fourier series over the reciprocal lattice vectors:

$$V(\mathbf{r}) = \sum_j V(\mathbf{G}_j)\exp(i\mathbf{r}\mathbf{G}_j). \tag{5.17}$$

Importantly, the coefficients $V(\mathbf{G}_j)$ are expressed in terms of the Fourier harmonics of the ion potential $V_0(\mathbf{r})$ alone:

$$V(\mathbf{G}_j) = 2V_0(\mathbf{G}_j)S(\mathbf{G}_j), \tag{5.18}$$

where Ω is the volume of the unit cell,

$$V_0(\mathbf{G}_j) = \frac{1}{\Omega}\int d\mathbf{r} V_0(\mathbf{r}) \exp(-i\mathbf{r}\mathbf{Q}_j),$$

and $S(\mathbf{G})$ is the structure factor of a diamond lattice:

$$S(\mathbf{G}_j) = \frac{1}{2}\left(\exp(i\boldsymbol{\tau} G_j) + \exp(-i\boldsymbol{\tau}\mathbf{G}_j)\right). \tag{5.19}$$

The crystal pseudopotential can also be easily constructed for III-V semiconductors with the zinc-blend lattice crystalline structure. In this case two types of pseudopotentials $V_{\mathrm{III}}(\mathbf{r})$ and $V_{\mathrm{V}}(\mathbf{r})$ for the two sorts of atoms must be introduced. The periodic potential can be again developed into the Fourier series in the form (5.17), where the coefficients $V(\mathbf{G})$ are:

$$V(\mathbf{G}) = \cos(\boldsymbol{\tau} G)V_s(\mathbf{G}) + i\,\sin(\boldsymbol{\tau} G)V_a(\mathbf{G}), \tag{5.20}$$

$V_s = V_{\mathrm{III}} + V_{\mathrm{V}}$ and $V_a = V_{\mathrm{III}} - V_{\mathrm{V}}$ are the symmetric and antisymmetric form factors. In the case of diamond lattice, when $V_{\mathrm{III}} = V_{\mathrm{V}}$, (5.20) coincides with (5.18, 5.19).

Because pseudopotentials do not have singularities within the core region and decay fast outside of the core, they can be considered as weak perturbations. Therefore, the plane wave basis $|\mathbf{k}\rangle = 1/\sqrt{\Omega}\exp(i\mathbf{k}\mathbf{r})$ is suitable to find the pseudo-wave function $\phi_{\mathbf{k}}(\mathbf{r})$:

$$\phi_{\mathbf{k}}(\mathbf{r}) = \sum_i a_i \left|\mathbf{k} + \mathbf{G}_i\right\rangle, \tag{5.21}$$

where the summation is over reciprocal lattice vectors $\mathbf{G}_j$. Substitution of (5.21) into (5.13) results in the following system of homogeneous equations:

$$\left[\frac{(\mathbf{k} + \mathbf{G}_i)^2}{2m_0} - E(\mathbf{k})\right] a_i + \sum_j V(\mathbf{G}_i - \mathbf{G}_j)a_j = 0. \tag{5.22}$$

In order the non-zero solution a_i exists the determinant of the matrix made of the coefficients in (5.22) must be set to zero. The corresponding secular equation determines the eigenenergies $E(\mathbf{k})$ in the crystal:

$$\det\left[\left[\frac{(\mathbf{k} + \mathbf{G}_i)^2}{2m_0} - E(\mathbf{k})\right]\delta_{\mathbf{G}_i,\mathbf{G}_j} + V(\mathbf{G}_i - \mathbf{G}_j)\right] = 0. \tag{5.23}$$

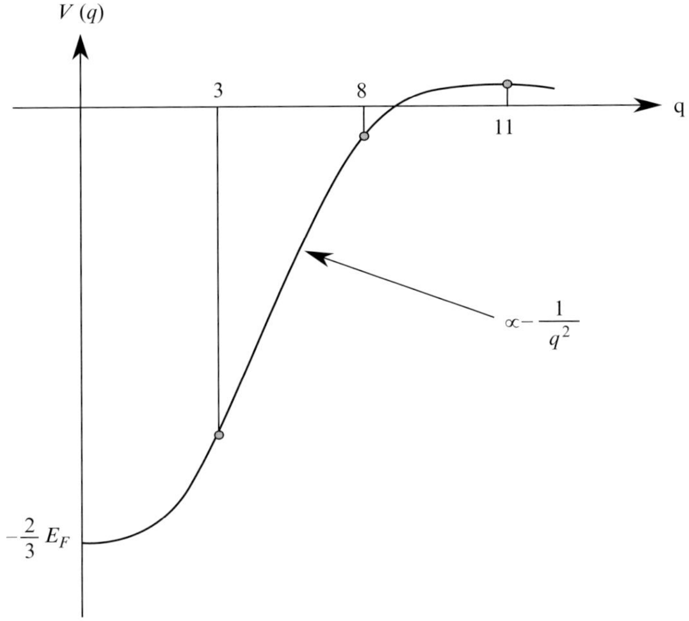

Fig. 5.2 Fourier-transform of the pseudopotential rapidly decreases at large q [31]. Due to its periodicity the pseudopotential is determined by the Fourier-harmonics at $|\mathbf{G}^2| = (2\pi/a)^2 \times (0,3,4,8,11,12,...)$

In the local pseudopotential method the ion pseudopotential is assumed to be spherically symmetric: $V_0(\mathbf{r}) = V_0(|\mathbf{r}|)$. In this case the form-factors depend on the absolute value of the reciprocal lattice vector $\mathbf{G}_i$. This assumption drastically reduces the number of form-factors needed in (5.23). These from-factors can be considered as free parameters which can be adjusted in order to accurately reproduce the band structure. Because of the periodicity of the reciprocal lattice the pseudopotentials are only needed at discrete G points. The square of the distance from the origin to each equivalent set of reciprocal lattice sites is equal to $|\mathbf{G}^2| = (2\pi/a)^2 \times (0, 3, 4, 8, 11, 12, \ldots)$ [31]. Because of the Fourier components $V_0(q)$ decrease fast (typically as $1/q^2$) with q increased, one can neglect the pseudopotential form factors for $G^2 > 11(2\pi/a)^2$, as shown in Fig. 5.2. This is the manifestation of the fact that the pseudopotential is a smooth function of $\mathbf{r}$, thus it can be approximated with only a few Fourier harmonics. For the diamond lattice structure there are only three relevant parameters $V(2\pi\sqrt{3}/a)$, $V(2\pi\sqrt{8}/a)$, and $V(2\pi\sqrt{11}/a)$. The value $V(0)$ gives only a rigid shift in energy and can be set to zero, while the value $V(4\pi/a) = 0$ because of the vanishing structure factor (5.19). The corresponding form-factors for silicon and germanium are listed in the Table 5.1 [24, 27].

As it was pointed out by Chelikowsky and Cohen, the local pseudopotential approximation is not sufficient when describing the properties of holes, namely

Table 5.1 Pseudopotential parameters used in calculations [24, 27] (1 Rydberg = 13.6 eV)

Parameter	Ge	Si	Units
V_3^S	−0.221	−0.2241	Ry
V_8^S	0.019	0.0520	Ry
V_{11}^S	0.056	0.0724	Ry
A_0	0.0	0.03	Ry
A_2	0.275	0.0	Ry
R_l	1.22	1.06	Å
μ_{so}	0.000965	0.00023	Ry
ξ	10.0911	8.0	Å^{-1}

the density of states in the valence band. To correct the behavior, an energy dependent non-local correction term V_{nl} is usually added to the local atomic pseudopotential [7]:

$$V_{nl}(\mathbf{r}, E) = \sum_{l=1}^{\infty} A_l(E) f_l(r) P_l, \tag{5.24}$$

where A_l is the energy-dependent constant, that can also be interpreted as the well depth [7], $f_l(\mathbf{r})$ is the function approximating the core wave function with l symmetry, and P_l is the projection operator for the l angular momentum. Using the series expansion of the plane wave

$$\exp(i\mathbf{qr}) = \sum_{l=0}^{\infty} i^l P_l(\cos\alpha) j_l(qr), \tag{5.25}$$

where $P_l(\cos\alpha)$ are the Legendre polynomials, α is the angle between $\mathbf{q}$ and $\mathbf{r}$, and $j_l(y)$ are the spherical Bessel functions of the first kind, the matrix elements of the non-local pseudopotentials can be written in the following form:

$$\langle \mathbf{k}+\mathbf{G}| |V_{nl}| |\mathbf{k}'+\mathbf{G}\rangle = \frac{4\pi}{\Omega} \sum_{l=0}^{\infty} (2l+1) P_l(\cos\beta) S(\mathbf{k}-\mathbf{k}') \tag{5.26}$$
$$\times \int_0^{\infty} drr^2 f_l(r) j_l(|\mathbf{k}+\mathbf{G}|r) j_l(|\mathbf{k}'+\mathbf{G}|r).$$

It turns out that only the components with $l = 0, 1, 2$ are important. For silicon, the core wave function is approximated by a Heaviside step function ($\Theta(x) = 1$ for $x \geq 0$, $\Theta(x) = 0$ for $x < 0$)

$$f_l(r) = \Theta(R_l - r), \tag{5.27}$$

while for germanium the core wave function is a Gaussian [24]:

$$f_l(r) = \exp\left(-\frac{r^2}{R_l^2}\right). \tag{5.28}$$

For the correct description of the valence band of silicon and germanium, the spin-orbit interaction responsible for the band splitting must be taken into account. The spin-orbit coupling is the result of the interaction of the electron spin $\boldsymbol{\sigma}$ with the magnetic field $\mathbf{B}$ produced by its orbital motion. The spin-orbit Hamiltonian H_{so} is in the form [31]:

$$H_{so} = \frac{\hbar}{4m_0^2c^2}[\nabla V \times \mathbf{p}]\boldsymbol{\sigma}, \tag{5.29}$$

where V is the potential energy, $\mathbf{p} = -i\hbar\nabla$ is the momentum operator, and $\boldsymbol{\sigma} = (\sigma_x, \sigma_y, \sigma_z)^T$ is the vector made out of the Pauli matrices:

$$\boldsymbol{\sigma}_x = \begin{pmatrix} 0 & 1 \\ 1 & 0 \end{pmatrix} \quad \text{and} \quad \boldsymbol{\sigma}_y = \begin{pmatrix} 0 & -i \\ i & 0 \end{pmatrix} \quad \text{and} \quad \boldsymbol{\sigma}_z = \begin{pmatrix} 1 & 0 \\ 0 & -1 \end{pmatrix}. \tag{5.30}$$

Substitution of the Bloch function (4.10) into (5.29) results in the following operator acting on the periodic amplitude of the Bloch function:

$$H_{so} = \frac{\hbar}{4m_0^2c^2}[\nabla V \times (\mathbf{p} + \mathbf{k})]\boldsymbol{\sigma}, \tag{5.31}$$

The term with the crystal momentum $\mathbf{k}$ is small as compared to the term with $\mathbf{p}$. This is because the momentum of the electron on a localized atomic orbital described by the term $< u_n|\mathbf{p}|u_n >$ is much larger than the typical crystal momentum $\mathbf{k}$. This also explains why the spin-orbit interaction is more important for valence electrons which are more localized than the conduction electrons and thus move faster on their orbitals. The matrix elements of the spin-orbit interaction can be presented in the form [24]:

$$V_{so}(\mathbf{K}s, \mathbf{K}'s') = -\frac{i\mu_{so}}{\hbar}\left(\frac{(\Omega/2)^{1/3}}{\pi}\right)^2 S(\mathbf{K} - \mathbf{K}')B(K)B(K')[\mathbf{K} \times \mathbf{K}']\left\langle s|\boldsymbol{\sigma}\left|s'\right.\right\rangle, \tag{5.32}$$

where

$$B(k) = \frac{5 - (K/\xi)^2}{5(1 + (K/\xi)^2)^4},$$

and μ_{so} is the spin-orbit parameter. The values of parameters used in the empirical pseudopotential band structure calculations are summarized in the Table 5.1. For germanium we have used the parameters from [24]. The spin-orbit interaction parameters μ_{so} and ξ for silicon are taken from [27].

In order to extend the empirical pseudopotential method on $Si_{1-x}Ge_x$ alloys the virtual crystal approximation is applied. The form-factors and the parameters of the non-local and spin-orbit interactions are interpolated linearly in the mole fraction x.

The expression for the lattice constant is taken in the following form [24]:

$$a(x) = a_0 + 0.0200326x(1 - x) + (a_{Ge} - a_0)x^2,$$

where $a_{Ge} = 0.5658\,\text{nm}$.

5.4 Semi-Empirical Tight Binding Method

The pseudopotential method is based on the separation of all the electrons into two groups: strongly localized core electrons and nearly or completely delocalized valence or conducting electrons. This allows to use the plane wave basis to describe the band structure by the empirical pseudopotential method. On the contrary, the tight-binding approach explores the band structure from another limiting case and uses the localized atomic orbitals as the basis function. Thus, this approach is expected to give good results for the valence band structure and somewhat less accurate results for the conduction bands. The reason that the two opposite approaches do work reasonably well is that the electrons in silicon are delocalized in the conduction band and are thus well described by the delocalized basis functions while the valence electrons are mostly localized in the bonds and may be described using the localized basis states. The atomic wave functions are usually classified by their orbital momenta. The states with the orbital momentum $l = 0, 1, 2$ are usually designated as s, p, d, correspondingly. Symmetrically orthogonalized atomic orbitals $\phi_l(\mathbf{r}-\mathbf{R})$ centered at the atom position $\mathbf{R}$ called Löwdin orbitals [18] are conveniently used as the pseudo-atomic basis states. They satisfy the one-particle Schrödinger equation with the Hamiltonian $h(\mathbf{r} - \mathbf{R}_i)$:

$$h(\mathbf{r} - \mathbf{R}_i)\phi_l(\mathbf{r} - \mathbf{R}_i) = E_l\phi_l(\mathbf{r} - \mathbf{R}_i). \tag{5.33}$$

The introduction of the Löwdin orbitals allows to get rid of the troublesome overlap integrals between the orbital states $a_l(\mathbf{r}-\mathbf{R})$ of the two atoms located at the positions $\mathbf{R}_i$ and $\mathbf{R}_j$

$$S_{ij}^{(mn)} = \int d\mathbf{r} a_m(\mathbf{r} - \mathbf{R}_i)a_n(\mathbf{r} - \mathbf{R}_j) \tag{5.34}$$

by the following transformation [8]:

$$\phi_l(\mathbf{r} - \mathbf{R}_i) = \sum_{mj} \left(S^{-1/2}\right)_{ij}^{(lm)} a_m(\mathbf{r} - \mathbf{R}_j). \tag{5.35}$$

When the atoms are assembled into the crystal lattice, the total Hamiltonian is the sum of the single atom Hamiltonians $h(\mathbf{r} - \mathbf{R}_i)$ and the term due to the inter-atomic interaction $\delta H(\mathbf{r})$:

$$H(\mathbf{r}) = \sum_i h(\mathbf{r} - \mathbf{R}_i) + \delta H(\mathbf{r}). \tag{5.36}$$

The inter-atomic interaction $\delta H(\mathbf{r})$ is small and can be taken as perturbation. By neglecting $\delta H(\mathbf{r})$ the Bloch function of the crystal is written via the Löwdin orbitals in the following form:

$$\Phi_{m\mathbf{k}}(\mathbf{r}) = |m\mathbf{k}\rangle = \frac{1}{\sqrt{N}} \sum_j \exp(i\mathbf{k}\mathbf{R}_j)\phi_m(\mathbf{r} - \mathbf{R}_j), \tag{5.37}$$

where N is the number of primitive cells. The wave functions (5.37) satisfy the properties

$$\Phi_{m\mathbf{k}}(\mathbf{r} + \mathbf{a}) = e^{i\mathbf{k}\mathbf{a}}\Phi_{m\mathbf{k}}(\mathbf{r}).$$

Using the orthonormality of the pseudoatomic Löwdin orbitals, the orthogonality of the Bloch functions (5.36) is ensured.

The functions (5.37) are used to expand the solution $\Psi_\mathbf{k}$ of the Schrödinger equation with the total Hamiltonian (5.36):

$$\Psi_\mathbf{k}(\mathbf{r}) = \sum_m C_m \Phi_{m\mathbf{k}}(\mathbf{r}). \tag{5.38}$$

This leads to the following equation for the energy dispersion $E(\mathbf{k})$

$$\det[\mathbf{H} - E(\mathbf{k})] = 0. \tag{5.39}$$

The matrix elements of the Hamiltonian from (4.8) computed with the basis functions (5.36) are [26]:

$$H_{ln}(\mathbf{k}) = \langle l\mathbf{k}|\, H \,|n\mathbf{k}\rangle = \sum_j \exp(i(\mathbf{R}_j\mathbf{k}) \int d\mathbf{r}\phi_l^*(\mathbf{r})H\phi_n(\mathbf{r} - \mathbf{R}_j). \tag{5.40}$$

The overlap integrals

$$I_{ln}(\mathbf{R}_j) = \int d\mathbf{r}\phi_l^*(\mathbf{r})H\phi_n(\mathbf{r} - \mathbf{R}_j) \tag{5.41}$$

are the parameters of the method. At $\mathbf{R}_j = 0$ they describe the on-site orbital energies ($l = n$) and hopping parameters between different orbitals at the same atom $l \neq n$, while for $\mathbf{R}_j \neq 0$ they represent hopping parameters for electrons between different sites. In order to reduce the number of parameters, several approximations are typically applied.

First, only a few orbital functions are included. The name of the model is typically related to the number of orbital functions preserved. The most commonly used nowadays $sp^3d^5s^*$ model employs one s, three p, five d, and one “excited” s^* orbital. The dimensionality of the matrix $\mathbf{H}$ in (5.36) is determined by the number of orbitals included. In case of several different basis atoms in the primitive cell the dimension of the matrix is correspondingly augmented [30].

The second simplification is based on a strong localization of the Löwdin orbitals. The overlap integrals decay rapidly with the distance between the atoms increased, which allows to restrict the summation in (5.38) only to the nearest neighbors or the next nearest neighbors.

Several semi-empirical tight-binding models are available in the literature, with different number of orbitals and order of neighbors included: sp^3s^* with the spin-orbit coupling [4], the second-near-neighbor spin-orbit model [2, 3], second- and third-near-neighbor without spin-orbit sp^3 model [11,17]. The next-nearest neighbor $sp^3d^5s^*$ model [5, 14] accurately reproduces the effective masses and gaps at Γ, L, and X symmetry points of the Brillouin zone. The next-nearest-neighbor model is easier adapted to strained nanostructures [15, 21]. The model was recently calibrated to accurately reproduce the valence band of silicon and germanium [5]. The parameters used are from [5] and are summarized in Table 5.2.

5.5 Comparison Between Different Numerical Methods

We briefly compare the numerical methods of band structure calculations. Fig. 5.3 demonstrates the band structure results obtained with the empirical pseudopotential method (EPM) and the $sp^3d^5s^*$ tight-binding method. For pseudopotentials we

Table 5.2 The parameters of the $sp^3d^5s^*$ model in Slater-Koster notation [26] for Si and Ge from [5] used in band structure calculations. λ is the strength of the spin-orbit coupling. All the values are in eV

Parameter	Si	Ge
E_s	−2.15168	−1.95617
E_p	4.22925	5.30970
E_{s^*}	19.11650	19.29600
E_d	13.78950	13.58060
λ	0.01989	0.10132
$ss\sigma$	−1.95933	−1.39456
$s^*s^*\sigma$	−4.2435	−3.56680
$ss^*\sigma$	−1.52230	−2.01830
$sp\sigma$	3.02562	2.73135
$s^*p\sigma$	3.15565	2.68638
$pp\sigma$	4.10364	4.28921
$pp\pi$	−1.51801	−1.73707
$pd\sigma$	−1.35554	−2.00115
$pd\pi$	2.38479	2.10953
$dd\sigma$	−1.68136	−1.32941
$dd\pi$	2.58880	2.56261
$dd\delta$	−1.81400	−1.95120

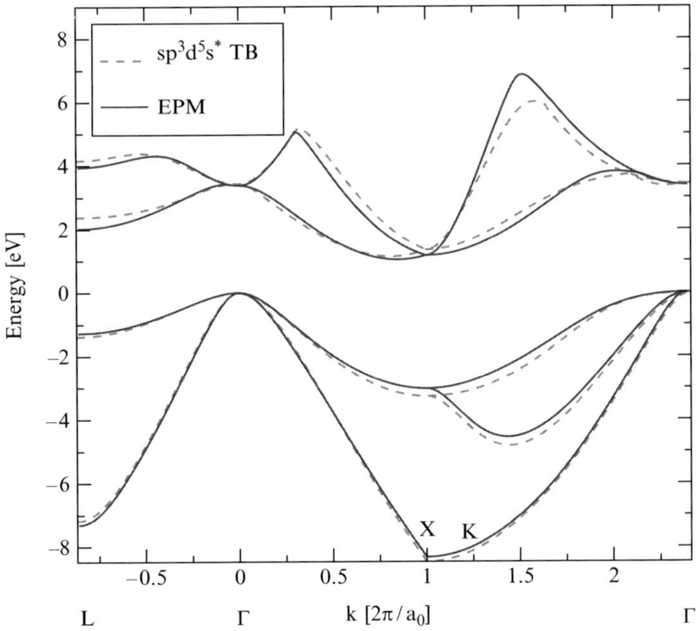

Fig. 5.3 Band structure from the EPM (*solid*) and from the $sp^3d^5s^*$ model (*dashed*)

use the parameters from [24]. The parameters of the pseudopotentials were chosen to reproduce the measurable quantities of silicon, like the energy gap and the effective masses. In addition, the spin-orbit coupling was included and calibrated to obtain the split-off band correctly [27] as explained in the subsection describing the pseudopotential method. Results of the band structure calculations obtained with the empirical pseudopotential method and $sp^3d^5s^*$ method with the parameters from [5] are compared in Fig. 5.3. It demonstrates that both methods agree reasonably well.

The valence band of silicon was reproduced reasonably well with a much simpler sp^3s^* tight-binding model [8]. Recent development of a more sophisticated $sp^3d^5s^*$ model [5] containing additional parameters has further improved the reproducibility of the valence band. It also allowed to obtain the correct effective masses for the conduction band of relaxed silicon. However, the conduction band minimum in the $sp^3d^5s^*$ model is further away from the X point than in the empirical pseudopotential calculations, where the valley minimum is located at the distance $k_0 = 0.15\frac{2\pi}{a_0}$ from the X point ($0.85\frac{2\pi}{a_0}$ from the Γ symmetry point). This leads to an almost two times larger gap between the two lowest conduction bands at the valley minima (Fig. 5.4) compared to $\Delta = 0.53$ eV found from the EPM. Because of this the $sp^3d^5s^*$ tight-binding model slightly underestimates the anisotropy of the conduction band. The effect is clearly visible in Fig. 5.5, which demonstrates that the EPM gives a more pronounced conduction band warping than

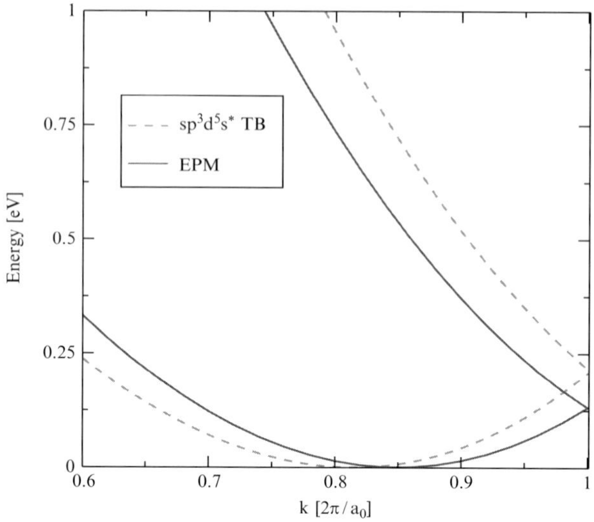

Fig. 5.4 Conduction bands close to the valley minimum from the EPM (*solid*) and from the $sp^3d^5s^*$ model (*dashed*)

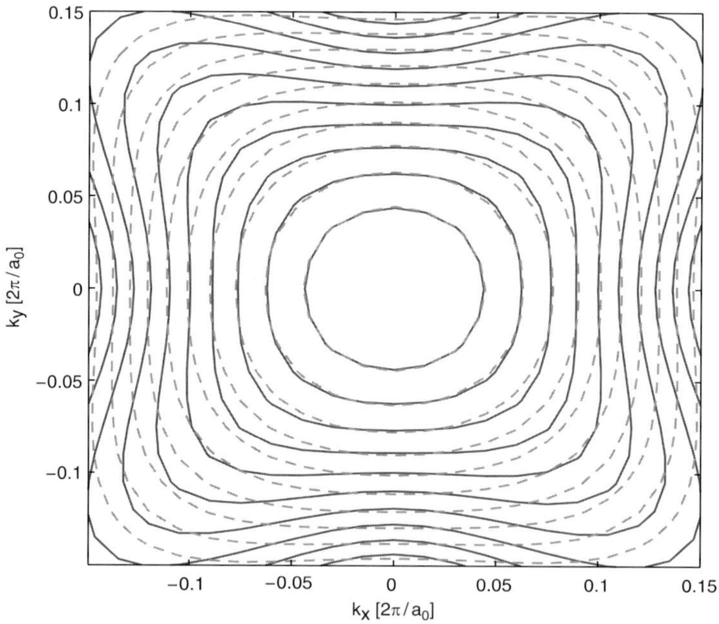

Fig. 5.5 Comparison between the k_x, k_y energy dispersion relations at the minimum k_0. The contour lines are spaced every 50 meV. *Solid lines* correspond to the EPM and the *dashed lines* to the $sp^3d^5s^*$ model

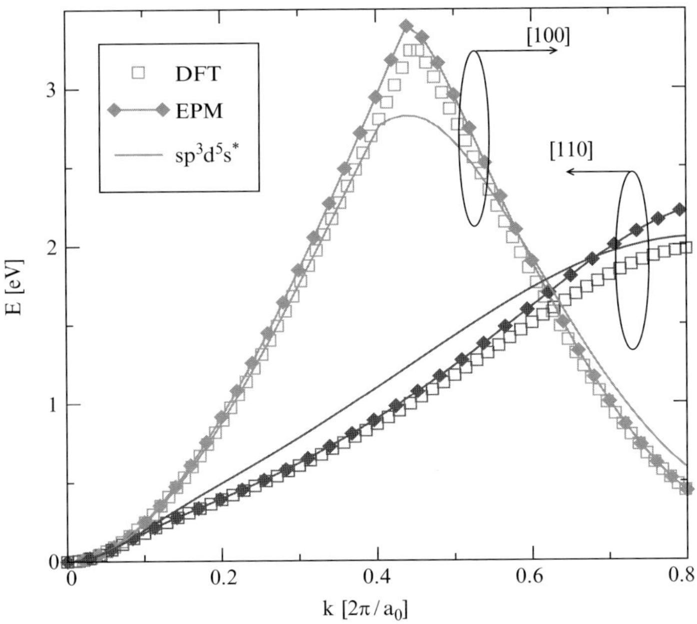

Fig. 5.6 Comparison of bulk dispersion relations close at the minimum of the [001] valleys of the conduction band in [100] and [110] directions. DFT [29] and EPM [28] results are similar, while the $sp^3d^5s^*$ tight-binding model [5] slightly underestimates anisotropy

the $sp^3d^5s^*$ tight-binding model. As follows from Fig. 5.6, the EPM results are the most accurate, when compared to the first principle density-functional band structure calculations obtained with the VASP [29].

References

1. Ando, T., Fowler, A.B., Stern, F.: Electronic properties of two-dimensional systems. Rev. Mod. Phys. **54**(2), 437–672 (1982)
2. Boykin, T.B.: Improved fits of the effective masses at γ in the spin-orbit, second-nearest-neighbor sp^3s^* model: Results from analytic expressions. Phys. Rev. B **56**(15), 9613–9618 (1997)
3. Boykin, T.B., Gamble, L.J., Klimeck, G., Bowen, R.C.: Valence-band warping in tight-binding models. Phys. Rev. B **59**(11), 7301–7304 (1999)
4. Boykin, T.B., Klimeck, G., Bowen, R.C., Lake, R.: Effective-mass reproducibility of the nearest-neighbor sp^3s^* models: Analytic results. Phys. Rev. B **56**(7), 4102–4107 (1997)
5. Boykin, T.B., Klimeck, G., Oyafuso, F.: Valence band effective-mass expressions in the $sp^3d^5s^*$ empirical tight-binding model applied to a Si and Ge parametrization. Phys. Rev. B **69**(11), 115201 (2004)
6. Casida, M.E.: Generalization of the optimized-effective-potential model to include electron correlation: A variational derivation of the sham-schlüter equation for the exact exchange-correlation potential. Phys. Rev. A **51**(3), 2005–2013 (1995)

7. Chelikowsky, J.R., Cohen, M.L.: Nonlocal pseudopotential calculations for the electronic structure of eleven diamond and zinc-blende semiconductors. Phys. Rev. B **14**(2), 556–582 (1976)
8. Datta, S.: Quantum transport: Atom to transistor. Cambridge University Press, Cambridge (2005)
9. Fermi, E.: Sullo spostamento per pressionedei termini elevati delle serie spettrali. Nuovo Cimento **11**, 157–166 (1934)
10. Hedin, L.: New method for calculating the one-particle green's function with application to the electron-gas problem. Phys. Rev. **139**(3A), A796–A823 (1965)
11. Helmholz, D., Lew Yan Voon, L.C.: Warping in the valence band of silicon. Phys. Rev. B **65**(23), 233,204 (2002). DOI 10.1103/PhysRevB.65.233204
12. Herring, C.: A new method for calculating wave functions in crystals. Phys. Rev. **57**(12), 1169–1177 (1940). DOI 10.1103/PhysRev.57.1169
13. Hohenberg, P., Kohn, W.: Inhomogeneous electron gas. Phys. Rev. **136**(3B), B864–B871 (1964)
14. Jancu, J.M., Scholz, R., beltram, F., Bassani, F.: Empirical $spds^*$ tight-binding calculation for cubic semiconductors: General method and material parameters. Phys. Rev. B **57**(11), 6493–6507 (1998)
15. Jancu, J.M., Voisin, P.: Tetragonal and trigonal deformations in zinc-blende semiconductors: A tight-binding point of view. Phys. Rev. B **76**(11), 115202 (2007)
16. Kohn, W., Sham, L.J.: Self-consistent equations including exchange and correlation effects. Phys. Rev. **140**(4A), 1133–1138 (1965)
17. Loehr, J.P., Talwar, D.N.: Exact parameter relations and effective masses within sp^3s zinc-blende tight-binding models. Phys. Rev. B **55**(7), 4353–4359 (1997)
18. Löwdin, P.O., Shull, H.: Natural orbitals in the quantum theory of two-electron systems. Phys. Rev. **101**(6), 1730–1739 (1956)
19. Luttinger, J.M., Kohn, W.: Motion of electrons and holes in perturbed periodic fields. Phys. Rev. **97**(4), 869–883 (1955)
20. Niquet, Y.M., Fuchs, M., Gonze, X.: Asymptotic behavior of the exchange-correlation potentials from the linear-response Sham–Schlüter equation. J. Chem. Phys. **118**(21), 9504–9518 (2003)
21. Niquet, Y.M., Rideau, D., Tavernier, C., Jaouen, H., Blase, X.: Onsite matrix elements of the tight-binding hamiltonian of a strained crystal: Application to silicon, germanium, and their alloys. Phys. Rev. B **79**(24), 245,201 (2009)
22. Perdew, J.P., Zunger, A.: Self-interaction correction to density-functional approximations for many-electron systems. Phys. Rev. B **23**(10), 5048–5079 (1981)
23. Phillips, J.C., Kleinman, L.: New method for calculating wave functions in crystals and molecules. Phys. Rev. **116**(2), 287–294 (1959)
24. Rieger, M.M., Vogl, P.: Electronic-band parameters in strained $Si_{1-x}Ge_x$ alloys on $Si_{1-y}Ge_y$ substrates. Phys. Rev. B **48**(19), 14,276–14,287 (1993)
25. Rinke, P., Qteish, A., Neugebauer, J., Scheffler, M.: Exciting prospects for solids: Exact-exchange based functionals meet quasiparticle energy calculations. Physica Status Solidy (B) **245**(5), 929–945 (2008)
26. Slater, J.C., Koster, G.F.: Simplified LCAO method for the periodic potential problem. Phys. Rev. **94**(6), 1498–1524 (1954)
27. Ungersboeck, E.: Numerische Berechnung der Bandstruktur von Halbleitern. Master's Thesis, Institute for Microelectronics, TU Wien (2002)
28. Ungersboeck, E., Dhar, S., Karlowatz, G., Sverdlov, V., Kosina, H., Selberherr, S.: The effect of general strain on band structure and electron mobility of silicon. IEEE Trans. Electron Devices **54**(9), 2183–2190 (2007)
29. VASP: Vienna Ab-initio Simulation Program. Kresse, G., Hafner, J.: Phys. Rev. B **47**, 558 (1993); *ibid.* B **49**, 14251 (1994); Kresse, G., Fertmueller, J.: Phys. Rev. B **54**, 11169 (1996); Computs. Mat. Sci. **6**, 15 (1996)
30. Vogl, P., Hjalmarson, H.P., Dow, J.D.: A semi-empirical tight-binding theory of the electronic structure of semiconductors. J. Phys. Chem. Solids **44**(5), 365–378 (1983)
31. Yu, P., Cardona, M.: Fundamentals of Semiconductors. Springer, Berlin (2003)

Chapter 6
Perturbative Methods for Band Structure Calculations in Silicon

6.1 The k·p Method for a Non-Degenerate Band

The methods of computing the band structure considered in the previous chapter 5 are intrinsically numerical. In some cases the analytical expressions for the band structure close to the extremum points are needed. The **k·p** method is based on the perturbative approach and allows to obtain the analytical band structure close to a chosen point provided the eigenenergies and eigenfunctions at this point are known. The idea of the method was formulated in the fundamental work by Luttinger and Kohn [11].

According to the Bloch theorem, the solution of the Schrödinger equation with the periodic potential $V(\mathbf{r})$

$$H\phi(\mathbf{r}) = \left(\frac{p^2}{2m_0} + V(\mathbf{r})\right)\phi(\mathbf{r}) = E\phi(\mathbf{r}) \tag{6.1}$$

is in the following form:

$$\psi_{n\mathbf{k}}(\mathbf{r}) = e^{i\mathbf{k}\cdot\mathbf{r}}u_{n\mathbf{k}}(\mathbf{r}). \tag{6.2}$$

where n is the band index, $\mathbf{k}$ is the wave vector in the first Brillouin zone, and $u_{n\mathbf{k}}(\mathbf{r})$ is the periodic Bloch amplitude. The wave functions (6.2) form the complete orthogonal basis set. This set is not unique. As it has been demonstrated by Luttinger and Kohn in their seminal work [11], the functions

$$\xi_{n\mathbf{k}}(\mathbf{r}) = e^{i(\mathbf{k}-\mathbf{k}_0)\cdot\mathbf{r}}\psi_{n\mathbf{k}_0}(\mathbf{r}). \tag{6.3}$$

also represent a complete orthonormal basis set provided the set of the functions $\psi_{n\mathbf{k}}(\mathbf{r})$ is orthonormal. Substituting the wave functions (6.3) into (6.1), one arrives to the following equation:

$$\left(\frac{p^2}{2m_0} + V(\mathbf{r}) + \frac{\hbar(\mathbf{k}-\mathbf{k}_0)\cdot\mathbf{p}}{m_0}\right)\psi_{n\mathbf{k}_0} = \left(E_{n\mathbf{k}} - \frac{\hbar^2(\mathbf{k}-\mathbf{k}_0)^2}{2m_0}\right)\psi_{n\mathbf{k}_0}. \tag{6.4}$$

V. Sverdlov, *Strain-Induced Effects in Advanced MOSFETs*, Computational Microelectronics, DOI 10.1007/978-3-7091-0382-1_6, © Springer-Verlag/Wien 2011

Assuming that the solution of the (6.4) at $\mathbf{k} = \mathbf{k}_0$ is known,

$$H_0 \psi_{n\mathbf{k}_0} = \left(\frac{p^2}{2m} + V(\mathbf{r}) \right) \psi_{n\mathbf{k}_0} = E_{n\mathbf{k}_0} \psi_{n\mathbf{k}_0}, \tag{6.5}$$

the eigenenergy $E_{n\mathbf{k}}$ in the vicinity of $E_{n\mathbf{k}_0}$ can be found by treating the term

$$H_1 = \frac{\hbar(\mathbf{k} - \mathbf{k}_0) \cdot \mathbf{p}}{m_0}$$

as a small perturbation to the unperturbed Hamiltonian H_0 defined in (6.5).

At the Γ symmetry point $\mathbf{k}_0 = (0, 0, 0)$, and (6.4) is recast into a more familiar form

$$\left(\frac{p^2}{2m_0} + \frac{\hbar \mathbf{k} \cdot \mathbf{p}}{m_0} + V(\mathbf{r}) \right) u_{n\mathbf{k}_0} = \left(E_{n\mathbf{k}} - \frac{\hbar^2 \mathbf{k}^2}{2m_0} \right) u_{n\mathbf{k}_0}. \tag{6.6}$$

Assuming that the solution of the (6.4) for $\mathbf{k}_0 = (0, 0, 0)$

$$\left(\frac{p^2}{2m} + V(\mathbf{r}) \right) u_{n0} = E_{n0} u_{n0}, \tag{6.7}$$

is known, the solution of (6.6) can be obtained by developing a perturbation expansion over $\frac{\hbar \mathbf{k} \cdot \mathbf{p}}{m_0}$. It is due to this term that the method has the name $\mathbf{k} \cdot \mathbf{p}$, or "k dot p" theory. The type of the quantum mechanical perturbation theory to apply depends on whether the bands for which one would like to find the expressions are degenerate at the $\mathbf{k}_0 = (0, 0, 0)$ point or not.

6.2 Effective Mass Theory for Non-Degenerate Bands

Let us assume that the band n has an extremum (a minimum or a maximum) at the point $\mathbf{k} = 0$. A generalization to the case when the extremum is at $\mathbf{k} = \mathbf{k}_0$ is straightforward [11]. Let us also assume that the E_{n0} is non-degenerate and is well separated by energy gaps from other bands at the point $\mathbf{k} = 0$. Treating the $\frac{\hbar \mathbf{k} \cdot \mathbf{p}}{m}$ term as a perturbation, we find the energy in the first order of the perturbation theory:

$$E_{n\mathbf{k}} = E_{n0} + \frac{\hbar}{m_0} \langle u_{n0} | \, \mathbf{k} \cdot \mathbf{p} \, | u_{n0} \rangle \,. \tag{6.8}$$

If the point $\mathbf{k} = \mathbf{k}_0$ is an extremum (a maximum or a minimum), then the first order correction (6.8) to the energy is zero,

$$E_{n\mathbf{k}} = E_{n0}, \tag{6.9}$$

and one has to proceed to the second order perturbation theory. To do so, the first order perturbation theory for the wave function is applied:

$$u_{n\mathbf{k}} = u_{n0} + \frac{\hbar}{m_0} \sum_{n' \neq n} \frac{\langle u_{n0} | \, \mathbf{k} \cdot \mathbf{p} \, | u_{n'0} \rangle}{E_{n0} - E_{n'0}} u_{n'0}. \tag{6.10}$$

The expression for the wave function (6.10) is used to obtain the energy correction due to the $\frac{\hbar \mathbf{k} \cdot \mathbf{p}}{m}$ term in the second order perturbation theory:

$$E_{n\mathbf{k}} = E_{n0} + \frac{\hbar^2 k^2}{2m_0} + \frac{\hbar^2}{m_0^2} \sum_{n' \neq n} \frac{| \langle u_{n0} | \, \mathbf{k} \cdot \mathbf{p} \, | u_{n'0} \rangle |^2}{E_{n0} - E_{n'0}} u_{n'0}. \tag{6.11}$$

By defining an inverse effective mass tensor m_{ij}^{-1} as the derivative

$$m_{ij}^{-1} = \frac{\partial^2 E_{n\mathbf{k}}}{\partial k_i \partial k_j}$$

one gets

$$m_{ij}^{-1} = \frac{1}{m_0} \delta_{ij} + \frac{2}{m_0^2} \sum_{n \neq n'} \frac{\langle u_{n0} | \, p_i \, | u_{n'0} \rangle \langle u_{n'0} | \, p_j \, | u_{n0} \rangle}{E_{n0} - E_{n'0}}. \tag{6.12}$$

The energy dispersion (6.11) can be recast into

$$E_{n\mathbf{k}} = E_{n0} + \frac{1}{2} \sum_{ij} \hbar^2 k_i m_{ij}^{-1} k_j = \frac{\hbar^2}{2} \mathbf{k}^T \hat{m}^{-1} \mathbf{k} \tag{6.13}$$

where $\hat{m}^{-1}$ is the effective mass tensor. It follows from the expression for the effective mass that the bigger the energy gap between the bands the smaller is their contribution into the effective mass. It should be pointed out that all non-zero matrix elements $\langle u_{n\mathbf{k}_0} | \mathbf{k} \cdot \mathbf{p} | u_{n'\mathbf{k}_0} \rangle$ contributing to the effective mass can be found from the matrix element theorem [16] by symmetry group considerations. The dispersion relation (6.13) is valid for the energies

$$|E_{n\mathbf{k}} - E_{n0}| \ll \min |E_{n0} - E_{n'0}|, \quad n \neq n'. \tag{6.14}$$

If the condition (6.14) is violated, the non-degenerate perturbation theory fails and must be replaced by its degenerate version. An example of the description of the band structure in silicon based on the degenerate perturbation theory will be presented below.

6.2.1 Electron Effective Mass in Relaxed Silicon

The electron band structure consists of six equivalent valleys shown in Fig. 4.3. The conduction band minima of silicon reside on the $\langle 001 \rangle$ axes at a distance of $0.15\frac{2\pi}{a_0}$ from the X symmetry points of the first Brillouin zone. The lowest conduction band Δ_1 is well separated from the nearest second conduction band $\Delta_{2'}$ by the energy $\Delta = 0.53\,eV$. Thus, the non-degenerate perturbation theory can be applied to find the dispersion of the lowest conduction band. Close to the minima the dispersion is parabolic. The eigenvalues $E_{n\mathbf{k}}$ are:

$$E_{n\mathbf{k}} = E_{n\mathbf{k}_0} + \frac{\hbar^2}{2m_0}\sum_{i,j} k_i\,\delta_{ij}\,k_j + \frac{\hbar^2}{m_0^2}\sum_{ij}\sum_{n'\neq n} k_i \frac{\langle u_{n\mathbf{k}_0}|\,p_i\,|u_{n'\mathbf{k}_0}\rangle\langle u_{n'\mathbf{k}_0}|\,p_j\,|u_{n\mathbf{k}_0}\rangle}{E_{n\mathbf{k}_0} - E_{n'\mathbf{k}_0}} k_j, \quad (6.15)$$

where

$$\langle u_{n\mathbf{k}_0}|\,p_j\,|u_{n'\mathbf{k}_0}\rangle = \frac{1}{\Omega}\int_{\Omega} u_{n\mathbf{k}_0}\frac{\hbar}{i}\frac{\partial}{\partial x_j}u_{n'\mathbf{k}_0}\,d\mathbf{r}\ . \quad (6.16)$$

Because $E_{n,\mathbf{k}_0}$ is the minimum, the linear terms in k_i is zero. The expression for the effective mass tensor $m^*_{n,ij}$ is:

$$\frac{1}{m^*_{n,ij}} = \frac{1}{m_0} + \frac{2}{m_0^2}\sum_{n'\neq n}\frac{\langle u_{n\mathbf{k}_0}|\,p_i\,|u_{n'\mathbf{k}_0}\rangle\langle u_{n'\mathbf{k}_0}|\,p_j\,|u_{n\mathbf{k}_0}\rangle}{E_{n\mathbf{k}_0} - E_{n'\mathbf{k}_0}}. \quad (6.17)$$

The effective mass tensor for the lowest conduction band Δ_1 is symmetric and can be diagonalized. In the principal coordinate system the tensor is diagonal and is characterized by the two transversal masses m_t and the longitudinal mass m_l. For the [001] valleys, or the valleys along the z axis the expressions for the effective masses are:

$$\frac{1}{m_l} = \frac{1}{m_0} + \frac{2}{m_0^2}\sum_{n'\neq\Delta_1}\frac{|\langle u_{\Delta_1\mathbf{k}_0}|p_z|u_{n'\mathbf{k}_0}\rangle|^2}{E_{\Delta_1\mathbf{k}_0} - E_{n'\mathbf{k}_0}} \quad (6.18)$$

and

$$\frac{1}{m_t} = \frac{1}{m_0} + \frac{2}{m_0^2}\sum_{n'\neq\Delta_1}\frac{|\langle u_{\Delta_1\mathbf{k}_0}|p_x|u_{n'\mathbf{k}_0}\rangle|^2}{E_{\Delta_1\mathbf{k}_0} - E_{n'\mathbf{k}_0}}. \quad (6.19)$$

Here Δ_1 denotes the band index n of the lowest conduction band. Therfore, the energy dispersion can be written as:

$$E(\mathbf{k}) = \frac{\hbar^2\,(k_z - k_{\min})^2}{2m_l} + \frac{\hbar^2\,(k_x^2 + k_y^2)}{2m_t}. \quad (6.20)$$

The relation (6.20) coincides with the well-known effective mass approximation for the energy dispersion of electrons in the conduction band of silicon (5.1) around the minimum located at $k_{min} = k_0$ To evaluate the values of the effective masses, the knowledge of the Bloch functions and the band energies at the minimum point is required. These values, and thus the masses, can be computed by the band structure calculation methods described in the previous chapter. The values of the effective masses in silicon are well known experimentally. Therefore the masses $m_l = 0.91m_0$ and $m_l = 0.196m_0$ can be used to calibrate the parameters of semi-empirical band structure calculation methods.

6.2.2 Approximations for the Conduction Band Dispersion at Higher Energies

At higher energies the deviation from the pure parabolic dispersion appears. In order to take into account deviations in the density of states from that given by the parabolic dispersion a more general dispersion similar to (5.2) including band non-parabolicity is typically used for the [001] valleys [9]:

$$E(\mathbf{k})(1 + \alpha E(\mathbf{k})) = \frac{\hbar^2 (k_z - k_{\min})^2}{2m_l} + \frac{\hbar^2 \left(k_x^2 + k_y^2\right)}{2m_t}. \tag{6.21}$$

The value of the non-parabolicity parameter $\alpha = 0.5\,\mathrm{eV}^{-1}$ is determined experimentally. However, the non-parabolic dispersion (6.21) is isotropic in (001) plane. For this reason it can only predict the modification of the density of states and can not describe the conduction band warping. As it was recently pointed out in [17], a more general description is needed to describe the subband structure correctly in (110) oriented ultra-thin silicon films. The conduction band model must take into account an anisotropy of the dispersion of the [001] valleys in the (001) plane. To derive a more accurate dispersion relation for the conduction band, the second conduction band $\Delta_{2'}$ ($i = 2$) closest to the first conduction band Δ_1 ($i = 1$) must be taken into account. We consider the pair of equivalent conduction band valleys along the [001] direction. Other valleys can be analyzed analogously.

The two conduction bands Δ_1 and $\Delta_{2'}$ become degenerate exactly at the X point. Since the minimum of the conduction band is only $k_0 = 0.15\frac{2\pi}{a_0}$ away from the X point, the dispersion around the minimum can be well described by the degenerate perturbation theory, which includes the two bands degenerate at the X point. Diagonal elements of the Hamiltonian $H_{ii}, i = 1, 2$ at the X point include the contribution of all the remote bands and can be easily obtained using the standard **k·p** theory [2]:

$$H_{ii}^0(k) = (-1)^{i-1}\frac{\hbar}{m_0}k_z p + \frac{\hbar^2 k_z^2}{2m_l} + \frac{\hbar^2 k_x^2}{2m_t} + \frac{\hbar^2 k_y^2}{2m_t}, \tag{6.22}$$

where m_0 is the free electron mass, m_t is the transversal, and m_l is the longitudinal effective mass. Let us point out that because the X point is not an extremum, the slope of the dispersion is non-zero. The sign of the matrix elements $(p_z)_{ii}$ is different for the band $i = 1$ and $i = 2$: $p = (p_z)_{11} = -(p_z)_{22}$. The values of k_z are counted from the X point.

In contrast to the 14 band model [7] or the recently developed 30 bands **k·p** model [14], our perturbation analysis allows to get excellent results with only two bands. This is because our approach is based on the perturbation theory around the X symmetry point, contrary to the perturbation methods [7, 14] developed around the Γ symmetry point which is far away from the conduction band minimum.

By taking into account the diagonal elements (6.22) alone, we recover the commonly used effective mass approximation for the conduction band (the linear term vanishes at the minimum $k_z = -k_0$). Indeed, by using the identity

$$\frac{p}{m_0} = \frac{k_0}{m_l}, \tag{6.23}$$

the diagonal elements (6.22) describe the two parabolic bands with the minima at the distance $\pm k_0$ symmetrically situated around the X point. The bands are degenerate at the X point provided the coupling between them is zero.

Within the second-order perturbation theory the coupling between the bands is described by the off-diagonal terms:

$$H_{12}^0(k) = \frac{\hbar^2 k_x k_y}{M}. \tag{6.24}$$

The parameter M is evaluated from the **k·p** perturbation theory as [8]:

$$\frac{1}{M} = \frac{2}{m_0^2} \left| \sum_{l \neq 1,2} \frac{(p_x)_{1l}(p_y)_{l2}}{E_k(X) - E_{\Delta_1}(X)} \right|.$$

Its value is close (but not equal) to

$$\frac{1}{M} \approx \frac{1}{m_t} - \frac{1}{m_0}.$$

Assuming the coupling is small and taking into account that the decoupled bands are degenerate, we use degenerate perturbation theory to obtain the following dispersion relation close to the minimum at $k_z = -k_0$:

$$\begin{aligned} E_0(\mathbf{k}) = {} & \frac{\hbar^2(\delta k_z)^2}{2m_l} + \frac{\hbar^2(k_x^2 + k_y^2)}{2m_t} \\ & - \frac{\Delta}{2}\left(\left[1 + \left(\frac{2\hbar^2 k_x k_y}{M\Delta}\right)^2\right]^{1/2} - 1\right). \end{aligned} \tag{6.25}$$

Here $\delta k_z = k_z + k_0$, $\Delta = 2\hbar k_0 p/m_0$ is the gap between the Δ_1 and the $\Delta_{2'}$ conduction bands at $k_z = -k_0$.

In Fig. 6.1 this analytical expression (dotted contour lines) is compared to the numerical band structure obtained from the empirical pseudo-potential method (EPM) for $k_z = -k_0$. Excellent agreement is found up to an energy of 0.5 eV. Figures 6.1–6.2 demonstrate that in [100] or [010] directions, when either k_x or k_y are zero, the dispersion remains parabolic. However, as soon as both k_x and k_y are non-zero, coupling between the bands appears which leads to non-parabolic correction in the dispersion. The non-parabolicity is strongest along the [110] and $[\bar{1}10]$ directions as anticipated in [17].

Let us now evaluate the value of the non-parabolicity parameter α in (6.21) based on the dispersion (6.25). Proceeding exactly as in [9], Appendix B, we arrive to the value $\alpha = 0.64\,\text{eV}^{-1}$, which is in reasonable agreement with the experimental value $\alpha = 0.5\,\text{eV}^{-1}$.

Finally, we would like to demonstrate that an underestimation of the conduction band warping in $sp^3d^5s^*$ tight-binding model [3] is related to the gap between the conduction bands at the minimum position k_0. As indicated in Fig. 5.4, the gap

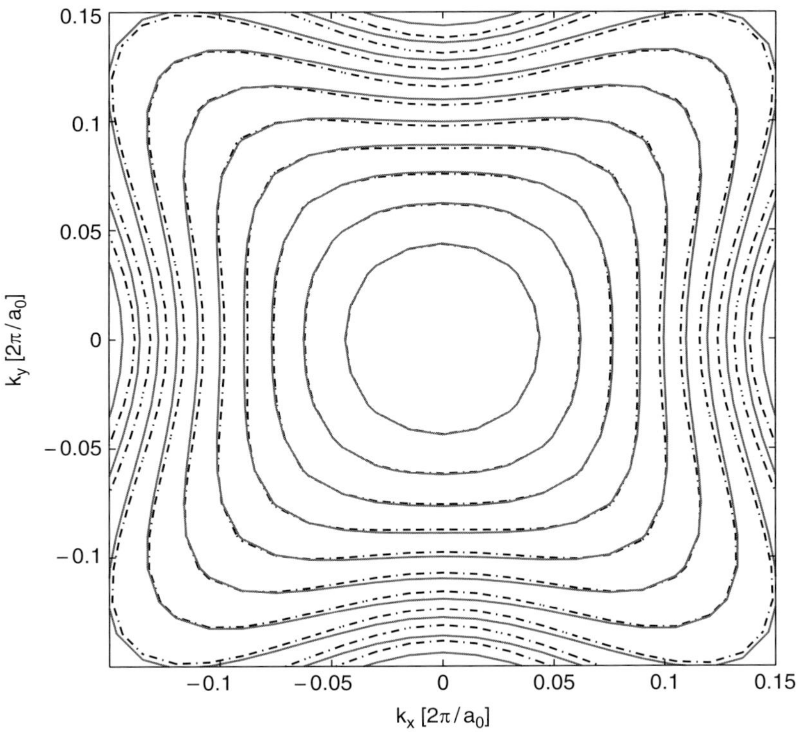

Fig. 6.1 Comparison of the dispersion relation (6.25) at the valley minimum (*dashed-dotted contour lines*) with the EPM results (*solid lines*). The distance between the equi-energy contour lines is 50 meV

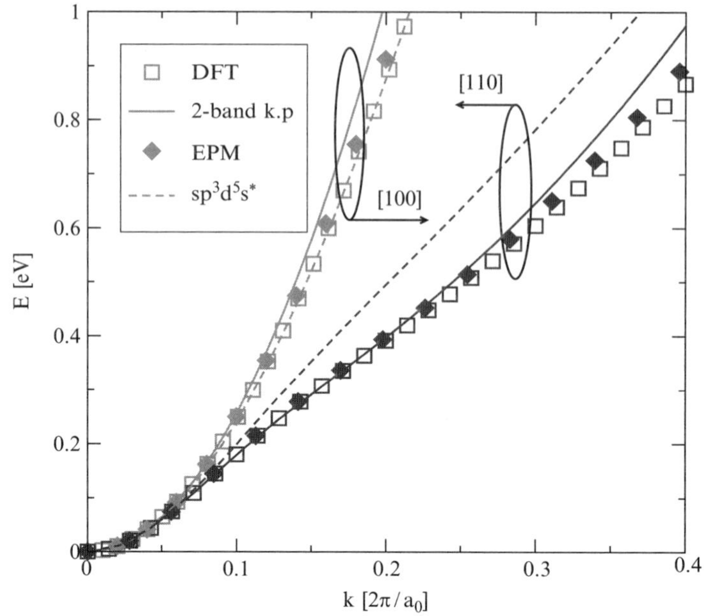

Fig. 6.2 Comparison of bulk dispersion relations close at the minimum of the [001] valleys of the conduction band in [100] and [110] directions. DFT [19], EPM [18], and **k·p** results are in good agreement. The $sp^3d^5s^*$ tight-binding model [3] slightly underestimates anisotropy

between the two lowest conduction bands predicted by the $sp^3d^5s^*$ model [3] is unrealistically large. This results in a smaller coupling between the bands. The solid contour lines shown in Fig. 6.3 obtained from (6.25) with an unrealistic value of $\Delta = 1.2\,\text{eV}$ reproduce the results of the tight-binding model. This confirms the observation that the larger gap between the two bands at the valley minimum results in less anisotropy of the conduction band.

6.3 Valence Band

In the **k·p** theory of the valence band in silicon the three top valence bands are typically considered. These are the heavy hole, light hole, and the split-off valence bands, which will be numbered by the index $n = 1, 2, 3$. Without spin–orbit interaction taken into account the three bands are degenerate at the Γ symmetry point. Therefore, the degenerate version of the **k·p** theory analogous to that for the conduction band at the X point must be applied. The Bloch function at the point **k** is built as a linear combination of the Bloch amplitudes of the three valence bands at the Γ point $\mathbf{k} = 0$:

$$\exp(i\mathbf{kr})u_{n\mathbf{k}} = \exp(i\mathbf{kr}) \sum_{n=1}^{3} C_n^{(0)} u_{n0} \tag{6.26}$$

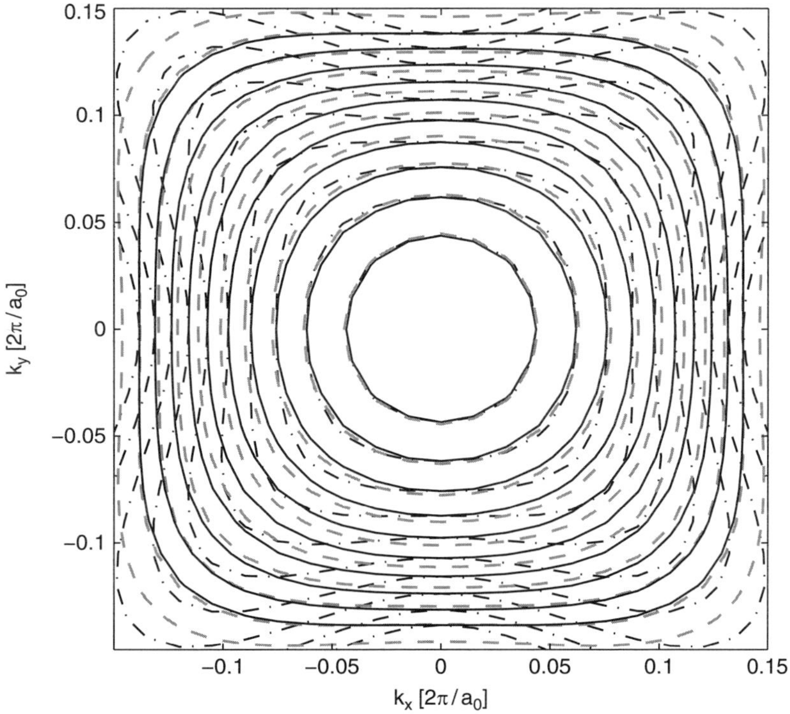

Fig. 6.3 Dispersion at the valley minimum obtained from the $sp^3d^5s^*$ model (*dashed contour lines*), from (6.25) with the correct EPM value $\Delta = 0.53\,\text{eV}$ (*dotted-dashed contours*), and from (6.25) with $\Delta = 1.2\,\text{eV}$ (*solid lines*). The distance between the equi-energy contour lines is 50 meV

Considering the term $H_1 = (\hbar/m_0)\mathbf{k}\cdot\mathbf{p}$ as a perturbation, we obtain the following equation for the coefficients $C_n^{(0)}$:

$$\sum_{n'=1}^{3}[D_{nn'}^{\alpha\beta}k_\alpha k_\beta - E_{\mathbf{k}}^{(2)}\delta_{nn'}]C_{n'}^{(0)} = 0 \qquad n = 1, 2, 3. \tag{6.27}$$

The equation has a non-trivial solution only if

$$\det|D_{nn'}^{\alpha\beta}k_\alpha k_\beta - E_{\mathbf{k}}^{(2)}\delta_{nn'}| = 0. \tag{6.28}$$

Here

$$D_{nn'}^{\alpha\beta} = \frac{\hbar^2}{m_0^2}\sum_{l\neq n}\frac{\langle u_{n0}|\, p_\alpha \,|u_{l0}\rangle\,\langle u_{l0}|\, p_\beta \,|u_{n'0}\rangle}{E_{n0} - E_{l0}}, \tag{6.29}$$

where the summation over remote bands $l \neq n$ is performed. The Greek indeces α and $\beta = 1, 2, 3$ stand for the coordinates x, y, z in the principal coordinate system.

In cubic crystals $D_{nn'}^{\alpha\beta}$ can be parameterized by the three parameters usually called L, M, N:

$$L = \frac{\hbar^2}{m_0} \sum_l \frac{\langle u_{10}| p_x |l_0\rangle \langle l_0| p_x |u_{10}\rangle}{E_{n0} - E_{l0}}, \tag{6.30}$$

$$M = \frac{\hbar^2}{m_0} \sum_l \frac{\langle u_{10}| p_y |l_0\rangle \langle l_0| p_y |u_{10}\rangle}{E_{n0} - E_{l0}}, \tag{6.31}$$

$$N = \frac{\hbar^2}{m_0} \sum_l \frac{\langle u_{10}| p_x |l_0\rangle \langle l_0| p_y |u_{20}\rangle + \langle u_{10}| p_y |l_0\rangle \langle l_0| p_x |u_{20}\rangle}{E_{n0} - E_{l0}}. \tag{6.32}$$

Using these parameters the **k·p** Hamiltonian is written in the form:

$$H_{3\times3} = \begin{pmatrix} \frac{\hbar^2}{2m_0} + Lk_x^2 + M(k_y^2+k_z^2) & Nk_xk_y & Nk_xk_z \\ Nk_xk_y & \frac{\hbar^2}{2m_0} + Lk_y^2 + M(k_x^2+k_z^2) & Nk_yk_z \\ Nk_xk_z & Nk_yk_z & \frac{\hbar^2}{2m_0} + Lk_z^2 + M(k_x^2+k_y^2) \end{pmatrix}. \tag{6.33}$$

The symmetry of the three Bloch amplitudes u_{n0} in diamond structures is of p-type [20]. This means that these wave functions are made by the p-type atomic orbitals with the eigenvalue $L = 1$ of the angular momentum operator **L**. The three Bloch wave functions u_{n0} at the Γ symmetry point are called in the next as $|e_1\rangle$, $|e_2\rangle$ and $|e_3\rangle$. Their symmetry properties are used to construct the spin–orbit Hamiltonian discussed in the next section.

6.3.1 Spin–Orbit Coupling in the Valence Band

The spin–orbit interaction is a relativistic effect important for carriers with high velocities. Since the conduction band states are constructed from the outer atomic orbitals where electron velocities are not large, the spin–orbit interaction is usually neglected for the conducting electrons. In contrast, the valence band is made out of internal orbitals positioned closer to the nucleus and thus characterized by relatively high carriers velocity. In order to describe the valence band correctly, the spin–orbit interaction must be taken into account properly.

The Hamiltonian of spin–orbit interaction is usually expressed as [20]

$$H_{SO} = \lambda \mathbf{LS}, \tag{6.34}$$

where **S** is the spin and **L** is the angular momentum operator, and λ is the constant of the spin–orbit coupling. Defining the operator of the total angular moment $\mathbf{J} = \mathbf{L} + \mathbf{S}$, one can demonstrate that

$$\mathbf{J}^2 = (\mathbf{L} + \mathbf{S})^2 = \mathbf{L}^2 + \mathbf{S}^2 + 2\mathbf{L} \cdot \mathbf{S},$$

from where we find the operator $\mathbf{L} \cdot \mathbf{S}$ as

$$\mathbf{L} \cdot \mathbf{S} = \frac{1}{2}\left(\mathbf{J}^2 - \mathbf{L}^2 - \mathbf{S}^2\right) \tag{6.35}$$

The eigenfunctions of the spin–orbit operator are the eigenfunctions of L^2, S^2, and the total angular momentum square J^2. These functions are different from the wave functions $|e_i\rangle$, which are the eigenfunctions of the orbital angular momentum.

One can extend the basis set $|e_i\rangle$ by including the spin degree of freedom $\sigma = \uparrow, \downarrow$. The Hamiltonian (6.33) written in the basis $|e_j, \sigma\rangle$ including spin is block-diagonal:

$$H_{6\times 6} = \begin{pmatrix} H_{3\times 3} & 0 \\ 0 & H_{3\times 3} \end{pmatrix}. \tag{6.36}$$

However, the functions of the extended set $|e_i, \sigma\rangle$ are the eigenfunctions of the L^2, L_z, S^2, and S_z operators. These functions are related to the eigenfunctions $|J, J_z\rangle$ of L^2, S^2, the total angular momentum square J^2, and its projection J_z by a linear transformation. In our case $s = 1/2$, $l = 1$, and the quantum number of the total angular momentum takes two values: $j = l + s$ and $j = l - s$. The eigenvalues of the projection operator J_z are equal to $j_z = -j, -j+1, ..., j-1, j$. The eigenvalue of the operator of spin–orbit interaction in the basis $|J, J_z\rangle$ are:

$$\lambda \mathbf{L} \cdot \mathbf{S}\,|J, J_z\rangle = \frac{\lambda \hbar^2}{2}\left[j(j+1) - l(l+1) - s(s+1)\right]|J, J_z\rangle\,. \tag{6.37}$$

Thus the Hamiltonian of the spin–orbit interaction is diagonal, with the elements

$$\begin{aligned} \left\langle \frac{3}{2}, J_i \middle| H_{so} \middle| \frac{3}{2}, J_j \right\rangle &= \frac{\Delta_{so}}{3}\delta_{ij}, \\ \left\langle \frac{1}{2}, J_i \middle| H_{so} \middle| \frac{1}{2}, J_j \right\rangle &= -\frac{2\Delta_{so}}{3}\delta_{ij}, \end{aligned} \tag{6.38}$$

where the value of the spin–orbit splitting $\Delta_{so} = 3\lambda$ was introduced. The values of the spin–orbit splitting for silicon, germanium, and several III-V semiconductors are collected in Table 6.1.

The coefficients of the linear transformation from the basis $|e_j, \sigma\rangle$ to the basis $|J, J_z\rangle$ are the Clebsch and Gordon coefficients with $s = 1/2$ and $l = 1$. Thus the transformation is in the form [12]:

Table 6.1 Spin–orbit splitting parameters [20]

Semiconductor	$\Delta_{so}(eV)$
Si	0.044
Ge	0.295
GaAs	0.341
InP	0.11
InAs	0.38

$$\begin{aligned}
\left|\tfrac{3}{2},\tfrac{3}{2}\right\rangle &= \tfrac{-1}{\sqrt{2}}\left(|e_1,\uparrow\rangle + i\,|e_2,\uparrow\rangle\right),\\
\left|\tfrac{3}{2},\tfrac{1}{2}\right\rangle &= = \tfrac{1}{\sqrt{6}}\left(-|e_1,\downarrow\rangle + i\,|e_2,\downarrow\rangle + 2\,|e_3,\uparrow\rangle\right),\\
\left|\tfrac{3}{2},-\tfrac{1}{2}\right\rangle &= \tfrac{1}{\sqrt{6}}\left(|e_1,\uparrow\rangle - i\,|e_2,\uparrow\rangle + 2\,|e_3,\downarrow\rangle\right)\\
\left|\tfrac{3}{2},-\tfrac{3}{2}\right\rangle &= \tfrac{1}{\sqrt{2}}\left(|e_1,\downarrow\rangle - i\,|e_2,\downarrow\rangle\right),\\
\left|\tfrac{1}{2},\tfrac{1}{2}\right\rangle &= \tfrac{-1}{\sqrt{3}}\left(|e_1,\downarrow\rangle + i\,|e_2,\downarrow\rangle + |e_3,\uparrow\rangle\right),\\
\left|\tfrac{1}{2},-\tfrac{1}{2}\right\rangle &= \tfrac{-1}{\sqrt{3}}\left(|e_1,\uparrow\rangle - i\,|e_2,\uparrow\rangle\;|e_3\downarrow\rangle\right).
\end{aligned} \tag{6.39}$$

This allows to write the Hamiltonian of the spin–orbit interaction in the basis $|e_i,\sigma\rangle$ [12]:

$$H_{so} = -\frac{\Delta_{so}}{3} - \begin{array}{c} \\ |e_1,\uparrow\rangle \\ |e_1,\downarrow\rangle \\ |e_2,\uparrow\rangle \\ |e_2,\downarrow\rangle \\ |e_3,\uparrow\rangle \\ |e_3,\downarrow\rangle \end{array} \begin{array}{c} \begin{array}{cccccc} |e_1,\uparrow\rangle & |e_1,\downarrow\rangle & |e_2,\uparrow\rangle & |e_2,\downarrow\rangle & |e_3,\uparrow\rangle & |e_3,\downarrow\rangle \end{array} \\ \begin{pmatrix} 0 & i & 0 & 0 & 0 & -1 \\ -i & 0 & 0 & 0 & 0 & i \\ 0 & 0 & 0 & 1 & -i & 0 \\ 0 & 0 & 1 & 0 & -i & 0 \\ 0 & 0 & i & i & 0 & 0 \\ -1 & -i & 0 & 0 & 0 & 0 \end{pmatrix} \end{array} \tag{6.40}$$

There exist different conventions for the phase factors in defining the coefficients in $|J, J_z\rangle$ as compared to (6.39). The convention for the phase factors used by Luttinger and Kohn [10, 11] and Rodriguez [15] and Chao [5, 6] relative to [12] are summarized in Table 6.2. Different phase factors do not of course affect the physical results.

The inverse transformation expressing $|e_j,\sigma\rangle$ via $|J, J_z\rangle$ is:

$$\begin{aligned}
|e_1,\uparrow\rangle &= \frac{1}{\sqrt{2}}\left(-\left|\frac{3}{2},\frac{3}{2}\right\rangle + \frac{1}{\sqrt{3}}\left|\frac{3}{2},-\frac{1}{2}\right\rangle - \sqrt{\frac{2}{3}}\left|\frac{1}{2},-\frac{1}{2}\right\rangle\right)\\
|e_1,\downarrow\rangle &= \frac{1}{\sqrt{2}}\left(-\frac{1}{\sqrt{3}}\left|\frac{3}{2},\frac{1}{2}\right\rangle - \sqrt{\frac{2}{3}}\left|\frac{1}{2},\frac{1}{2}\right\rangle + \left|\frac{3}{2},-\frac{3}{2}\right\rangle\right)\\
|e_2,\uparrow\rangle &= \frac{i}{\sqrt{2}}\left(\left|\frac{3}{2},\frac{3}{2}\right\rangle + \frac{1}{\sqrt{3}}\left|\frac{3}{2},-\frac{1}{2}\right\rangle - \sqrt{\frac{2}{3}}\left|\frac{1}{2},-\frac{1}{2}\right\rangle\right)
\end{aligned}$$

Table 6.2 The convention for the phase factors used by Luttinger and Kohn [10, 11] and Rodriguez [15] and Chao [5, 6]

Wave function	$\left\|\frac{3}{2},\frac{3}{2}\right\rangle$	$\left\|\frac{3}{2},\frac{1}{2}\right\rangle$	$\left\|\frac{3}{2},\frac{-1}{2}\right\rangle$	$\left\|\frac{3}{2},\frac{-3}{2}\right\rangle$	$\left\|\frac{1}{2},\frac{1}{2}\right\rangle$	$\left\|\frac{1}{2},\frac{-1}{2}\right\rangle$
Luttinger-Kohn	-1	$-i$	1	i	-1	i
Rodriguez-Chao	1	1	1	1	-1	-1

$$|e_2,\downarrow\rangle = \frac{i}{\sqrt{2}}\left(\frac{1}{\sqrt{3}}\left|\frac{3}{2},\frac{1}{2}\right\rangle + \sqrt{\frac{2}{3}}\left|\frac{1}{2},\frac{1}{2}\right\rangle + \left|\frac{3}{2},-\frac{3}{2}\right\rangle\right)$$
$$|e_3,\uparrow\rangle = \sqrt{\frac{2}{3}}\left|\frac{3}{2},\frac{1}{2}\right\rangle - \frac{1}{\sqrt{3}}\left|\frac{1}{2},\frac{1}{2}\right\rangle$$
$$|e_3,\downarrow\rangle = \sqrt{\frac{2}{3}}\left|\frac{3}{2},-\frac{1}{2}\right\rangle + \frac{1}{\sqrt{3}}\left|\frac{1}{2},-\frac{1}{2}\right\rangle \tag{6.41}$$

The total **k** · **p** Hamiltonian including spin–orbit interaction is usually written in the basis $|J, J_z\rangle$ [12]:

$$H = -\begin{array}{c} \\ \left|\frac{3}{2},\frac{3}{2}\right\rangle \\ \left|\frac{3}{2},\frac{1}{2}\right\rangle \\ \left|\frac{3}{2},-\frac{1}{2}\right\rangle \\ \left|\frac{3}{2},-\frac{3}{2}\right\rangle \\ \left|\frac{1}{2},\frac{1}{2}\right\rangle \\ \left|\frac{1}{2},-\frac{1}{2}\right\rangle \end{array}
\begin{array}{c} \begin{array}{cccccc} \left|\frac{3}{2},\frac{3}{2}\right\rangle & \left|\frac{3}{2},\frac{1}{2}\right\rangle & \left|\frac{3}{2},-\frac{1}{2}\right\rangle & \left|\frac{3}{2},-\frac{3}{2}\right\rangle & \left|\frac{1}{2},\frac{1}{2}\right\rangle & \left|\frac{1}{2},-\frac{1}{2}\right\rangle \end{array} \\
\begin{pmatrix}
P+Q & -S & R & 0 & \frac{-1}{\sqrt{2}}S & \sqrt{2}R \\
-S^\dagger & P-Q & 0 & R & -\sqrt{2}Q & \sqrt{\frac{3}{2}}S \\
R^\dagger & 0 & P-Q & S & \sqrt{\frac{3}{2}}S^\dagger & \sqrt{2}Q \\
0 & R^\dagger & S^\dagger & P+Q & -\sqrt{2}R^\dagger & \frac{-1}{\sqrt{2}}S^\dagger \\
\frac{-1}{\sqrt{2}}S^\dagger & -\sqrt{2}Q & \sqrt{\frac{3}{2}}S & -\sqrt{2}R & P+\Delta & 0 \\
\sqrt{2}R^\dagger & \sqrt{\frac{3}{2}}S^\dagger & \sqrt{2}Q & \frac{-1}{\sqrt{2}}S & 0 & P+\Delta
\end{pmatrix} \end{array} \tag{6.42}$$

Here

$$P(\mathbf{k}) = \frac{k^2}{3}\left[(L+2M) + \frac{3\hbar^2}{2m_0}\right] \tag{6.43}$$
$$Q(\mathbf{k}) = \frac{1}{6}\left[(L-M)(k_x^2+k_y^2-2k_z^2)\right] \tag{6.44}$$
$$S(\mathbf{k}) = \frac{-N}{\sqrt{3}}(k_x - ik_y)k_z \tag{6.45}$$
$$R(\mathbf{k}) = \frac{-1}{2\sqrt{3}}\left[(M-L)(k_x^2-k_y^2) + i2Nk_xk_y\right] \tag{6.46}$$

6.3.2 Dispersion of the Valence Band in Silicon

Because the point symmetry group of the diamond lattice contains inversion, the Hamiltonian (6.42) is doubly degenerate, and the dispersion relations do not depend on the spin projection. The Hamiltonian (6.42) can be presented in the form [12]:

$$H = \begin{pmatrix} B^+ & 0 \\ 0 & B^- \end{pmatrix}, \tag{6.47}$$

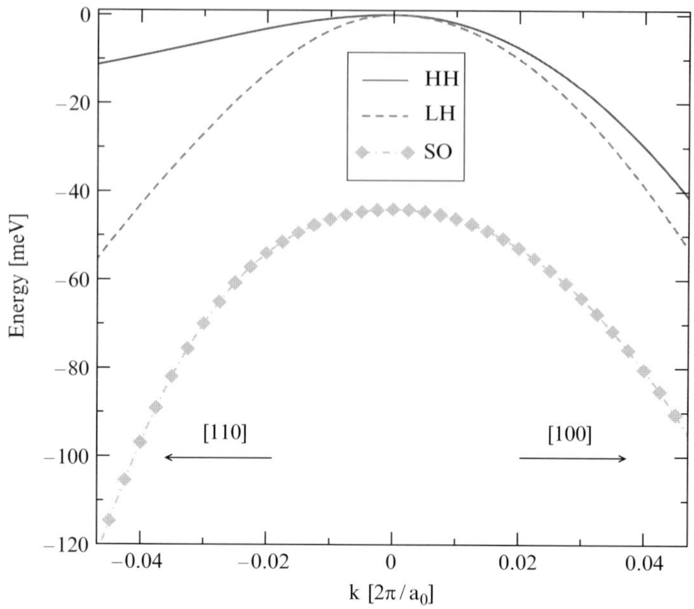

Fig. 6.4 Energy dispersions of the heavy hole, light hole, and split-off bands in unstraned silicon along [110] and [100] directions. The heavy hole and the light hole bands are degenerate at the Γ-point

where the 3×3 matrices $B^{\pm}$ reveal the valence band spectrum. The dispersion relations can be found analytically [12], however, they are quite cumbersome, and we address the reader to the original paper [12].

Dispersion of the heavy hole (HH), light hole (LH), and split-off (SO) bands in the [100] and [110] directions are shown in Fig. 6.4. The HH and LH bands are degenerate at the Γ point, while due to the spin–orbit coupling the energy of the maximum of the split-off band is 44 meV lower. The surface plots of energy dispersions are shown in Fig. 6.5, Fig. 6.6, and Fig. 6.7, for HH, LH, and split-off bands, respectively. The figures demonstrate that the energy dispersions are not isotropic. In order to produce the plots we have used the parameters listed in Table 6.3.

6.3.3 Luttinger Parameters

In case when the spin–orbit coupling Δ_{so} is large, the coupling between the split-off band and the heavy hole and light hole bands can be ignored, and a simpler analytical dispersion for the HH and LH bands can be obtained. In this case the dispersion of the HH and LH bands is determined by the following 4×4 Hamiltonian conveniently written as [11]:

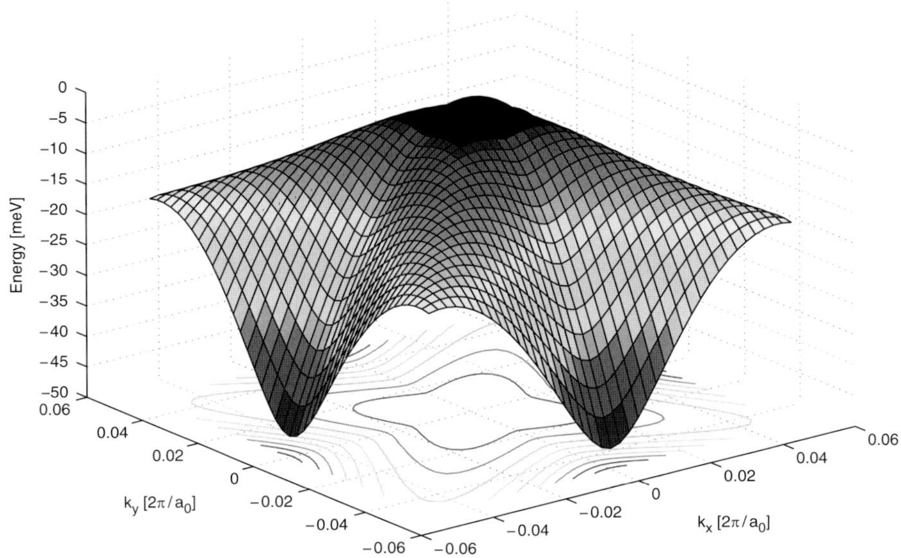

Fig. 6.5 Energy dispersion of the heavy hole band in relaxed silicon

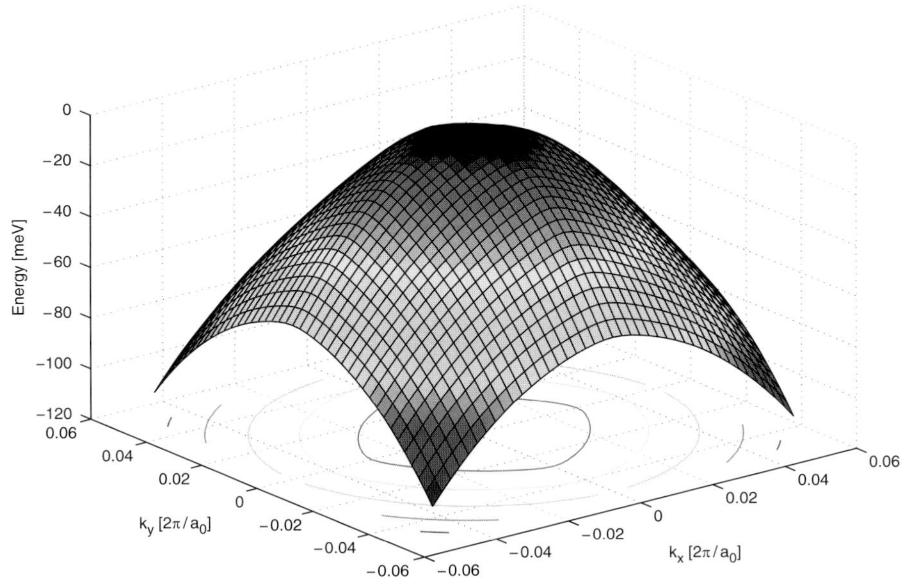

Fig. 6.6 Energy dispersion of the light hole band in relaxed silicon

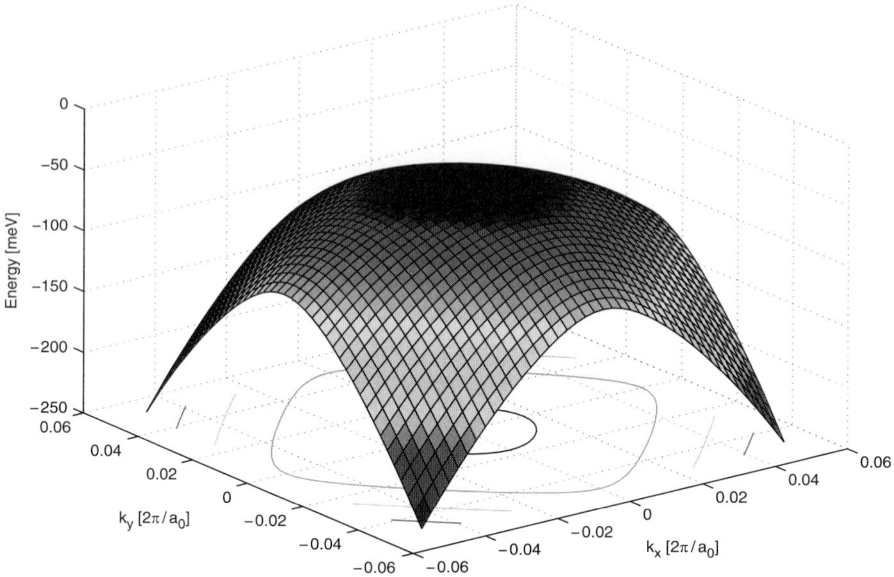

Fig. 6.7 Energy dispersion of the split-off hole band in silicon

Table 6.3 The parameters used for the silicon valence band

Parameter	Value	Units
L	−6.53	$\frac{\hbar^2}{2m_0}$
M	−4.64	$\frac{\hbar^2}{2m_0}$
N	−8.75	$\frac{\hbar^2}{2m_0}$
Δ_{so}	44	meV

$$H = \frac{\hbar^2}{2m_0}\left[(\gamma_1 + \frac{5\gamma_2}{2})k^2 - 2\gamma_2(k_x^2 J_x^2 + k_y^2 J_y^2 + k_z^2 J_z^2) \right.$$
$$\left. -4\gamma_3(\{k_x k_y\}\{J_x J_y\} + \{k_y k_z\}\{J_y J_z\} + \{k_z k_x\}\{J_z J_x\})\right]. \quad (6.48)$$

Here $\{\}$ stands for an anti-commutator:

$$\{J_\alpha J_\beta\} = \frac{1}{2}(J_\alpha J_\beta + J_\beta J_\alpha), \quad (6.49)$$

and J_x, J_y and J_z are the projection operators of the total angular momentum operator **J**. They satisfy the standard commutation relations:

$$[J_x J_y] = J_x J_y - J_y J_x = i J_z, \qquad \text{etc.} \quad (6.50)$$

With the split-off band decoupled, the HH and LH bands correspond to $j = 3/2$. In the basis $\left|\frac{3}{2}, J_z\right\rangle$, these momentum projection operators are represented by the following matrices:

$$J_x = \frac{i}{2}\begin{pmatrix} 0 & \sqrt{3} & 0 & 0 \\ -\sqrt{3} & 0 & 1 & 0 \\ 0 & -1 & 0 & \sqrt{3} \\ 0 & 0 & -\sqrt{3} & 0 \end{pmatrix} \quad J_y = \frac{1}{2}\begin{pmatrix} 0 & \sqrt{3} & 0 & 0 \\ \sqrt{3} & 0 & 1 & 0 \\ 0 & 1 & 0 & \sqrt{3} \\ 0 & 0 & \sqrt{3} & 0 \end{pmatrix}$$

$$J_z = \frac{1}{2}\begin{pmatrix} 3 & 0 & 0 & 0 \\ 0 & 1 & 0 & 0 \\ 0 & 0 & -1 & 0 \\ 0 & 0 & 0 & -3 \end{pmatrix} \tag{6.51}$$

Introducing the following dimensionless Luttinger parameters:

$$\begin{aligned} \gamma_1 &= -\frac{2m_0}{3\hbar^2}(L+2M) - 1, \\ \gamma_2 &= -\frac{m_0}{3\hbar^2}(L-M), \\ \gamma_3 &= -\frac{m_0}{3\hbar^2}N, \end{aligned} \tag{6.52}$$

the dispersion relations for the HH band (E_+) and LH band (E_-) are conveniently written in the form [4]:

$$E_\pm(k) = \gamma_1 \frac{\hbar^2 k^2}{2m_0} \pm \frac{\hbar^2}{m_0}[\gamma_2^2 k^4 + 3(\gamma_3^2 - \gamma_2^2)(k_x^2 k_y^2 + k_y^2 k_z^2 + k_z^2 k_x^2)]^{1/2}. \tag{6.53}$$

Only in case $\gamma_2 = \gamma_3$ the dispersions (6.53) are parabolic:

Table 6.4 The Luttinger parameters for several semiconductors with zincblende and diamond lattice structures [1, 13]

	γ_1	γ_2	γ_3
Si	4.22	0.39	1.44
Ge	13.35	4.25	5.69
GaP	4.2	0.98	1.66
GaAs	7.65	2.41	3.28
GaSb	11.8	4.03	5.26
InP	6.28	2.08	2.76
InAs	19.67	8.37	9.29
InSb	35.08	15.64	16.91
ZnS	2.54	0.75	1.09
ZnSe	3.77	1.24	1.67
ZnTe	3.74	1.07	1.64
CdTe	5.29	1.89	2.46

$$E_{\pm}(k) = \frac{\hbar^2}{2m_0}(\gamma_1 \pm 2\gamma_2)k^2. \tag{6.54}$$

In a general situation the square root in (6.53) remains, and the HH and LH band dispersions can not be described by the effective masses. The values of the Luttinger parameters $\gamma_1, \gamma_2, \gamma_3$ for several semiconductors are summarized in Table 6.4.

We stress again that the separation of the total 6×6 Hamiltonian into the 4×4 Hamiltonian for the HH and LH bands and the 2×2 Hamiltonian for the split-off band is possible only, when the spin–orbit splitting Δ_{so} is much larger than the coupling terms between the states $\left|\frac{3}{2}, J_z\right\rangle$ and $\left|\frac{1}{2}, J_z\right\rangle$ in the Hamiltonian (6.42). In silicon, however, the spin–orbit splitting is only 44meV, which is roughly $2k_BT$, and a more general approach of Manku [12] to diagonalize the full Hamiltonian (6.42) must be applied.

References

1. Baldereschi, A., Lipari, N.O.: Spherical model of shallow acceptor states in semiconductors. Phys. Rev. B **8**(6), 2697–2709 (1973)
2. Bir, G.L., Pikus, G.E.: Symmetry and Strain-Induced Effects in Semiconductors. Willey, New York - Toronto (1974)
3. Boykin, T.B., Klimeck, G., Oyafuso, F.: Valence band effective-mass expressions in the $sp^3d^5s^*$ empirical tight-binding model applied to a Si and Ge parametrization. Phys. Rev. B **69**(11), 115201 (2004)
4. Broido, D.A., Sham, L.J.: Effective masses of holes at gaas-algaas heterojunctions. Phys. Rev. B **31**(2), 888–892 (1985)
5. Chao, C.Y.P., Chuang, S.L.: Resonant tunneling of holes in the multiband effective-mass approximation. Phys. Rev. B **43**(9), 7027–7039 (1991)
6. Chao, C.Y.P., Chuang, S.L.: Spin-orbit-coupling effects on the valence-band structure of strained semiconductor quantum wells. Phys. Rev. B **46**(7), 4110–4122 (1992). DOI 10.1103/PhysRevB.46.4110
7. El Kurdi, M., Fishman, G., Sauvage, S., Boucaud, P.: Comparison between 6-band and 14-band **k·p** formalisms in SiGe/Si heterostructures. Phys. Rev. B **68**(16), 165, 333 (2003)
8. Hensel, J.C., Hasegawa, H., Nakayama, M.: Cyclotron resonance in uniaxially stressed silicon. II. Nature of the covalent bond. Phys. Rev. **138**(1A), A225–A238 (1965)
9. Jacoboni, C., Reggiani, L.: The Monte Carlo method for the solution of charge transport in semiconductors with applications to covalent materials. Rev. Mod. Phys. **55**(3), 645–705 (1983)
10. Luttinger, J.M.: Quantum theory of cyclotron resonance in semiconductors: General theory. Phys. Rev. **102**(4), 1030–1041 (1956)
11. Luttinger, J.M., Kohn, W.: Motion of electrons and holes in perturbed periodic fields. Phys. Rev. **97**(4), 869–883 (1955)
12. Manku, T., McGregor, J.M., Nathan, A., Roulston, D.J., Noel, J.P., Houghton, D.C.: Drift hole mobility in strained and unstrained doped $Si_{1-x}Ge_x$ alloys. IEEE Trans. Electron Devices **40**(11), 1990–1996 (1993)
13. Mayer, H., Rössler, U.: Nonparabolicity in the conduction band of II-VI semiconductors. Solid State Commun. **87**(2), 81–84 (1993)
14. Rideau, D., Feraille, M., Ciampolini, L., Minondo, M., Tavernier, C., Jaouen, H., Ghetti, A.: Strained Si, Ge, and $Si_{1-x}Ge_x$ alloys modeled with a first-principles-optimized full-zone **k·p** method. Phys. Rev. B **74**(19), 195,208 (2006)

15. Rodríguez, S., López-Villanueva, J.A., Melchor, I., Carceller, J.E.: Hole confinement and energy subbands in a silicon inversion layer using the effective mass theory. J. Appl. Phys. **86**(1), 438–444 (1999)
16. Tinkham, M.: Group Theory and Quantum Mechanics. McGraw-Hill, New York (1964)
17. Uchida, K., Kinoshita, A., Saitoh, M.: Carrier transport in (110) nMOSFETs: Subband structure, non-parabolicity, mobility characteristics, and uniaxial stress engineering. In: Intl. Electron Devices Meeting, pp. 1019–1021 (2006)
18. Ungersboeck, E., Dhar, S., Karlowatz, G., Sverdlov, V., Kosina, H., Selberherr, S.: The effect of general strain on band structure and electron mobility of silicon. IEEE Trans. Electron Devices **54**(9), 2183–2190 (2007)
19. VASP: Vienna Ab-initio Simulation Program. Kresse, G., Hafner, J.: Phys. Rev. B **47**, 558 (1993); *ibid.* B **49**, 14251 (1994); Kresse, G., Fertmueller, J.: Phys. Rev. B **54**, 11169 (1996); Comput. Mat. Sci. **6**, 15 (1996)
20. Yu, P., Cardona, M.: Fundamentals of Semiconductors. Springer, Berlin (2003)

Chapter 7
Strain Effects on the Silicon Crystal Structure

7.1 Strain-Induced Symmetry Reduction of Silicon Crystal Lattice

When homogeneous strain is applied to a crystal, the Bravais lattice basis vectors $\mathbf{a}_i{}'$ transforms as

$$\mathbf{a}_i{}' = (\mathbb{1} + \hat{\varepsilon}) \cdot \mathbf{a}_i\,, \tag{7.1}$$

where $\hat{\hat{\varepsilon}}$ is the strain tensor [1] and $\mathbb{1}$ is the unit matrix. The application of strain modifies the volume Ω_0' of the primitive unit cell:

$$\Omega_0' = \Omega_0(1 + \varepsilon_{xx} + \varepsilon_{yy} + \varepsilon_{zz}), \tag{7.2}$$

Thus, the volume of the new primitive cell Ω_0' changes, when the hydrostatic strain component, $\varepsilon_{xx} + \varepsilon_{yy} + \varepsilon_{zz}$, is non-zero.

This is not the only modification of the crystal lattice, however. Depending on strain conditions, the type of the Bravais lattice may change, and the symmetry of the crystal lattice may be altered. The set of the symmetry transformation under which the crystal lattice remains invariant is in general modified by applying stress.

Relaxed silicon belongs to O_h symmetry group. The point symmetry group O_h possesses the highest symmetry from all the groups of crystal lattices.

7.1.1 O_h Symmetry

The point group O_h contains 48 symmetry elements:

- The unity operation,
- Clockwise rotation of 180° about the principal axes $\mathbf{e}_i$ (three operations),
- Clockwise and counter-clockwise rotation of 120° about the four space diagonals (eight operations),
- Clockwise rotation of 180° about the axes $\mathbf{e}_p$ representing the diagonals on the faces of a unit cube (six operations),

V. Sverdlov, *Strain-Induced Effects in Advanced MOSFETs*, Computational Microelectronics, DOI 10.1007/978-3-7091-0382-1_7, © Springer-Verlag/Wien 2011

Table 7.1 Point group and symmetry elements of strained lattices that originate, when stress is applied along various high symmetry directions to an initially cubic lattice O_h. The Schönflies symbols are used to specify the point group. $|P(\Gamma)|$ denotes the number of elements of the point group

Point group	Symmetry elements	$\lvert P(\Gamma)\rvert$	Stress direction
O_h	$\mathrm{E}\ 2_i\ 3_j^+\ 3_j^-\ 2_p\ 4_i^+\ 4_i^-\ \mathrm{I}\ \bar{2}_i\ \bar{3}_j^+\ \bar{3}_j^-\ \bar{2}_p\ \bar{4}_i^+\ \bar{4}_i^-$	48	Relaxed
D_{4h}	$\mathrm{E}\ 4_z^+\ 4_z^-\ 2_i\ 2_s\ \mathrm{I}\ \bar{4}_z^+\ \bar{4}_z^-\ \bar{2}_i\ \bar{2}_s$	16	Stress along ⟨100⟩
D_{3d}	$\mathrm{E}\ 3_z^+\ 3_z^-\ 2_{i''}\ \mathrm{I}\ \bar{3}_z^+\ \bar{3}_z^-\ \bar{2}_{i''}$	12	Stress along ⟨111⟩
D_{2h}	$\mathrm{E}\ 2_i\ \mathrm{I}\ \bar{2}_i$	8	Stress along ⟨110⟩
C_{2h}	$\mathrm{E}\ 2_z\ \mathrm{I}\ \bar{2}_z$	4	Stress along ⟨120⟩
S_2	E I	2	Other directions

- Clockwise and counter-clockwise rotations of 90° about the principal axes $\mathbf{e}_i$ (six operations).

Since the symmetry is preserved for any of these 24 operations followed by the inversion operation, there are 48 symmetry operations in total.

The symmetry of diamond structure is lowered by distortion of the crystal. Because stress applied to a diamond structure reduces the symmetry, the point symmetry group of the strained crystal is a subgroup of O_h: it contains only those symmetry elements which are preserved under strain. The effect of a homogeneous strain on the symmetry of the Bravais lattice depends on the direction along which stress is applied.

There exist two different paths to lower the symmetry of the point group O_h to the point group S_2 [1]:

$$O_h \rightarrow D_{4h} \rightarrow D_{2h} \rightarrow C_{2h} \rightarrow S_2 \tag{7.3}$$

$$O_h \rightarrow D_{3d} \rightarrow C_{2h} \rightarrow S_2. \tag{7.4}$$

In Table 7.1 five directions of uniaxial stress are given that yield a symmetry reduction to the five point subgroups D_{4h}, D_{3d}, D_{2h}, C_{2h}, S_2 of the group O_h.

7.1.2 D_{4h} Symmetry

When the silicon lattice is stressed in the ⟨001⟩ direction, or along a fourfold axis $\mathbf{e}_i$, the cubic lattice of symmetry class O_h becomes a square cuboid (rectangular parallelepiped on a square base), representing the Bravais parallelepiped belonging to class D_{4h} [1]. A similar symmetry reduction is observed, if biaxial strain is applied in a {001} plane. The D_{4h} class is a member of the tetragonal crystal system. The point group D_{4h} has 16 symmetry elements listed in Table 7.1. Note that there are no threefold symmetry axes, and only one fourfold symmetry axis remains.

The strain tensor which yields the symmetry reduction $O_h \rightarrow D_{4h}$ is in the form $\hat{\varepsilon}_{(001)}$ from (3.34) or $\hat{\varepsilon}_{[001]}$ from (3.36). It has non-zero elements in the diagonal (e.g. $\varepsilon_{11} = \varepsilon_{22} \neq \varepsilon_{33}$), while all off-diagonal elements are zero.

7.1.3 D_{3d} Symmetry

If uniaxial stress is applied along the $\langle 111 \rangle$ directions, or parallel to the cube diagonals corresponding to the threefold symmetry axes $\mathbf{e}_j$ the cubic face-centered lattice transforms to a primitive rhombohedral lattice belonging to the crystal class D_{3d} of the trigonal (= rhombohedral) system [1].

The symmetry class D_{3d} includes, apart from the trivial unity transformation and inversion, the threefold rotations about the direction of stress and twofold rotations around three axes perpendicular to the threefold axis, resulting in twelve symmetry operations.

The strain tensor resulting from stress along $\langle 111 \rangle$, which yields a $O_h \rightarrow D_{3d}$ symmetry reduction, is in the form $\hat{\varepsilon}_{(111)}$ from (3.34) or $\hat{\varepsilon}_{[111]}$ from (3.36). It contains equal off-diagonal components $\varepsilon_{12} = \varepsilon_{13} = \varepsilon_{23}$ and equal diagonal components $\varepsilon_{11} = \varepsilon_{22} = \varepsilon_{33}$. Because for biaxial strain in $\{111\}$ plane the strain tensor has got a similar form, the same symmetry reduction by straining the crystal biaxially in the $\{111\}$ plane is achieved.

7.1.4 D_{2h} Symmetry

The Bravais lattice of the crystal class O_h is converted to a parallelepiped of the orthorhombic system belonging to D_{2h} in two ways [1, 7]:

1. Dilatation or compression of different strength along two of the three fourfold axes $\mathbf{e}_i$. This results in a parallelepiped with rectangular faces (cuboid). Of the five twofold axes $\mathbf{e}_i$ and $\mathbf{e}_s$ of D_{4h}, only the three $\mathbf{e}_i$ along the edges of the parallelepiped remain. This symmetry reduction can be achieved by applying stress of different magnitude along two of the three equivalent $\langle 100 \rangle$ directions, simultaneously. In this case, the strain tensor is given by $\varepsilon_{11} \neq \varepsilon_{22} \neq \varepsilon_{33}$ and contains vanishing off-diagonal components.
2. The deformation originates from shearing the unit cube, thus altering the angles between the basis vectors. The result is a rectangular parallelepiped with rhombic base, which is also invariant under D_{2h}. Of the original five twofold axes $\mathbf{e}_i$ and $\mathbf{e}_s$ only two (diagonals of the base) remain. This type of lattice results, when uniaxial stress is applied along $\langle 110 \rangle$ or from biaxial strain in a $\{110\}$ plane. The strain tensor has the form

$$\hat{\varepsilon} = \begin{pmatrix} \varepsilon_{11} & \varepsilon_{12} & 0 \\ \varepsilon_{12} & \varepsilon_{11} & 0 \\ 0 & 0 & \varepsilon_{33} \end{pmatrix}, \tag{7.5}$$

where the components of the strain tensor are related to stress according to (3.36). This group has only eight symmetry elements given in Table 7.1.

7.1.5 C_{2h} Symmetry

Higher symmetry reduction results from deforming the base of the Bravais parallelepiped of the orthorhombic system so that the angle between its edges is changed. In this way the invariant parallelepiped of the system C_{2h} is obtained from the cubic lattice O_h [1]. It contains four symmetry operations given in Table 7.1 with only one twofold symmetry axis.

Uniaxial stress in [120] direction can achieve this kind of symmetry reduction. The strain tensor has three different non-zero diagonal components and one off-diagonal component

$$\hat{\varepsilon} = \begin{pmatrix} \varepsilon_{11} & \varepsilon_{12} & 0 \\ \varepsilon_{12} & \varepsilon_{22} & 0 \\ 0 & 0 & \varepsilon_{33} \end{pmatrix}. \tag{7.6}$$

7.2 Internal Strain Parameter

The strain tensor (3.4) determines only the deformation of the primitive cell as a whole. In strained lattice, the displacements of the basis atoms relative to their positions in the relaxed lattice appears. This internal displacement has no impact on the lattice symmetry, however, it affects the band dispersions [3]. The internal displacement is not described by the strain tensor, and the information about the positions of the atoms must be obtained from the lattice structure calculations in a strained crystal.

The displacement of the atoms due to stress in each fcc sublattice is determined by the strain tensor, however, the relative displacement of the sublattices determined by the position of the second basis atom in the primitive cell of the strained lattice remains undefined. To obtain the exact position of the central atom an additional parameter for this displacement has to be introduced.

To define this internal strain parameter let us consider the change of atomic positions in the primitive unit cell under strain shown in Fig. 7.1. In the relaxed lattice, the central atom is situated exactly in the center of the tetrahedron. In the stressed lattice the vertex atoms move into their new positions defined by the strain tensor. If the central atom is assumed to move into its position determined by the strain tensor, it may get displaced from the center of the distorted tetrahedron formed by the four vertex atoms as shown in Fig. 7.1. However, in this case the central atom will be closer to some of the vertex atoms of the strained tetrahedron than to others, and the total energy of the system will rise. In order to minimize the lattice energy, the central atom will be displaced towards the center of the deformed tetrahedron

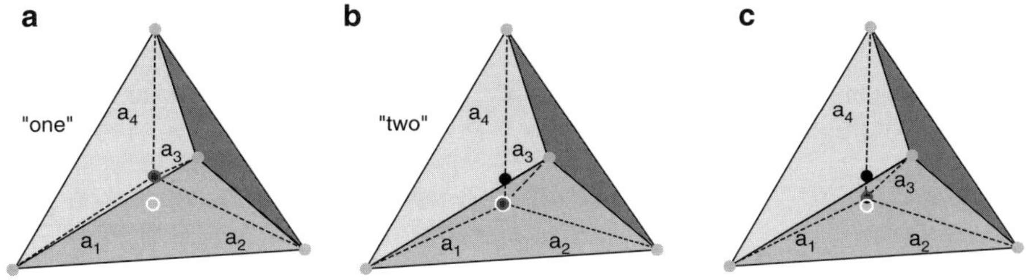

Fig. 7.1 Unit cell of strained diamond structure with the central atom being at: (**a**) The position "one" defined by the strain tensor; (**b**) The position "two" with equal bonds to the four vertex atoms; (**c**) The real position determined by the internal strain parameter ξ

defined as the point with equal distances to each of the four vertex atoms. However, because of an increase of the energy due to non-central forces between the surrounding atoms [4], the central atom does not completely relax to the center of the strained tetrahedron as demonstrated in Fig. 7.1.

For general stress conditions, this additional displacement of the central atom, or the displacement of one of the inter-penetrating fcc lattice with respect to another, is described by the displacement factor called the internal strain parameter ξ. To define it, let us first determine the positions of the four vertex atoms of the tetrahedron defined by the strain tensor. Next, let us determine the position of the central atom in the deformed tetrahedron, called the position "one", by assuming that there was no additional displacements between the two sublattices, and the positions of atoms in the second sublattice are also determined by the strain tensor. Second, let us find the center of the deformed tetrahedron as the point situated at equal distances from the four vertex atoms and call it the position "two". The actual position of the central atom in the strained lattice does not coincide with either of the two limiting positions and will be located on a line between the two defined positions. The internal strain parameter ξ denotes the fraction of the distance between these two specified positions at which the central atom is actually located. The internal strain parameter ξ is set to zero, if the central atom remains at the position "one" determined by the strain tensor, and $\xi = 1$ if the central atom moves to the position "two" characterized by all four bonds between the atom and the vertexes being of the same length.

In a real crystal the value of the internal strain parameter $0 \leq \xi \leq 1$. The value of the internal strain parameter $\xi = 0.54 \pm 0.04$ was extracted from the experiment [2]. The experimental value coincides with the results of theoretical calculations [5], which give $\xi = 0.53$. Recently, advanced theoretical calculations based on the first-principle ab-initio program VASP (Vienna ab-initio simulation program) [10] yield the value $\xi = 0.5$ [9], which is very close to the previously obtained values.

The position of the central atom can be easily determined for practically relevant stress conditions. Because in case of uniaxial stress applied along $\langle 100 \rangle$ direction the center of the deformed primitive unit cell, the position "two", coincides with the position "one" defined by the strain tensor, no internal displacement occurs.

For stress in [110] direction the additional displacement along [001] axis occurs. The value of this additional displacement is [8]

$$u_z = -\frac{\xi}{2} \frac{(1+\varepsilon_{xx})\varepsilon_{xy}}{1+\varepsilon_{zz}} a_0. \tag{7.7}$$

If now uniaxial stress is along [111] direction, the additional displacement is in the same [111] direction, with the value equal to

$$u = -\frac{\xi}{2} \varepsilon_{xy} a_0 (1, 1, 1). \tag{7.8}$$

7.3 Strain and Symmetry of the Brillouin Zone

The Brillouin zone of the strained crystal can be obtained by an appropriate deformation of the Brillouin zone of the relaxed crystal [1]. Because the symmetry of the strained crystal is reduced, the Brillouin zone of the strained crystal possesses less symmetry.

A possible choice for the irreducible wedge of relaxed Si is depicted in Fig. 7.2. It should be noted that this specific choice is not unique, and other shapes for the irreducible wedge, reflecting the symmetries of O_h, can be found [6]. The volume of the specified irreducible wedge is $\Omega_{\mathrm{BZ}}/48$ in accordance with the relation (4.7).

From (4.7) it can be concluded that the higher the point symmetry of the crystal lattice, the smaller is the volume of the irreducible wedge. The symmetries of the Brillouin zone and the size of the irreducible wedges for the crystal systems D_{4h}, D_{3d}, D_{2h}, C_{2h}, and S_2 are briefly discussed below.

The symmetry operations of the D_{4h} class enforce the following invariance of the energy dispersions under reflections

$$E_n(k_x, k_y, k_z) = E_n(|k_x|, |k_y|, |k_z|). \tag{7.9}$$

Therefore, in order to obtain the dispersion relations in the whole Brillouin zone, it is enough to compute them only for the first octant. The additional symmetry

$$E_n(k_x, k_y, k_z) = E_n(k_x, k_z, k_y) \tag{7.10}$$

for stress along [100] direction allows additionally reducing the volume of irreducible wedge by a factor of two. The irreducible wedge is shown in Fig. 7.2.

The invariants of the energy band structure for the D_{3d} class are listed in Table 7.2. In this table the Miller index notation of the cubic lattice is used to specify

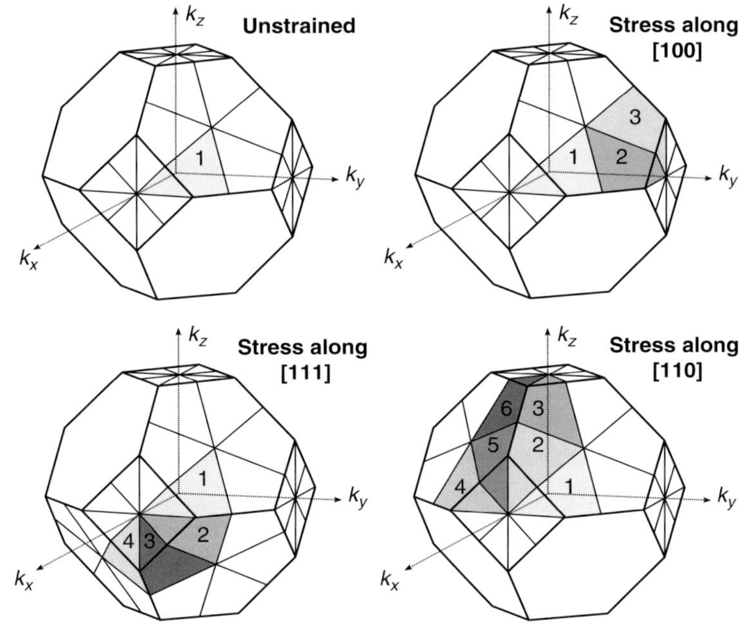

Fig. 7.2 Irreducible wedges for relaxed silicon and stresses along [100], [110], and [111] directions

Table 7.2 Symmetry operations leaving the band structure invariant for the D_{3d} class. Directions of the rotation axes are given in Miller index notation for the cubic lattice

Symmetry operation	Energy	Symmetry operation	Energy
E	$E_n(k_x, k_y, k_z)$	I	$E_n(-k_x, -k_y, -k_z)$
$3^{-}_{[111]}$	$E_n(k_z, k_x, k_y)$	$\bar{3}^{-}_{[111]}$	$E_n(-k_y, -k_z, -k_x)$
$3^{+}_{[111]}$	$E_n(k_y, k_z, k_x)$	$\bar{3}^{+}_{[111]}$	$E_n(-k_y, -k_z, -k_x)$
$2_{[\bar{1}10]}$	$E_n(-k_y, -k_x, -k_z)$	$\bar{2}_{[\bar{1}10]}$	$E_n(k_y, k_x, k_z)$
$2_{[\bar{1}01]}$	$E_n(-k_z, -k_y, -k_x)$	$\bar{2}_{[\bar{1}01]}$	$E_n(k_z, k_y, k_x)$
$2_{[0\bar{1}1]}$	$E_n(-k_x, -k_z, -k_y)$	$\bar{2}_{[0\bar{1}1]}$	$E_n(k_x, k_z, k_y)$

the directions of the rotation axes. The twelve involved symmetry operations give rise to a volume of the irreducible wedge of $\Omega_{\mathrm{BZ}}/12$ shown in Fig. 7.2.

The D_{2h} class contains eight symmetry elements, thus the volume of the irreducible wedge is $\Omega_{\mathrm{BZ}}/8$. When non-equal stress is applied along two of the three fourfold axes $\mathbf{e}_i$, any octant of the Brillouin zone can be chosen as the irreducible wedge. Certain care must be taken, when choosing the wedge in presence of uniaxial stress along [110] direction [7]. The irreducible wedge is depicted in Fig. 7.2.

For stress in [120] direction the symmetry class of the lattice reduces to C_{2h}, and the volume of the irreducible wedge is $\Omega_{\mathrm{BZ}}/4$. Finally, for stress along directions different from those listed in Table 7.1 the crystal loses all rotational symmetries and is invariant only under inversion. The resulting crystal class is S_2, and half of the first Brillouin zone must be chosen as the irreducible wedge for band structure calculation.

References

1. Bir, G.L., Pikus, G.E.: Symmetry and Strain-Induced Effects in Semiconductors. Wiley, New York (1974)
2. Cousins, C.S.G., Gerward, L., Olsen, J.S., Selsmark, B., Sheldon, B.J.: Surface effects in uniaxially stressed crystals: the internal-strain parameters of silicon and germanium revised. J. Phys. C Solid State Phys. **20**, 29–37 (1987)
3. Fischetti, M.V., Laux, S.E.: Monte Carlo simulation of electron transport in Si: The first 20 years. In: 26th European Solid State Device Research Conference, pp. 813–820 (1996)
4. Kleinman, L.: Deformation potentials in silicon. I. Uniaxial strain. Phys. Rev. **128**(6), 2614–2621 (1962)
5. Nielsen, O.H., Martin, R.M.: Quantum-mechanical theory of stress and force. Phys. Rev. B **32**(6), 3780–3791 (1985)
6. Stanley, J., Goldsman, N.: New irreducible wedge for scattering rate calculations in full-zone Monte Carlo simulations. VLSI Des. **8**(1–4), 413–417 (1998)
7. Ungersboeck, E.: Advanced modeling of strained CMOS technology. Dissertation, Institute for Microelectronics, TU Wien (2007)
8. Ungersboeck, E., Dhar, S., Karlowatz, G., Kosina, H., Selberherr, S.: Physical modeling of electron mobility enhancement for arbitrarily strained silicon. In: International Workshop on Computational Electroncis, pp. 141–142 (2006)
9. Ungersboeck, E., Gös, W., Dhar, S., Kosina, H., Selberherr, S.: The effect of uniaxial stress on band structure and electron mobility of silicon. Math. Comput. Simul. **79**(4), 1071–1077 (2008)
10. VASP: Vienna Ab-initio Simulation Program. Kresse, G., Hafner, J.: Phys. Rev. B **47**, 558 (1993); *ibid.* B **49**, 14251 (1994); Kresse, G., Fertmueller, J.: Phys. Rev. B **54**, 11169 (1996); Computs. Mat. Sci. **6**, 15 (1996)

Chapter 8
Strain Effects on the Silicon Band Structure

8.1 Linear Deformation Potential Theory

8.1.1 Conduction Band

The deformation potential theory to describe the influence of strain on the band structure was developed by Bardeen and Shockley [2] and later generalized by Herring and Vogt [7]. Within this theory the energy is represented as a Taylor series in powers of lattice strain, and the expansion is truncated after the terms linear in strain. The theory thus relates the shifts of the energy bands to small deformations of the crystal as:

$$\Delta E(\mathbf{k}) = \sum_{ij} D_{ij}^{(\mathbf{k})} \varepsilon_{ji}. \tag{8.1}$$

It follows from (8.1) that the band shift is linear in strain. The coefficients of proportionality $D_{ij}^{(\mathbf{k})}$ form a second rank tensor. This tensor called the deformation potential tensor is a characteristic of a given non-degenerate band at a chosen point **k**. The tensor is symmetric and therefore has only six independent components. In cubic semiconductors the number of components is further reduced to three [7].

Based on the linear deformation potential theory, the shift of the conduction band minima with stress in silicon and germanium can be evaluated. The shift depends on the magnitude of forces applied and their directions with respect to the valley orientations. For arbitrary stress conditions, the degenerate minima in silicon are split. The value of the valley splitting, which is linear in strain within the linear deformation potential theory [7] is completely determined by the only two deformation potentials D_d and D_u [1]. The general form of the linear energy shift (8.1) for one of the six degenerate valleys $i = 1, 2, .., 6$ in silicon for an arbitrary homogeneous deformation can be written in the following form

$$\Delta E_c^{(i)} = D_d^{\Delta} \operatorname{Tr}(\hat{\varepsilon}) + D_u^{\Delta} \mathbf{a}_i^T \hat{\varepsilon} \mathbf{a}_i, \tag{8.2}$$

where $\mathbf{a}_i$ is a unit vector parallel to the $\mathbf{K_0}$ vector determining the minimum position of the valley i. It follows from (8.2) that the shift of the mean energy of the

V. Sverdlov, *Strain-Induced Effects in Advanced MOSFETs*, Computational Microelectronics, DOI 10.1007/978-3-7091-0382-1_8, © Springer-Verlag/Wien 2011

conduction band depends on the hydrostatic pressure:

$$\Delta E_{c,av}^{(k)} = (D_d^{\Delta} + \frac{1}{3} D_u^{\Delta}) Tr(\hat{\varepsilon}). \tag{8.3}$$

Because the deformation potentials D_d and D_u have different values at different locations in the Brillouin zone, the average energy shifts are different for different valley types. This difference has to be taken into account, when more than one type of valleys is considered, since the relative shift of the mean energy causes a repopulation of carriers between these types of valleys. For example, the energy shifts of the conduction valleys at the L-point is determined by the same expression (8.2), with the deformation potentials D_d^L and D_u^L computed at the L-point [1]. However, if the valley were at the Γ symmetry point, it is enough to know a single deformation potential to determine the valley shift:

$$\delta E_0^{\Gamma} = D^{\Gamma}\,\mathrm{Tr}(\hat{\varepsilon}). \tag{8.4}$$

Using the above relations the valley splitting due to stress along any direction can be obtained once the strain tensor is known. Depending on stress conditions, the energy shifts of different valleys can be different. Uniaxial [001] stress or biaxial strain in (001) plane can be used to partly remove the degeneracy of the six equivalent valleys in silicon as demonstrated in Fig. 8.1.

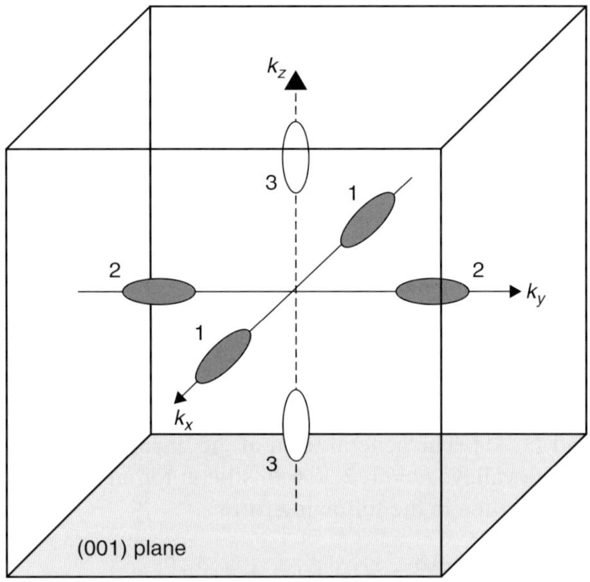

Fig. 8.1 Uniaxial [001] stress or biaxial strain in (001) plane partly lifts the degeneracy of the six equivalent valleys in the conduction band of silicon by shifting the k_z valleys "3" relative to the k_x and k_y valleys. The valleys "1" and "2" remain degenerate

Table 8.1 Strain-induced energy shifts of the conduction band valleys of cubic semiconductors, when uniaxial stress is applied along a given direction. It is assumed that the bands are not degenerate at the position of their minima, although the equivalent valleys located at different points in the Brillouin zone can be degenerate with each other

Stress direction	Valley	Valley direction	$\delta E/P$
[100]	Δ	[100]	$D_d^{\Delta}(s_{11}+2s_{12})+D_u^{\Delta}s_{11}$
	Δ	[010][001]	$D_d^{\Delta}(s_{11}+2s_{12})+D_u^{\Delta}s_{12}$
	L	$[111][11\bar{1}][1\bar{1}\bar{1}][\bar{1}11]$	$D_d^{L}(s_{11}+2s_{12})+D_u^{L}/3(s_{11}+2s_{12})$
	Γ	[000]	$D_d^{\Gamma}(s_{11}+2s_{12})$
[110]	Δ	[100][010]	$D_d^{\Delta}(s_{11}+2s_{12})+D_u^{\Delta}/2(s_{11}+s_{12})$
	Δ	[001]	$D_d^{\Delta}(s_{11}+2s_{12})+D_u^{\Delta}s_{12}$
	L	$[111][11\bar{1}]$	$D_d^{L}(s_{11}+2s_{12})+D_u^{L}/3(s_{11}+2s_{12}+s_{44})$
	L	$[\bar{1}11][1\bar{1}1]$	$D_d^{L}(s_{11}+2s_{12})+D_u^{L}/3(s_{11}+2s_{12}-s_{44})$
	Γ	[000]	$D_d^{\Gamma}(s_{11}+2s_{12})$
[111]	Δ	[100][010][001]	$D_d^{\Delta}(s_{11}+2s_{12})+D_u^{\Delta}/3(s_{11}+2s_{12})$
	L	[111]	$D_d^{L}(s_{11}+2s_{12})+D_u^{L}/3(s_{11}+2s_{12}+2s_{44})$
	L	$[\bar{1}11][1\bar{1}1][11\bar{1}]$	$D_d^{L}(s_{11}+2s_{12})+D_u^{L}/3(s_{11}+2s_{12}-2/3s_{44})$
	Γ	[000]	$D_d^{\Gamma}(s_{11}+2s_{12})$

The analytical expressions for the energy shifts of the conduction band valleys for three stress directions [100], [110], and [111] are summarized in Table 8.1 [14].

8.1.2 Valence Band

In case of bands degenerate at a certain symmetry point of the Brillouin zone strain causes not only relative shifts between the bands but also modifies their shapes. In order to describe the band warping, an additional Hamiltonian linear in strain is added to a **k·p** Hamiltonian of the relaxed crystal. The most general form of this additional strain Hamiltonian is given by its matrix element

$$\delta H(\hat{\varepsilon})_{mm'} = \sum_{i,j} D_{mm'}^{ji}\varepsilon_{ij}, \tag{8.5}$$

where m, m' denote the corresponding components of the **k·p** Hamiltonian, and D_{ij} is the deformation potential operator which transforms under symmetry operations as a second rank tensor [8]. From the symmetry of the strain tensor with respect to i, j one concludes that $D_{ij} = D_{ji}$, therefore maximum six independent deformation potential operators exist. In cubic semiconductors the minima of the conduction and the valence bands are situated at the symmetry lines of the Brillouin zone which further reduces the number of independent constants determining the deformation potential operators. Due to the symmetry of the states at these points the deformation potential operators of a particular state can be described in terms of two or three deformation potential constants [7, 12]. The values of the deformation potentials are usually determined using electrical, optical, microwave techniques or by analyzing stress-induced indirect absorption edges. The deformation potential constants can also be computed numerically by comparing the band structure of strained and relaxed systems. The values of deformation potentials for silicon and germanium obtained by different methods are summarized in [5]. The formal expressions for the deformation potentials obtained from the perturbative expansion of the Hamiltonian of the strained crystal for the deformation potentials will be obtained in Sect. 8.2 below.

8.1.3 Stress-Induced Band Splitting of the Valence Bands

Strain induced splitting and warping of the bands can be calculated within the $\mathbf{k}\cdot\mathbf{p}$ model after adding the strain-dependent perturbation Hamiltonian δH. Within the deformation potential theory the energy dispersions of the valence bands are obtained from a perturbation Hamiltonian $\delta H(\hat{\varepsilon})$ added to the $\mathbf{k}\cdot\mathbf{p}$ Hamiltonian (6.33):

$$\boldsymbol{\delta}\mathbf{H}(\hat{\boldsymbol{\varepsilon}}) = \begin{pmatrix} l\varepsilon_{xx} + m(\varepsilon_{yy} + \varepsilon_{zz}) & n\varepsilon_{xy} & n\varepsilon_{zx} \\ n\varepsilon_{xy} & l\varepsilon_{yy} + m(\varepsilon_{zz} + \varepsilon_{xx}) & n\varepsilon_{yx} \\ n\varepsilon_{zx} & n\varepsilon_{yz} & l\varepsilon_{zz} + m(\varepsilon_{xx} + \varepsilon_{yy}) \end{pmatrix} \begin{matrix} |e_1\rangle \\ |e_2\rangle . \\ |e_3\rangle \end{matrix} \tag{8.6}$$

In the basis $|e_i, \sigma\rangle$, where $\sigma = \uparrow, \downarrow$, the matrix of the perturbation to the valence band Hamiltonian is

$$\boldsymbol{\delta}\mathbf{H}_{strain} = \begin{pmatrix} \boldsymbol{\delta}\mathbf{H}(\hat{\boldsymbol{\varepsilon}}) & \mathbf{0}_{3\times 3} \\ \mathbf{0}_{3\times 3} & \boldsymbol{\delta}\mathbf{H}(\hat{\boldsymbol{\varepsilon}}) \end{pmatrix} \begin{matrix} |\uparrow\rangle \\ |\downarrow\rangle \end{matrix}, \tag{8.7}$$

where $\boldsymbol{\delta}\mathbf{H}$ is determined by (8.6). The three parameters l, m, n denote valence band deformation potentials. The effect of spin-orbit coupling is taken into account by introducing a spin-orbit interaction term, H_{so}. The total 6×6 Hamiltonian of the valence band in strained silicon in the basis $|e_i, \sigma\rangle$ is the sum of (6.36), (8.7) and (6.40):

$$H(\hat{\varepsilon}) = H + \boldsymbol{\delta}\mathbf{H}_{strain} + H_{so}. \tag{8.8}$$

The Hamiltonian (8.8) can be diagonalized by performing a suitable unitary transformation. After some mathematical manipulations, Manku [9] has arrived at the following form describing the valence band of strained silicon:

$$\begin{aligned} &h'_{11}h'_{22}h'_{33} + 2h'_{12}h'_{23}h'_{13} - h'_{11}h^2_{23} - h'_{22}h^2_{13} - h'_{33}h^2_{12} \\ -&\frac{\Delta_{so}}{3}\left(h'_{11}h'_{22} + h'_{11}h'_{33} + h'_{22}h'_{33} - h^2_{12} - h^2_{13} - h^2_{23}\right) = 0. \end{aligned} \tag{8.9}$$

Here the h_{ij} and the h'_{ii} are determined by the matrices $H_{3\times 3}$ $\delta H(\hat{\varepsilon})$ defined by (6.33) and (8.6), respectively, through the relation

$$h_{ij} = H_{ij} + \delta H_{ij}(\hat{\varepsilon}) \qquad h'_{ii} = h_{ii} + \frac{\hbar^2 \mathbf{k}^2}{2m_0} - E(\mathbf{k}), \tag{8.10}$$

where $E(\mathbf{k})$ is the energy to be determined. Equation (8.10) can be simplified to [9]

$$\sum_{i=0}^{3} \sum_{j=0}^{3-i} a_{ij} E(\mathbf{k})^j k^{2i} = 0, \tag{8.11}$$

which is a cubic equation in $E(\mathbf{k})$. Its solutions give the energies for the HH, LH and the split-off bands for a particular value of $\mathbf{k}$ and arbitrary strain. The components a_{ij} depend on the spherical angles determining the direction $\mathbf{k}$ in the crystallographic system and are functions of the strain tensor. The strain-induced splitting of the valence bands can be obtained by setting $\mathbf{k} = 0$ in (8.11) to give

$$\sum_{i=0}^{3} a_i E^i = 0, \tag{8.12}$$

where the coefficients a_i are defined as [9]

$$a_0 = \frac{\Delta_{so}}{3}(pq + pr + qr - n^2\varepsilon_T^2 + n^2(2n\varepsilon_{xy}\varepsilon_{yz}\varepsilon_{xz} - p\varepsilon_{yz}^2 - q\varepsilon_{xz}^2 - r\varepsilon_{xy}^2) + pqr \tag{8.13}$$

$$a_1 = \frac{2\Delta_{so}}{3}(p + q + r) + (pq + pr + qr - n^2\varepsilon_T^2) \tag{8.14}$$

$$a_2 = p + q + r - \Delta_{so} \tag{8.15}$$

$$a_3 = -1 \tag{8.16}$$

$$\varepsilon_T^2 = \varepsilon_{xy}^2 + \varepsilon_{yz}^2 + \varepsilon_{xz}^2 \tag{8.17}$$

and

$$\begin{pmatrix} p \\ q \\ r \end{pmatrix} = \begin{pmatrix} l & m & m \\ m & l & m \\ m & m & l \end{pmatrix} \begin{pmatrix} \varepsilon_{xx} \\ \varepsilon_{yy} \\ \varepsilon_{zz} \end{pmatrix}. \tag{8.18}$$

Modification of the valence band due to uniaxial or biaxial stress is well understood within the six-band $\mathbf{k}\cdot\mathbf{p}$ model [9, 13, 17]. Here we show some results for the technologically relevant compressive stress along [110] direction. The values of the deformation potentials l, m, n used are listed in Table 8.2.

Table 8.2 The valence band parameters used

Parameter	Value	Units
l	−2.44	eV
m	4.37	eV
n	−6.8	eV

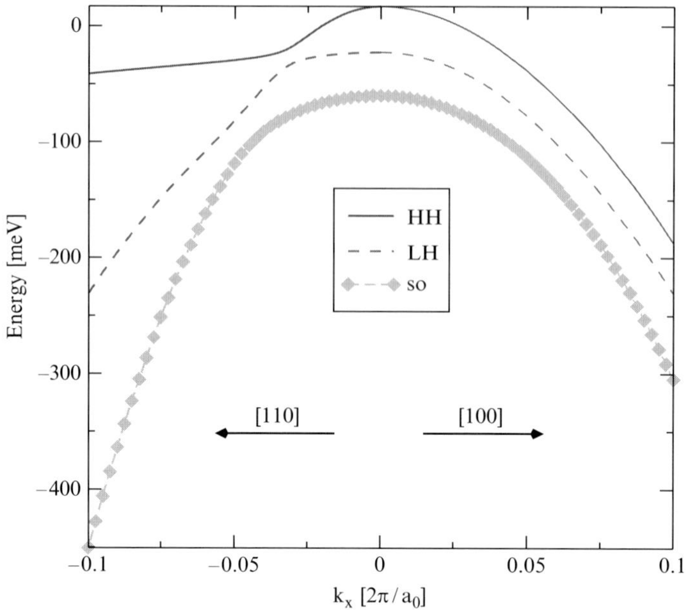

Fig. 8.2 Energy dispersions of the heavy hole, light hole, and split-off bands along [110] and [100] directions in compressively stressed silicon. Stress of 1 GPa is applied in [110] directions. The degeneracy between the heavy hole and light hole bands at the Γ-point is lifted

Figure 8.2 displays the dispersion of the HH, LH, and split-off valence bands, when compressive stress of 1 GPa along the [110] direction is applied. A clear splitting between the bands appears. The dependence of the band splitting on the value of the compressive stress along [110] direction is shown in Fig. 8.3. At high stress values only the HH band is occupied and thus contributing to transport.

Apart from splitting, a substantial valence band warping appears in strained silicon. Figure 8.4 shows the in-plane dispersion of the HH valence band at $k_z = 0$, with and without stress. In comparison to the dispersion of the HH band in relaxed silicon a substantial modification of the dispersion of the in-plane is observed. The heavy-mass contribution from the wings along the [110] direction clearly seen in Fig. 8.4 is substantially reduced in silicon compressively stressed along [110] [17]. The effective mass decrease along the [110] stress direction is also visible in Fig. 8.2. Stress also modifies the dispersion of the LH and SO bands. Figure 8.5 displays stress-induced modification of the LH band dispersion. Stress-induced band splitting and warping is the main reason of the hole mobility enhancement (or degradation) in strained silicon.

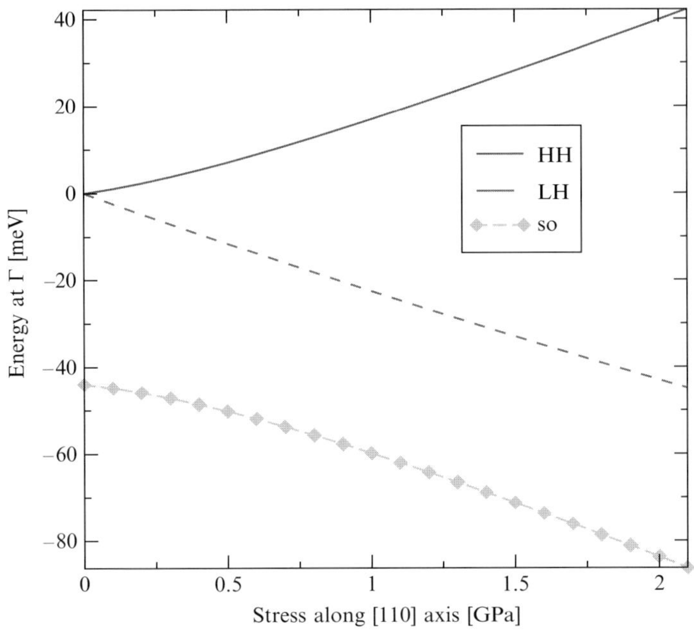

Fig. 8.3 Stress induced splitting of the heavy hole, light hole, and split-off bands in silicon at the Γ-point, according to (8.12), (8.17) and (8.18). Compressive stress is applied along [110] direction

8.2 Inclusion of Strain into Perturbative Band Structure Calculations

Until now the deformation potentials describing the perturbation of the Hamiltonian in strained lattices were the constants introduced phenomenologically. Here, following [3], we outline the procedure how the deformation potentials can in principle be computed.

The Hamiltonian describing the electron spectrum in a crystal subject to homogeneous strain is:

$$H(\epsilon) = \frac{p^2}{2m_0} + V_S(\mathbf{r}) + \frac{\hbar}{4m_0^2c^2}(\nabla V_S[\mathbf{p}\boldsymbol{\sigma}]), \tag{8.19}$$

where $V_S(\mathbf{r})$ is the potential in the strained crystal. In the last term describing the spin-orbit interaction $\boldsymbol{\sigma} = (\sigma_x, \sigma_y, \sigma_z)$ is the vector assembled from the Pauli matrices. The Hamiltonian (8.19) should be contrasted with the Hamiltonian of the relaxed crystal

$$H_0 = \frac{p^2}{2m_0} + V(\mathbf{r}) + \frac{\hbar}{4m_0^2c^2}(\nabla V[\mathbf{p}\boldsymbol{\sigma}]), \tag{8.20}$$

with V_S replaced by V.

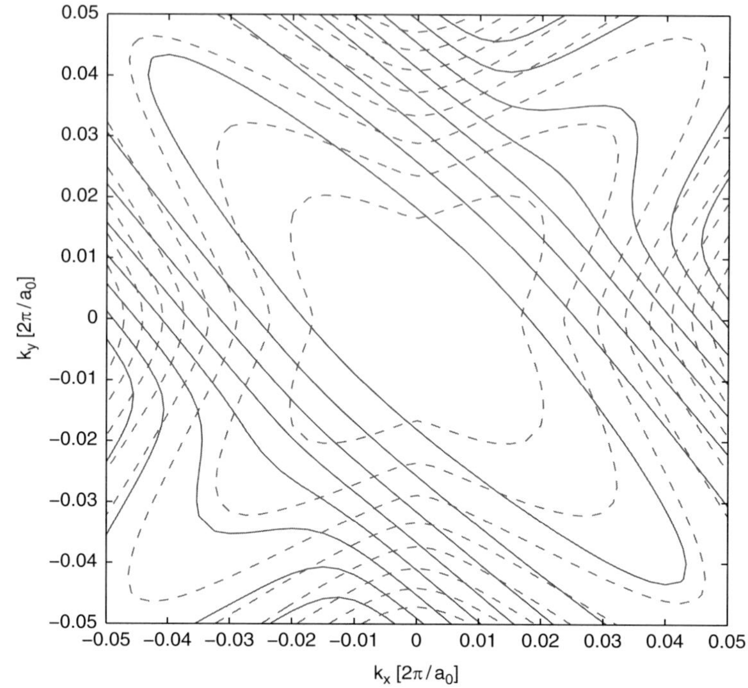

Fig. 8.4 Energy dispersion of the heavy hole band in silicon compressively strained along [110] direction by applying 1 GPa stress. The dashed contour lines are for the heavy hole band dispersion of relaxed silicon. Contour lines are every 5 meV.

Let us assume that strain is small. In this case only the terms linear in strain can be kept in the Hamiltonian of the strained crystals. However, we can not simply develop the difference between the crystal potentials $V_S - V$ into the series in strain because, due to mismatch in the lattice constants, the difference between the potential is not small. To demonstrate it, let us put the first atom of strained and relaxed Bravais lattices at the origin. Then the displacement in the strained lattice of a remote lattice point $\mathbf{r}_m$ is proportional to $\boldsymbol{\epsilon}\mathbf{r}_m$ and can become comparable to the lattice constant or even larger. It then follows that the difference $V_S(\mathbf{r}_m) - V(\mathbf{r}_m)$ is of the order of the crystal potential $V(\mathbf{r}_m)$ irrespective of the strain value.

In order to develop a perturbative expansion of the Hamiltonian (8.19) of a strained lattice, a coordinate transformation has to be first performed. This is done by choosing a new coordinate system in strained crystal in such a way that the positions of the Bravais lattice points $\mathbf{r}'_m$ coincide with the corresponding points in the relaxed crystal. The transformation can be written as [3]:

$$\mathbf{r}' = (1 + \hat{\varepsilon})^{-1}\mathbf{r} \approx (1 - \hat{\varepsilon})\mathbf{r}, \tag{8.21}$$

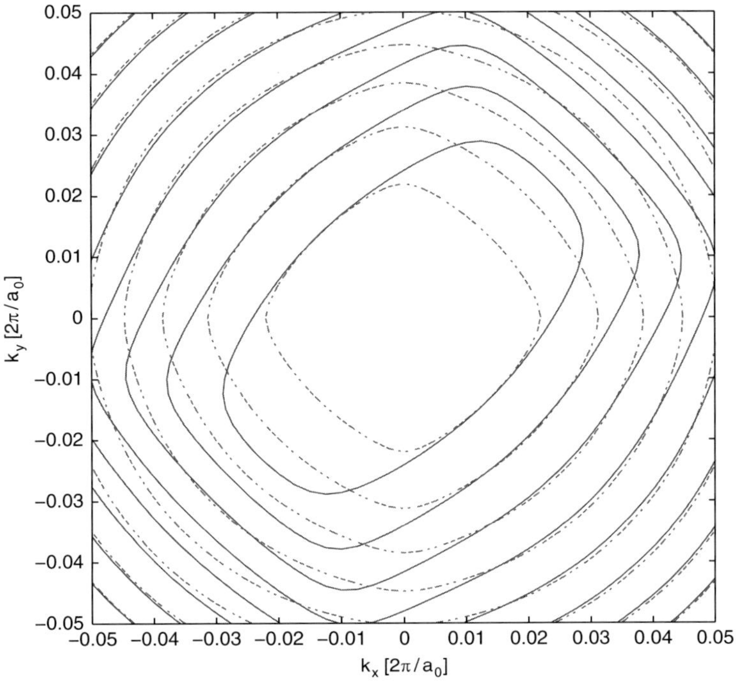

Fig. 8.5 Energy dispersion of the light hole band in silicon compressively strained along [110] direction by applying 1 GPa stress. The dashed contour lines are for the heavy hole band dispersion of relaxed silicon. Contour lines are every 10 meV.

or

$$\mathbf{r} = (1 + \hat{\varepsilon})\mathbf{r}', \tag{8.22}$$

where the j component of $(\hat{\varepsilon}\mathbf{r}')_j = \sum_i \varepsilon_{ji} r'_i$. Taking into account the coordinate transformation one can derive that the momentum operator $\mathbf{p} = -i\hbar\nabla_{\mathbf{r}}$ transforms as

$$\mathbf{p} = (1 - \hat{\varepsilon})\mathbf{p}', \tag{8.23}$$

where $\mathbf{p}' = -i\hbar\nabla_{\mathbf{r}'}$. The transformation (8.23) takes the operator p^2 in the kinetic energy part of the Hamiltonian in (8.19) into

$$p^2 = p'^2 - 2\sum_{ij} p'_i \varepsilon_{ij} p'_j \tag{8.24}$$

and the potential energy $V_S(\mathbf{r})$ into $V_S((1 + \hat{\varepsilon})\mathbf{r}')$:

$$V_S(\mathbf{r}) \rightarrow V_S((1 + \hat{\varepsilon})\mathbf{r}'). \tag{8.25}$$

Following [3], let us change the notation $\mathbf{r}' \to \mathbf{r}$. The potential $V_S((1+\hat{\varepsilon})\mathbf{r})$ is now having the same period as $V(\mathbf{r})$, and their difference may be expanded in a series in terms of strain $\hat{\varepsilon}$ as:

$$V_S((1+\hat{\varepsilon})\mathbf{r}) - V(\mathbf{r}) = \sum_{ij} V_{ij}(\mathbf{r})\varepsilon_{ji}, \tag{8.26}$$

where

$$V_{ij}(\mathbf{r}) = \frac{1}{2-\delta_{ij}} \lim_{\hat{\varepsilon}\to 0} \frac{V_S((1+\hat{\varepsilon})\mathbf{r}) - V(\mathbf{r})}{\varepsilon_{ij}}, \tag{8.27}$$

The factor one-half for $i \neq j$ is needed because in this case there are two identical terms $V_{ij}\varepsilon_{ji}$ and $V_{ji}\varepsilon_{ij}$ entering (8.26). The transformation (8.21) brings the Hamiltonian (8.19) into

$$H(\hat{\varepsilon}) = H_0 + H_{\hat{\varepsilon}} + H_{\hat{\varepsilon}so}, \tag{8.28}$$

where

$$H_{\hat{\varepsilon}} = \sum_{ij} \left(-\frac{p_i \varepsilon_{ij} p_j}{m_0} + V_{ij}\varepsilon_{ji} \right) \tag{8.29}$$

and

$$H_{\hat{\varepsilon}so} = \frac{\hbar}{4m_0^2c^2} \sum_{ij} \left(\varepsilon_{ij}\nabla_j V_s[\boldsymbol{\sigma}\mathbf{p}]_j - \nabla_i \varepsilon_{ij} V_s[\boldsymbol{\sigma}\, p]_j - [\boldsymbol{\sigma}\nabla V_s]_i \varepsilon_{ij} p_j \right) \tag{8.30}$$

Under the transformation (8.21) a Bloch function $\psi_{n\mathbf{k}'} = u_{n\mathbf{k}} \exp(i\mathbf{k}\mathbf{r})$ of the Hamiltonian of the strained system (8.19) which corresponds to the eigenenergy $E_n(\mathbf{k}')$ at the wave vector $\mathbf{k}'$ becomes

$$\psi'_{n\mathbf{k}} = u'_{n\mathbf{k}'}((1+\hat{\varepsilon})\mathbf{r}')e^{i\mathbf{k}'(1+\hat{\varepsilon})\mathbf{r}'} = u'_n((1+\hat{\varepsilon})\mathbf{r}')e^{i\mathbf{k}\mathbf{r}}, \tag{8.31}$$

where $\mathbf{k} = (1+\hat{\varepsilon})\mathbf{k}'$, or

$$k_i = (k'_i + \sum_j \varepsilon_{ij} k'_j).$$

The Bloch function (8.31) has the same periodicity as the Bloch functions in the relaxed lattice with the same wave vector $\mathbf{k}$ and can be expanded in terms of the wave functions $\xi_{n\mathbf{k}}$ (6.3):

$$\psi'_{n\mathbf{k}} = \sum_{n'} c_{nn'} \xi_{n'\mathbf{k}}(\mathbf{r}). \tag{8.32}$$

Let us now calculate the change in energy due to strain

$$\delta E(\mathbf{k}) = E(\hat{\varepsilon}, (1-\hat{\varepsilon})\mathbf{k})) - E_0(\mathbf{k}), \tag{8.33}$$

where $E_0(\mathbf{k})$ is the energy of the relaxed crystal and $E(\hat{\varepsilon}, (1-\hat{\varepsilon})\mathbf{k})$ is the energy of the strained crystal at the point $\mathbf{k}' = (1-\hat{\varepsilon})\mathbf{k}$ to which $\mathbf{k}$ is displaced under stress. We substitute the series expansion (8.32) for the wave function $\psi'_{\mathbf{k}'}$ into the Hamiltonian (8.28) and multiply by the conjugate function $\xi^*_{n'\mathbf{k}_0}(\mathbf{r})$. After the integration over $\mathbf{r}$ of the resulting expression one obtains the following system of equations:

$$\sum_{n'} \left((E_{n'}(\mathbf{k}_0) - E)\delta_{n'n} + H'_{nn'} \right) c_{nn'} = 0, \tag{8.34}$$

where the perturbation Hamiltonian H' is [3]

$$H' = H_{\mathbf{k}} + H_{\hat{\varepsilon}} + H_{\hat{\varepsilon}so} + H_{\hat{\varepsilon}\mathbf{k}}, \tag{8.35}$$

Here

$$H_{\mathbf{k}} = \frac{\hbar^2 k^2}{2m_0} + \frac{\hbar \mathbf{k}\pi}{m_0}, \tag{8.36}$$

$$\pi = \mathbf{p} + \frac{\hbar}{2m_0^2 c^2}[\sigma \nabla V].$$

Here $H_{\hat{\varepsilon}}$ and $H_{\hat{\varepsilon}so}$ are defined by (8.29) and (8.30), respectively, and

$$H_{\hat{\varepsilon}\mathbf{k}} = -2\frac{\hbar \sum_{ij} p_i \varepsilon_{ij} k_j}{m_0}. \tag{8.37}$$

In accordance with general degenerate perturbation theory described in Sect. 6, by treating the Hamiltonian H' from (8.35) as a perturbation the modification of energy $E - E_m(\mathbf{k}_0)$ due to strain in the vicinity of the point $\mathbf{k}_0$ is found from the solution of the following equation:

$$\det\left[H'_{mm'} - (E - E_m(\mathbf{k}_0))\right] = 0, \tag{8.38}$$

where the index m' runs through the all bands degenerate with the band $E_m(\mathbf{k}_0)$ at the point $(\mathbf{k}_0)$ in the relaxed crystal, and the matrix elements in the non-relativistic approximation ($c \to \infty$) are:

$$H'_{\mathbf{k},mm'} = \frac{\hbar^2 k^2}{2m_0}\delta_{mm'} + \frac{\hbar}{m_0}\mathbf{k}\mathbf{p}_{mm'} + \frac{\hbar^2}{m_0^2}\sum_s \frac{(\mathbf{k}\mathbf{p}_{m's})(\mathbf{k}\mathbf{p}_{sm})}{E_m - E_s}, \tag{8.39}$$

$$H'_{\hat{\varepsilon}\mathbf{k},mm'} = -2\frac{\hbar}{m_0}\sum_{ij}(p_i)_{mm'}\varepsilon_{ij}k_j + \frac{\hbar}{m_0}\sum_s \frac{(\mathbf{k}\mathbf{p}_{m's})(H_{\hat{\varepsilon}})_{sm} + (H_{\hat{\varepsilon}})_{m's}(\mathbf{k}\mathbf{p}_{sm})}{E_m - E_s}, \tag{8.40}$$

where $(p_i)_{mm'}$ is the matrix element of the i-th projection of the operator $\mathbf{p}$, and

$$H'_{\hat{\varepsilon},mm'} = \sum_{ij}\left(-\frac{(p_i p_j)_{mm'}}{m_0} + V^{ij}_{mm'} \right)\varepsilon_{ij}. \tag{8.41}$$

By introducing

$$D^{ij}_{mm'} = \left(-\frac{(p_i p_j)_{mm'}}{m_0} + V^{ij}_{mm'} \right) \tag{8.42}$$

the Hamiltonian $H'_{\hat{\varepsilon},mm'}$ (5.18) can be written in the form

$$H'_{\hat{\varepsilon},mm'} = \sum_{ij} D^{ij}_{mm'} \varepsilon_{ij}, \tag{8.43}$$

which is exactly the form (8.5). Therefore, (8.42) is the microscopic expression which in principle allows to calculate the values of the deformation potentials. In order to do so we need to know the Bloch functions at the point $\mathbf{k_0}$ of the relaxed crystal as well as the components V_{ij} defined by (8.27).

It is difficult, however, to determine the explicit form of the components V_{ij} because it requires the exact solution of the self-consistent problem in the strained crystal. It is therefore mandatory to have a numerical method that allows to obtain the deformation potential. Generalization of the numerical methods to compute the band structure described in Chap. 5 to include strain is necessary. Such a procedure for tight-binding $sp^3d^5s^*$ model was recently reported [10]. A generalization of the empirical pseudopotential method to include strain is outlined in the next section.

8.3 Empirical Pseudopotential Method with Strain

The empirical pseudopotential method can be easily adapted to incorporate strain effects. It has been already used to investigate the band structure of biaxially strained $Si_{1-x}Ge_x$ grown epitaxially on $Si_{1-y}Ge_y$ for various surface orientations [4,11,16]. In general, to handle arbitrary strain conditions, the following modifications in the band structure calculation have to be incorporated [14]. First, the lattice vectors $\mathbf{a}_i{}'$ of the strained crystal must be calculated by deforming the vectors $\mathbf{a}_i$ of the relaxed crystal according to (7.1). The change of the normalizing volume of the strained unit cell Ω'_0 is determined by (7.2).

Next, from the strained lattice basis vectors, the strained reciprocal lattice vectors $\mathbf{b}_i{}'$ are obtained. Strain-induced symmetry reduction gives rise to a change in shape and volume of the irreducible wedge of the first Brillouin zone [15].

The basis vectors of the reciprocal lattice are used to calculate the reciprocal lattice vectors $\mathbf{G}'_j$ of the strained lattice. Taking into account the periodicity of the problem, the following expansion of the pseudo wave function $\phi_\mathbf{k}$ in the plane wave basis is employed.

The form factors of the local pseudopotential are needed at the strained reciprocal lattice vectors. For this purpose an interpolation of the pseudopotential is required. Different methods have been proposed [6,11,14]. One of them is a cubic spline interpolation through the pseudopotential form factors, V_0, V_3, V_8, V_{11}, and V_{3k_F} [14]. Following [11], V_0 is set to $-2E_\mathrm{F}/3$, and $V_{3k_\mathrm{F}} = 0$, where k_F denotes the Fermi wave vector of the free electron gas. Next, the additional displacement of the two fcc sublattices in the diamond structure modeled by an internal strain parameter (displacement factor) ξ must be incorporated into the calculation.

We use the empirical pseudopotential method to demonstrate the non-trivial modifications of the conduction band structure under shear strain in the next section.

References

1. Balslev, I.: Influence of uniaxial stress on the indirect absorption edge in silicon and germanium. Phys. Rev. **143**, 636–647 (1966)
2. Bardeen, J., Shockley, W.: Deformation potentials and mobilities in non-polar crystals. Phys. Rev. **80**(1), 72–80 (1950)
3. Bir, G.L., Pikus, G.E.: Symmetry and Strain-Induced Effects in Semiconductors. Wiley, New York (1974)
4. Fischetti, M.V., Laux, S.E.: Band structure, deformation potentials, and carrier mobility in Si, Ge, and SiGe alloys. J.Appl.Phys. **80**(4), 2234–2252 (1996)
5. Fischetti, M.V., Laux, S.E.: Monte Carlo simulation of electron transport in Si: The first 20 years. In: 26th European Solid State Device Research Conference, pp. 813–820 (1996)
6. Friedel, P., Hybertsen, M.S., Schlüter, M.: Local empirical pseudopotential approach to the optical properties of Si/Ge superlattices. Phys. Rev. B **39**(11), 7974–7977 (1989)
7. Herring, C., Vogt, E.: Transport and deformation-potential theory for many-valley semiconductors with anisotropic scattering. Phys. Rev. **101**(3), 944–961 (1956)
8. Hinckley, J., Singh, J.: Influence of substrate composition and crystallographic orientation on the band structure of pseudomorphic Si-Ge alloy films. Phys. Rev. B **42**, 3546–3566 (1990)
9. Manku, T., McGregor, J.M., Nathan, A., Roulston, D.J., Noel, J.P., Houghton, D.C.: Drift hole mobility in strained and unstrained doped $Si_{1-x}Ge_x$ alloys. IEEE Trans. Electron Devices **40**(11), 1990–1996 (1993)
10. Niquet, Y.M., Rideau, D., Tavernier, C., Jaouen, H., Blase, X.: Onsite matrix elements of the tight-binding hamiltonian of a strained crystal: Application to silicon, germanium, and their alloys. Phys. Rev. B **79**(24), 245201 (2009)
11. Rieger, M.M., Vogl, P.: Electronic-band parameters in strained $Si_{1-x}Ge_x$ alloys on $Si_{1-y}Ge_y$ substrates. Phys. Rev. B **48**(19), 14,276–14,287 (1993)
12. Singh, J.: Physics of Semiconductors and their Heterostructures. McGraw-Hill, New York (1993)
13. Sun, G., Sun, Y., Nishida, T., Thompson, S.E.: Hole mobility in silicon inversion layers: Stress and surface orientation. J. Appl. Phys. **102**(8), 084501 (2007)
14. Ungersboeck, E.: Advanced modeling of strained CMOS technology. Dissertation, Institute for Microelectronics, TU Wien (2007)
15. Ungersboeck, E., Dhar, S., Karlowatz, G., Kosina, H., Selberherr, S.: Physical modeling of electron mobility enhancement for arbitrarily strained silicon. In: International Workshop on Computational Electroncis, pp. 141–142 (2006)
16. Van de Walle, C.G., Martin, R.M.: Theoretical calculations of heterojunction discontinuities in the Si/Ge system. Phys. Rev. B **34**(8), 5621–5634 (1986)
17. Wang, E., Matagne, P., Shifren, L., Obradovic, B., Kotlyar, R., Cea, S., Stettler, M., Giles, M.D.: Physics of hole transport in strained silicon MOSFET inversion layers. IEEE Trans. Electron Devices **53**(8), 1840–1851 (2006)

Chapter 9
Strain Effects on the Conduction Band of Silicon

9.1 Limitation of the Effective Mass Approximation for the Conduction Band of Silicon

The conduction band in silicon consists of six equivalent valleys with their energy minima located close to the corresponding X-points of the first Brillouin zone. Within the usually used parabolic approximation (5.1) each valley is characterized by two transversal and one longitudinal effective mass [1]. At higher energy a non-parabolic isotropic correction must be included to reproduce the density of states correctly [9].

The change in the conduction band of silicon under biaxial stress is well understood [7, 12]. Biaxial stress causes splitting between the six equivalent valleys. The splitting prompts re-populations between the valleys and also reduces inter-valley scattering. These effects lead to a substantial bulk mobility enhancement due to a biaxial tensile stress. Mobility in biaxially stressed silicon both in the bulk [7] and in the electron inversion layers [6, 17] was carefully investigated. The application of local stress techniques results in creating uniaxially stressed silicon in the channel. Since the usual channel orientation on a (001) wafer is along [110] direction, the local stress is typically aligned with the [110] axis. Although already being used in mass production, the stress along [110] direction has received surprisingly little attention within the research community. Only recently a systematic experimental study of the mobility modification due to [110] stress was performed [19]. It was shown that the electron mobility data on (001) substrate under [110] uniaxial stress is consistent with the conductivity mass depending on the stress value, in contrast to biaxially stressed silicon, where the conductivity mass was shown to be virtually independent of the stress value [7]. This effective mass dependence cannot be recovered within the effective mass approximation, and a generalization of the conduction band description in uniaxially stressed silicon is needed. Another shortcoming of the effective mass approximation for the conduction band becomes apparent in structures with thin silicon bodies. Confining carriers within thin films reduces the channel dimension in transversal direction, which further improves gate channel control. The quantization energy in ultra-thin silicon films may reach hundreds of meV. The parabolic band approximation usually employed for subband structure

V. Sverdlov, *Strain-Induced Effects in Advanced MOSFETs*, Computational Microelectronics, DOI 10.1007/978-3-7091-0382-1_9, © Springer-Verlag/Wien 2011

calculations of confined electrons in silicon inversion layers becomes insufficient in ultra-thin films. A recent study of subband energies and transport in (001) and (110) oriented thin films reveals that even a non-parabolic isotropic dispersion relation is not sufficient to describe experimental data, and a direction-dependent anisotropic non-parabolicity must be introduced [18].

In order to overcome these difficulties, the effective mass approximation usually applied for the conduction band must be generalized. An approach based on the full band structure computed with the empirical pseudo-potential method [5] is promising and has been recently generalized to include strain and spin-orbit interaction [14]. Although it uses a realistic band structure, the method is computationally demanding and needs to be improved to include the self-consistent solution of Poisson equation.

Another approach is based on the **k·p** theory. Recently, a 30-band **k·p** method [11] was employed to investigate subbands in thin films. The method gives an accuracy comparable with the 6-band **k·p** method for the valence band. Since the method is developed around the Γ-symmetry point, it requires all 30 bands to obtain results for the conduction band minima located close to the edge of the first Brillouin zone. The two-band **k·p** model [3, 8, 15, 20] is developed in the vicinity of the X-point and thus provides a natural framework to compute the subband structure, in particular the dependences of the electron effective masses on shear strain and thickness, in thin films. In the case of a square potential well with infinite walls, which is a good approximation for the confining potential in ultra-thin films, the subband structure can be obtained analytically [16]. This allows for an analysis of subband energies, effective masses, non-parabolicity, and the low-field mobility on film thickness for arbitrary stress conditions. The peculiarity of [110] uniaxial stress is that it produces an off-diagonal element of the strain tensor and results in a shear distortion of the crystal. Under shear deformation a significant change in the band structure appears. Namely, due to non-zero values of the shear deformation potential D, the degeneracy between the two lowest conduction bands at the X-points of the Brillouin zone along the [001] axis is lifted [3]. Thus, for non-zero values of an additional energy splitting between the two conduction bands appears at the X-point. Due to this splitting, the effective masses in the valleys along [001] direction are substantially modified and become functions of the shear stress value. It is worth noticing that uniaxial stress along [001] direction, which produces a biaxially stressed (001) substrate, does not contain the shear strain component. Therefore, the influence of [110] uniaxial stress on the band structure and transport must be carefully investigated. In the following we briefly review the main ideas behind the two-band **k·p** model for a valley in the conduction band of silicon. Then we shortly analyze the unprimed subband structure in (001) ultra-thin films, obtaining analytical expressions for the effective masses and non-parabolicity parameter. With these parameters the non-parabolic subband approximations for the subband dispersions are constructed. The non-parabolic subband dispersions can be embedded into a subband Monte Carlo code in order to compute the low-field mobility.

9.2 The Two-Band k·p Model

The energy dispersion of the conduction band in relaxed silicon computed with help of several methods in [100] and [110] directions is shown in Fig. 6.2. The method based on non-local empirical pseudo-potentials from [12, 20] is the most accurate one as compared to DFT band structure results obtained with VASP [21]. The $sp^3d^5s^*$ tight-binding model with parameters from [4] does not reproduce the anisotropy of the conduction band correctly. In addition, an accurate calibration of the parameters of the $sp^3d^5s^*$ model to describe the modification of the conduction band in strained silicon was performed only recently [10]. The **k·p** theory is a well established method to describe the band structure analytically. The **k·p** method reproduces the band structure accurately at energies below 0.5 eV, which is sufficient to describe the subband structure and transport properties of advanced MOSFETs.

In the diamond crystal structure, the lowest two conduction bands Δ_1 and $\Delta_{2'}$ are degenerate at the X-point due to a special symmetry of the diamond structure: the existence of three glide reflection planes, given by $x = a_0/8$, $y = a_0/8$, and $z = a_0/8$ [22]. The plane $z = a_0/8$ is called a glide plane, when it leaves the crystal structure invariant under a translation by $\frac{a_0}{4}(1,1,0)$ followed by a reflection with respect to this plane. If now the strain tensor contains a shear component ε_{xy}, which is a result from stressing the crystal along the [110] direction, the strained lattice belongs to an orthorhombic crystal system. The glide reflection plane $z = a_0/8$ is no longer a symmetry element, thus the degeneracy of the two lowest conduction bands Δ_1 and $\Delta_{2'}$ at the X-point can be lifted [3, 8]. However, in the case silicon biaxially strained in (001) plane the glide reflection symmetry is preserved. Therefore, the crucial difference between silicon grown epitaxially on (001) $Si_{1-y}Ge_y$ substrate or silicon uniaxially strained/stressed along an [100] from silicon stressed along [110] direction is the presence of the non-zero shear strain component ε_{xy}, which removes the glide reflection plane and lifts the degeneracy between the two conduction bands Δ_1 and $\Delta_{2'}$ of the [001] valleys at the X-pont.

In order to obtain the analytical expressions for energy dispersion within the **k·p** theory let us consider the valley pair along the [001] direction. Other pairs of valleys can be analyzed in a similar fashion. In relaxed silicon the two conduction bands Δ_1 ($i = 1$) and $\Delta_{2'}$ ($i = 2$) become degenerate exactly at the X-points. Since the minimum of the conduction band is only $k_0 = 0.15\frac{2\pi}{a}$ away from the X-point, the dispersion around the minimum is well described by the degenerate perturbation theory built at the X-point, which includes only these two bands. Diagonal elements of the Hamiltonian H_{ii} of the strained crystal at the X-point are:

$$H_{ii}^0(\mathbf{k}) = (-1)^{i-1}\frac{\hbar}{m_0}k_z p + \frac{\hbar^2 k_z^2}{2m_l} + \frac{\hbar^2 k_x^2}{2m_t} + \frac{\hbar^2 k_y^2}{2m_t} + \delta E_C, \tag{9.1}$$

where $i = 1, 2$, m_0 is the free electron mass, m_t is the transversal and m_l the longitudinal effective mass. The only difference with respect to the two-band Hamiltonian (6.22) already discussed in Chap. 6 is the presence of the strain-induced shift of both valleys [2]:

$$\delta E_C = D_d(\varepsilon_{xx} + \varepsilon_{yy} + \varepsilon_{zz}) + D_u \varepsilon_{zz}, \tag{9.2}$$

with D_d denoting the dilation and D_u the uniaxial deformation potentials for the conduction band. Here we took into account that the matrix elements $(p_z)_{ii}$ are different only in sign, which is positive for the lower band:

$$p = (p_z)_{11} = -(p_z)_{22}.$$

The values of k_z are counted from the X point. The coupling between the two bands is described by the off-diagonal terms including shear strain [8]:

$$H_{ij}(k) = H_{ij}^0 - D\varepsilon_{xy}, \tag{9.3}$$

where $D = 14\,\text{eV}$ denotes the shear deformation potential, and

$$H_{12}^0(\mathbf{k}) = \frac{\hbar^2 k_x k_y}{M}.$$

The parameter M can be evaluated from the $\mathbf{k}{\cdot}\mathbf{p}$ perturbation theory: We have computed M by the empirical pseudopotential method at the point $k_z = -k_0$, where the numerical value is close to $M \approx m_t/(1 - m_t/m_0)$ reported in [8]. With degenerate perturbation theory we obtain the following dispersion relations of the [001] valleys including the shear strain component for the two lowest conduction bands:

$$\begin{aligned} E(\mathbf{k}) = {} & \frac{\hbar^2 k_z^2}{2m_l} + \frac{\hbar^2(k_x^2 + k_y^2)}{2m_t} + \delta E_C \\ & \pm \sqrt{\left(\frac{\hbar}{m_0} k_z p\right)^2 + \left(D\varepsilon_{xy} - \frac{\hbar^2 k_x k_y}{M}\right)^2}. \end{aligned} \tag{9.4}$$

Below we briefly analyze the behavior of the dispersion relations (9.5) and compare it to the results of the empirical pseudopotential simulations.

9.2.1 Valley Shift Due to Shear Strain

It follows from (9.4) that shear strain ε_{xy} lifts the degeneracy between the conduction bands of the two [001] valleys at the X-point by opening a gap

$$\delta E_{X[001]\varepsilon_{xy}} = 2D|\varepsilon_{xy}| \tag{9.5}$$

The splitting is linear in shear strain. This linear splitting transforms into a nonlinear shift of the valley minimum. This is a consequence of two facts:

- In relaxed silicon the [001] valley minimum is located at the distance $k_0 = 0.15\,(2\pi/a_0)$ away from the corresponding X-point;
- The position of the valley minimum depends on the shear strain component.

In order to find the valley shift due to shear strain we first find the position of the minimum from (9.4). Introducing dimensionless strain as

$$\eta = \frac{2D\varepsilon_{xy}}{\Delta}, \tag{9.6}$$

the position of the minimum is determined as [15, 20]:

$$k_{\min} = -k_0\sqrt{1-\eta^2}, \quad |\eta| < 1. \tag{9.7}$$

It follows from (9.7) that the minimum moves closer to the X-point with the shear strain increased. Because it depends on ε_{xy}^2, the minimum approaches the X-point for tensile ($\varepsilon_{xy} > 0$) as well as for compressive ($\varepsilon_{xy} < 0$) stress along [110] direction. At high strain values $\eta > 1$ the minimum rests exactly at the $X-$*point*. To estimate strain value corresponding to $|\eta| = 1$ we use $\Delta = 0.53\,\text{eV}$, $D = 14\,\text{eV}$, so $\varepsilon_{xy} = 0.019$, or 1.9%. This value of strain is achieved at 6 GPa uniaxial stress applied along [110] direction. Such a value is far above the level of 2 GPa stress achieved so far by the semiconductor industry, however, in nanowires strain of up to 4% can be achieved [13].

The energy shift of the minimum due to shear strain can now be evaluated:

$$\Delta E_{\text{shear}} = \begin{cases} -\frac{\Delta}{4}\eta^2 & , \quad |\eta| < 1 \\ -(2|\eta| - 1)\Delta/4\,, & |\eta| > 1 \end{cases} \tag{9.8}$$

Figure 9.1 shows the dependence of the absolute value of the valley shift as the function of shear strain η. It is observed that for $|\eta| < 1$ the energy shift is proportional to a square of shear strain (9.8), regardless the fact that we have used the linear deformation potential theory in (9.3). This nonlinearity is the consequence of the fact that the valley minimum is located not at the X-point where the linear deformation potential theory was written. As soon as the minimum reaches the X-point, the linear dependence of the valley shift is restored.

Figure 9.2 demonstrates a perfect agreement between k_z energy dependence predicted by (9.4) and the results of numerical pseudopotential calculations. The gap linear in shear strain opening at the X-point is clearly seen for $\varepsilon \neq 0$ in Fig. 9.2. Because the dependence of the additional valley shift is proportional to the square of shear strain, the question naturally arises on how relevant this additional shift is and is it necessary to include higher terms beyond the deformation potential approximation to describe the shift correctly. Figure 9.3 compares the valley shift due to shear strain ε_{xy} and the diagonal component ε_{zz} computed from (9.8) and (9.2), correspondingly (solid line), with the shift obtained from the empirical pseudopotential method (dotted line). The agreement persists up to a high level of strain. The difference between the two curves remains small as compared to the valley shift due to the two contributions (9.8) and (9.2) indicating that the corrections due to the higher order deformation potentials are small.

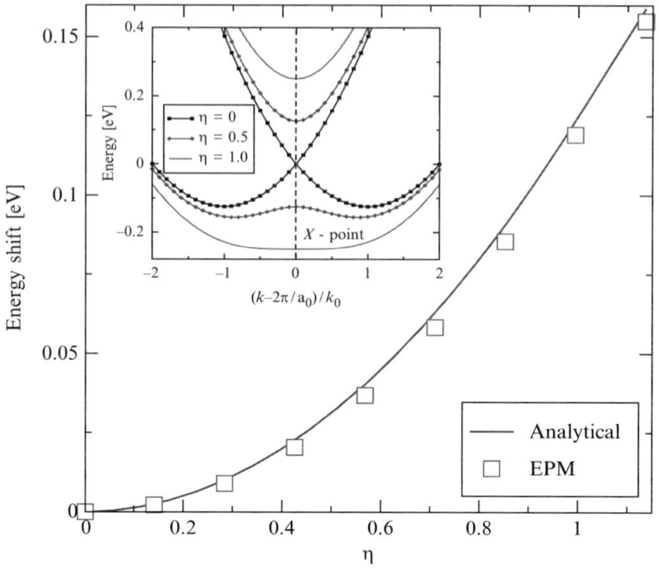

Fig. 9.1 [001] valley energy shift as function of the dimensionless off-diagonal component of the strain tensor, as predicted by (9.8) and by EPM calculations. *Inset*: conduction band profile along the [001] direction for different stress values

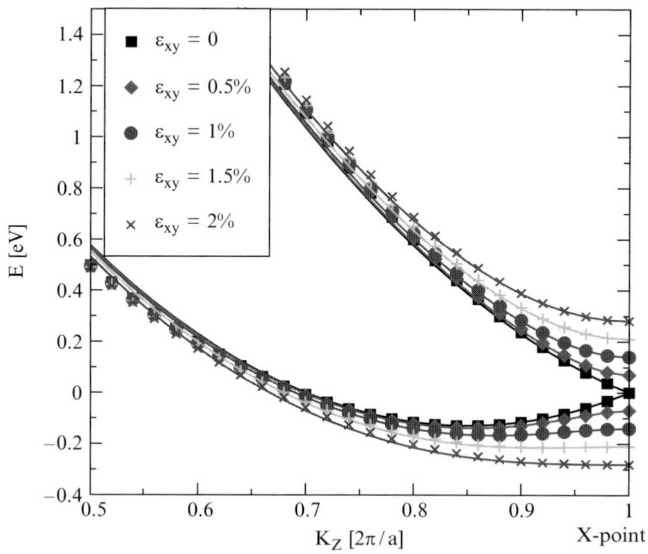

Fig. 9.2 Comparison of the k_z dispersions obtained from the analytical model (9.4) (*lines*) and the EPM calculations (*symbols*), for several values of shear strain. In the analytical dispersion (9.4) k_z is counted from the X-point ($k_z = 2\pi/a - K_z$), $D = 14\,\text{eV}$ and $m_l = 0.91m_0$

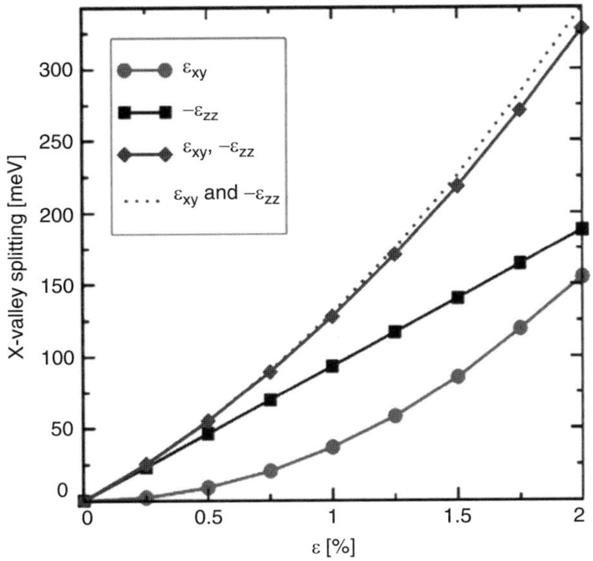

Fig. 9.3 Energy shifts of the [001] valleys due to ε_{xy} and ε_{zz} computed from (9.8) and (9.2), correspondingly, are compared to the shift computed with the empirical pseudopotential method. This figure demonstrates that the contribution due to higher order deformation potentials may be ignored up to high strain level. This check is necessary to prove the relevance of the energy shift (9.8) quadratic in shear strain

9.2.2 Stress-Dependent Transversal Effective Masses

A shear strain component ε_{xy} modifies the effective masses of the [001] valleys. Evaluating the second derivatives of (9.4) at the band minimum (9.8), we obtain two different branches for the effective mass across (m_{t1}) and along (m_{t2}) the stress direction [110] [15,20]:

$$\frac{m_t}{m_{t1}(\eta)} = \begin{cases} \left(1 - \eta \dfrac{m_t}{M}\right) & , \quad |\eta| < 1 \\ \left(1 - \mathrm{sgn}(\eta) \dfrac{m_t}{M}\right), & \quad |\eta| > 1 \end{cases} \tag{9.9}$$

$$\frac{m_t}{m_{t2}(\eta)} = \begin{cases} \left(1 + \eta \dfrac{m_t}{M}\right) & , \quad |\eta| < 1 \\ \left(1 + \mathrm{sgn}(\eta) \dfrac{m_t}{M}\right), & \quad |\eta| > 1 \end{cases} \tag{9.10}$$

Here, $\mathrm{sgn}(\eta)$ denotes the sign function. In Fig. 9.4 the analytical expressions for the transversal masses (9.9) and (9.10) are compared with the masses obtained from EPM calculations. To improve the agreement at high values of stress $\eta \approx 1$, the deformation potential D is set to be slightly stress dependent in the form $D(\eta) = D + \beta\eta^2$, with $\beta = 0.7$ eV.

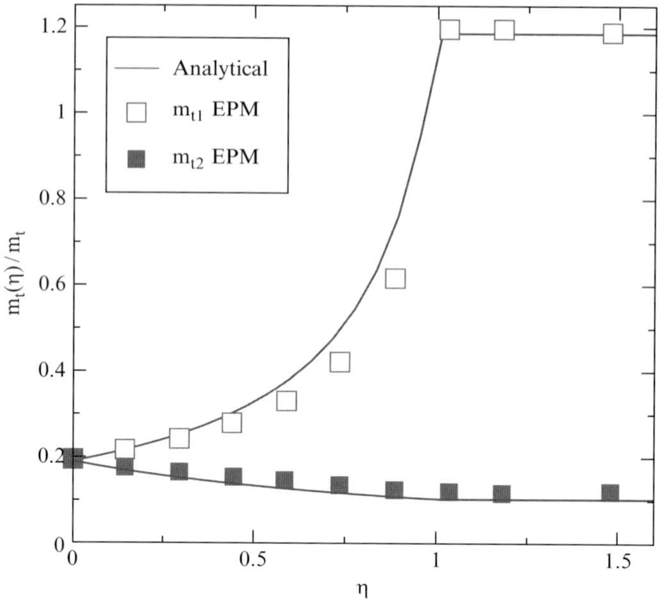

Fig. 9.4 Dependence of the [001] valley transversal effective mass on the dimensionless [110] uniaxial strain as predicted by (9.9) and (9.10) (*lines*) and EPM calculations (*symbols*). Shear stress generates strong anisotropy in the transversal mass

Interestingly, the effective mass depends "linearly" on the shear strain component, contrary to an assessment in [2] that within the linear deformation potential theory there should not be any warping of the conduction band existing. A comparison of the effective mass dependences on strain extracted from the mobility measurement in strained samples [19] with (9.9) and (9.10) shown in Fig. 9.5 displays an excellent agreement.

9.2.3 Dependence on Strain of the Longitudinal Effective Mass

The dependence of $m_l(\eta)$ on shear strain η can be found analogously [15, 20]:

$$m_l(\eta)/m_l = \begin{cases} (1-\eta^2)^{-1} & , \quad |\eta| < 1 \\ (1-1/|\eta|)^{-1} & , \quad |\eta| > 1 \end{cases} \tag{9.11}$$

The transversal mass becomes infinite, when the valley minimum touches the X-point (Fig. 9.6). At this moment the terms higher in order than k_z^2 must be considered in the dispersion.

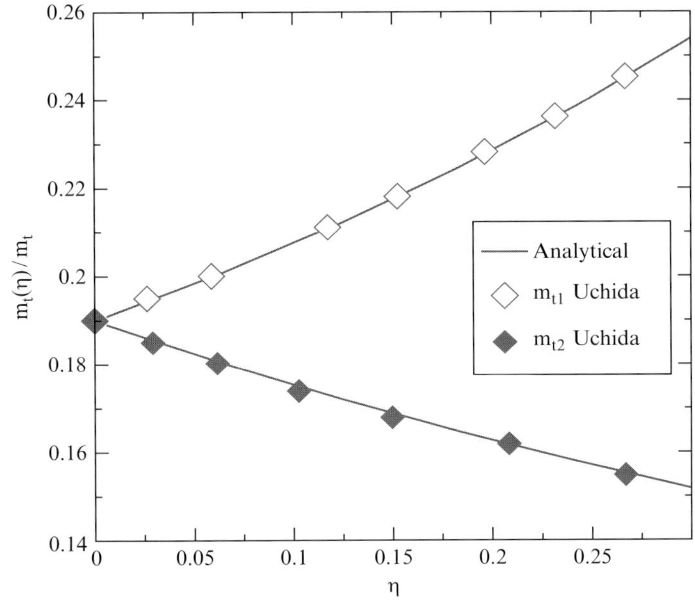

Fig. 9.5 Comparison of (9.9) and (9.10) to the transversal masses extracted from mobility measurements [19]

The cyclotron mass from the experiment [8] is shown in Fig. 9.7 by symbols. In this experiment, a uniaxial tensile stress was applied along the [110] direction. This stress produces the shear strain component ε_{xy}, which affects the masses of the [001] valleys. The direction of applied magnetic field within the plane (001) can be arbitrarily rotated. The angle ϕ determines the direction of the magnetic field with respect to the $[\bar{1}10]$ axis.

The cyclotron mass m_C is defined by the expression

$$m_C(\phi) = \sqrt{m_l m_t(\phi)}, \tag{9.12}$$

where

$$m_t(\phi) = \frac{m_{t1} m_{t2}}{m_{t1}\cos^2(\phi) + m_{t2}\sin^2(\phi)}. \tag{9.13}$$

Figure 9.7 indicates that, apart from the m_l increase described by (9.11), there might be an additional contribution into the longitudinal effective mass which results in the necessity to shift the theoretical curve upwards in order to match experimental data. However, the dependence of the cyclotron mass on angle ϕ is reproduced perfectly supporting the change of the transversal masses with shear strain described by (9.9) and (9.10).

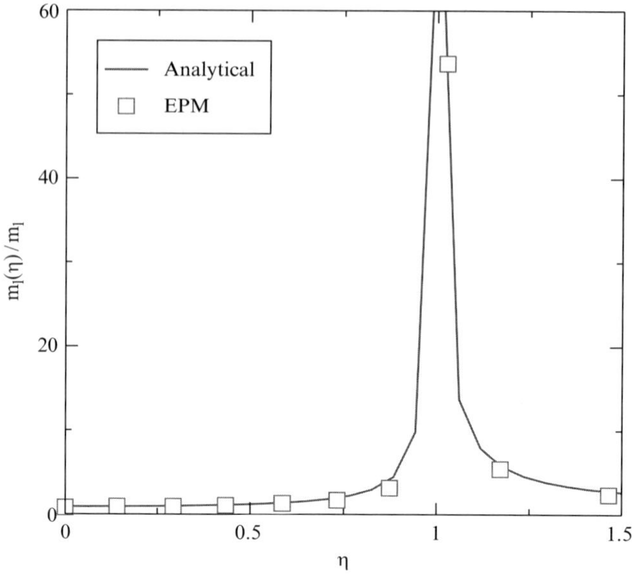

Fig. 9.6 Stress dependence of longitudinal effective mass in the [001] valleys due to [110] stress. Effective mass diverges at $\eta = 1$ suggesting that full-band theory must be used for such stress values

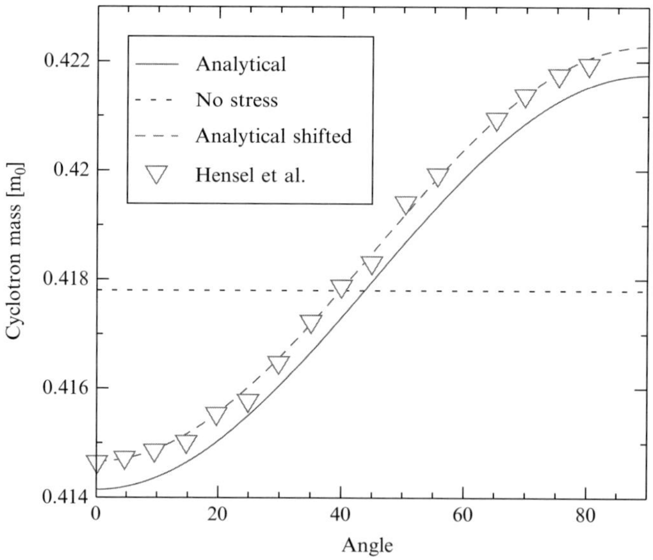

Fig. 9.7 Cyclotron mass as function of field direction obtained with (9.12). While the angular dependence described by (9.12) is accurate, an additional increase to m_l is introduced to reproduce the data from [8]

9.2.4 *Stress and Non-Parabolicity*

Shear strain affects the value of the non-parabolicity parameter α as well. In order to find the dependence, we rewrite the dispersion relation (9.4) in the vicinity of the valley minimum $k_{\min}$, which in the case of $k_x, k_y \neq 0$ is equal to

$$k_{\min} = -k_0 \sqrt{1 - 4\left(\frac{D\varepsilon_{xy} - \frac{\hbar^2 k_x k_y}{M}}{\Delta}\right)^2}, \tag{9.14}$$

and, therefore, depends on $k_x k_y$. Expanding (9.4) around the minimum (9.14) for small $p_z = k_z - k_{\min}$, one obtains:

$$\begin{aligned} E(\mathbf{k}) = & \frac{\hbar^2 p_z^2}{2m_l(\eta)} + \frac{\hbar^2 (k_x^2 + k_y^2)}{2m_t} - \frac{\Delta}{4} \\ & -\frac{1}{\Delta}\left(D\varepsilon_{xy} - \frac{\hbar^2 k_x k_y}{M}\right)^2, \end{aligned} \tag{9.15}$$

Taking into account the shifts of the valley minima (9.14) and the effective mass changes (9.9), (9.10), we rewrite (9.15) as

$$\begin{aligned} E(\mathbf{k}) = & \frac{\hbar^2 p_z^2}{2m_l(\eta)} + \frac{\hbar^2 \tilde{k}_x^2}{2m_{t2}(\eta)} + \frac{\hbar^2 \tilde{k}_y^2}{2m_{t1}(\eta)} \\ & -\frac{\hbar^4 \left(\tilde{k}_x^2 - \tilde{k}_y^2\right)^2}{2M^2 \Delta}, \end{aligned} \tag{9.16}$$

where $\tilde{k}_x = (k_x + k_y)/\sqrt{2}$ and $\tilde{k}_y = (k_x - k_y)/\sqrt{2}$ are the momentum projections in the rotated coordinate system, where the new x axis is along the [110] stress direction. The non-parabolic term written in the original xy coordinate system is similar to the corresponding term in [9]. However, the parabolic term contains the effective masses $m_{t1}(\eta), m_{t2}(\eta)$, which are modified due to [110] stress. Introducing new variables $\tilde{x} = p_x/\sqrt{m_{t2}(\eta)}$, $\tilde{y} = p_y/\sqrt{m_{t1}(\eta)}$, we separate the renormalization due to stress of the parabolic part of the density-of-states (DOS) from the non-parabolic contribution. One rewrites (9.16) at $k_z = -k_0(\eta)$ as

$$\begin{aligned} E(\tilde{x}, \tilde{y}) = & \frac{\hbar^2 \left(\tilde{x}^2 + \tilde{y}^2\right)}{2} \\ & - \left(m_{t2}(\eta)\tilde{x}^2 - m_{t1}(\eta)\tilde{y}^2\right)^2 \frac{\hbar^4}{2M^2 \Delta}. \end{aligned} \tag{9.17}$$

The last term gives the correction to the density of states of the parabolic bands. Assuming the last term to be small, we compute the correction to the DOS identifying it with the isotropic non-parabolicity parameter $\alpha(\eta)$. In order to find the dependence of $\alpha(\eta)$ on strain we write the following expression for the DOS:

$$D(E) = \int \frac{dk_x dk_y}{(2\pi)^2} \delta(E - E(\mathbf{k})) = \frac{\sqrt{m_{t1}(\eta) m_{t2}(\eta)}}{2(2\pi)^2} \int_{E=const} d\phi \frac{d\zeta^2(E,\phi)}{dE}, \tag{9.18}$$

where

$$\zeta^2 = \frac{k_x^2}{m_{t1}} + \frac{k_y^2}{m_{t2}}. \tag{9.19}$$

Taking into account the dependences (9.9) and (9.10) of the masses $m_{t1}(\eta), m_{t2}(\eta)$ on the stress η, we arrive at an expression for the strain dependent non-parabolicity parameter $\alpha(\eta)$:

$$\alpha(\eta) = \alpha_0 \frac{1 + 2(\eta m_t/M)^2}{(1 - (\eta m_t/M)^2)^2} \tag{9.20}$$

Expression (9.20) is plotted in Fig. 9.8. The non-parabolicity parameter depends on the square of the stress value and therefore does not depend on whether stress is tensile or compressive. This is expected, since tensile strain in [110] direction produces compression along $[1\bar{1}0]$, thus α can not depend on the sign of stress. Due to the square dependence, the relative increase of $\alpha(\eta)$ is more pronounced at relatively large stress values.

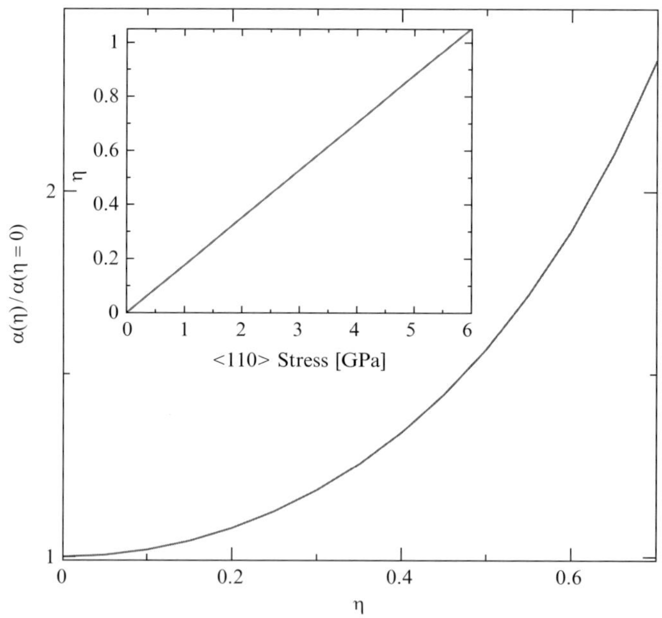

Fig. 9.8 Non-parabolicity parameter in the [001] valleys increases as function of [110] tensile stress, according to (9.20). *Inset*: Relation between dimensionless strain η and stress in GPa

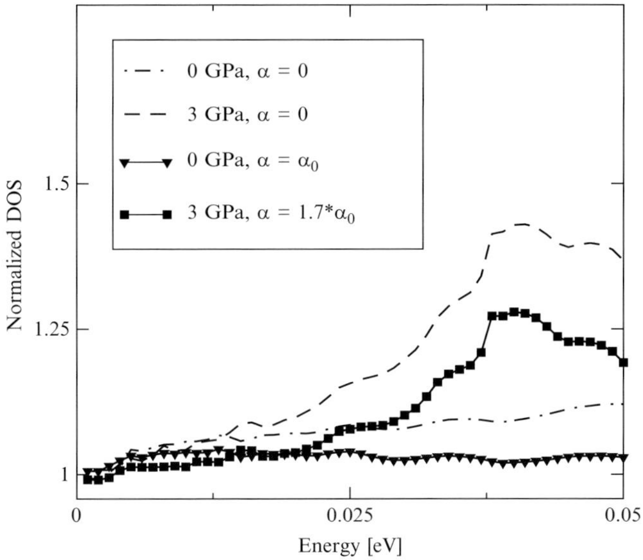

Fig. 9.9 Numerical DOS in the [001] valley normalized to the DOS analytical expressions obtained without (*dotted lines*) and with stress dependent non-parabolicity parameter (*solid lines*), for relaxed Si and Si under [110] uniaxial stress

Figure 9.9 shows the density-of-states of the [001] valley as a function of energy relative to the minimum, obtained numerically from empirical pseudo-potential calculations, for the unstressed case and stress equal to 3 GPa. The value of the non-parabolicity parameter at 3 GPa is 70% higher than the relaxed value α_0, according to (9.20). The DOS is normalized to the analytical DOS corresponding to the parabolic dispersion (dotted lines), with strain dependent transversal effective masses (9.9), (9.10) and strain dependent longitudinal mass $m_l(\eta)$. Without the non-parabolicity parameter included, considerable deviations in the ratio of the numerical DOS to its analytical value is observed at higher energies. The ratio becomes much closer to unity, if the non-parabolicity correction due to stress dependent $\alpha(\eta)$ is taken into account in the analytical DOS (Fig. 9.9, solid lines). However, a deviation of the numerical DOS from the analytical model with the stress dependent non-parabolicity parameter is observed for energies larger than $k_B T$ at high stress value. This happens due to the fact that the energy difference between the value at the minimum and the value at the X-point decreases with stress (see Fig. 9.2). A pronounced peak in the numerical DOS appears, which corresponds to the flat dispersion close to the X-point. The energy value of the peak is shown in Fig. 9.10 for different stress values. It is in agreement with theoretical predictions based on (9.4). For stress values larger than 3 GPa the energy difference from the minimum to the value at the X-point becomes smaller than $2 \times k_B T$, as seen from Fig. 9.10. In this case the non-parabolic approximation with the strain-dependent

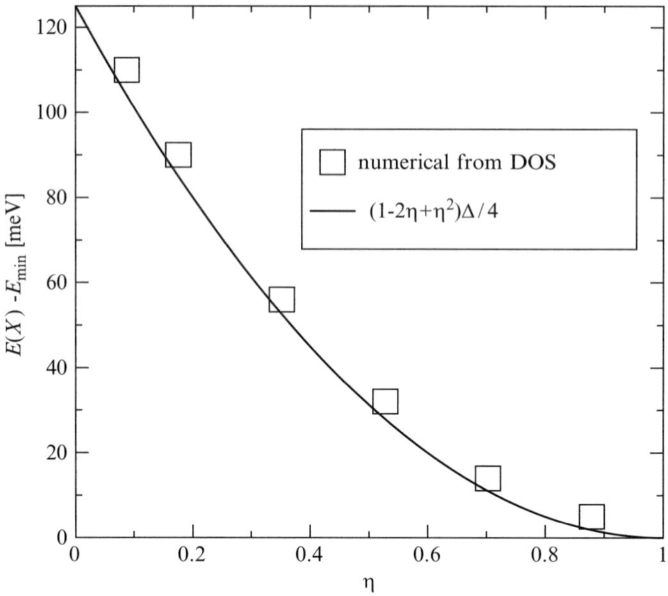

Fig. 9.10 Energy at the X-point relative to the energy value at the minimum. The *solid line* is a theoretical prediction based on (9.4)

effective masses and the non-parabolicity parameter becomes insufficient for the description of even low-field mobility, and a full-band description is required [20].

9.2.5 Comparison of the Two-Band k·p Model with Strain to the Empirical Pseudo-Potential Calculations

We now demonstrate that the two-band **k·p** model [3, 8, 15, 20] not only describes accurately the dependences of the valley shifts and the effective masses on the shear strain component. By comparing the model to results from the empirical pseudopotential method (EPM) we show that the two-band **k·p** model predicts the correct energy dispersion in a wider range of momenta in the Brillouin zone capturing non-parabolicity effects due to the interaction between the two lowest conduction bands.

Figure 9.11 demonstrates a good agreement between the analytical band structure described by (9.4) and the numerical one obtained by the empirical pseudopotential method in the case, when tensile stress of 150MPa in [110] is combined with compressive stress in $[\bar{1}10]$ direction. The chosen strain configuration allows to generate only the shear strain component. Comparison with the analytical band structure within the parabolic effective mass approximation with strain-dependent masses

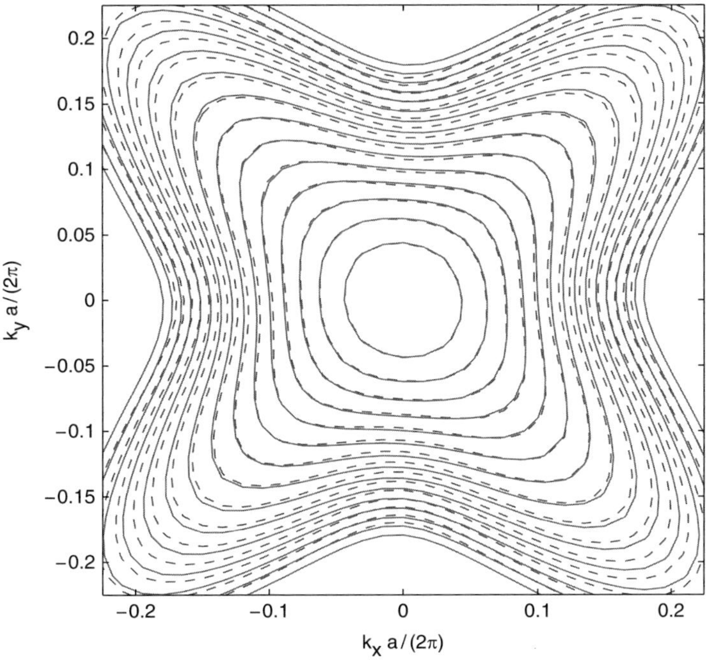

Fig. 9.11 Comparison between the analytical model (9.4) (*dashed lines*) and the EPM calculations (*solid lines*). The contour lines are spaced at 50 meV. Tensile stress in [110] and compressive stress in $[\bar{1}10]$ direction of 150 MPa in each direction is applied

(9.9), (9.10) is shown in Fig. 9.12. The agreement between all the three methods is good for energies close to the valley minimum. At energies larger than 40 meV the parabolic approximation becomes less accurate, while the analytical two-band **k·p** model (9.4) closely follows the numerical dispersion.

The reason for such a good agreement between the two-band **k·p** model (9.4) and the numerical band structure is revealed in [3], where it is shown that the Hamiltonian obeying symmetry of the Brillouin zone at the X-point must be in the form

$$H = A\mathbb{1} + B\sigma_z + C\sigma_x, \tag{9.21}$$

where σ_i are the Pauli matrices, $\mathbb{1}$ is the unit matrix, and A, B, and C are the parameters. The two-band **k·p** Hamiltonian for a [001] valley in silicon defined by (9.1) and (9.2) can be written as:

$$H = \left(\frac{\hbar^2 k_z^2}{2m_l} + \frac{\hbar^2 k_x^2}{2m_t} + \frac{\hbar^2 k_y^2}{2m_t} + \delta E_C\right) + \frac{\hbar}{m_0} k_z p \sigma_z + \left(D\varepsilon_{xy} - \frac{\hbar^2 k_x k_y}{M}\right)\sigma_x, \tag{9.22}$$

which is exactly the form required by symmetry considerations. Since the valley minimum is not far from the X-symmetry point and moves closer when stress is

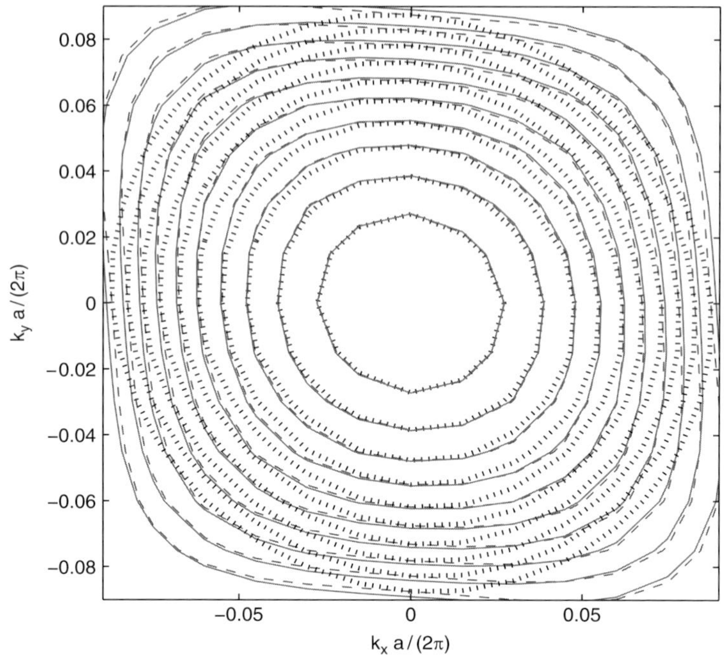

Fig. 9.12 Comparison between the analytical model (9.4) (*dashed lines*), the EPM calculations (*solid lines*), and the parabolic approximation with strain dependent effective masses (9.9), (9.10) (*dotted lines*). The contour lines are spaced at 20 meV. Tensile stress in [110] and compressive stress in $[\bar{1}10]$ direction of 150 MPa in each direction is applied

applied, the Hamiltonian (9.22) does excellent work in approximating the conduction band structure. Because the **k·p** model is valid in a larger portion of the Brillouin zone as compared to the parabolic approximation, it can be used to investigate the dispersion of the quantized subbands in MOSFET surface layers and in thin silicon films.

References

1. Ando, T., Fowler, A.B., Stern, F.: Electronic properties of two-dimensional systems. Rev. Mod. Phys. **54**(2), 437–672 (1982)
2. Balslev, I.: Influence of uniaxial stress on the indirect absorption edge in silicon and germanium. Phys. Rev. **143**, 636–647 (1966)
3. Bir, G.L., Pikus, G.E.: Symmetry and Strain-Induced Effects in Semiconductors. Wiley, New York (1974)
4. Boykin, T.B., Klimeck, G., Oyafuso, F.: Valence band effective-mass expressions in the $sp^3d^5s^*$ empirical tight-binding model applied to a Si and Ge parametrization. Phys. Rev. B **69**(11), 115201 (2004)

5. Esseni, D., Palestri, P.: Linear combination of bulk bands method for investigating the low-dimensional electron gas in nanostructured devices. Phys. Rev. B **72**(16), 165342 (2005)
6. Fischetti, M.V., Gámiz, F., Hänsch, W.: On the enhanced electron mobility in strained-silicon inversion layers. J. Appl. Phys. **92**(12), 7320–7324 (2002)
7. Fischetti, M.V., Laux, S.E.: Monte Carlo simulation of electron transport in Si: The first 20 years. In: 26th European Solid State Device Research Conference, pp. 813–820 (1996)
8. Hensel, J.C., Hasegawa, H., Nakayama, M.: Cyclotron resonance in uniaxially stressed silicon. II. Nature of the covalent bond. Phys. Rev. **138**(1A), A225–A238 (1965)
9. Jacoboni, C., Reggiani, L.: The Monte Carlo method for the solution of charge transport in semiconductors with applications to covalent materials. Rev. Mod. Phys. **55**(3), 645–705 (1983)
10. Niquet, Y.M., Rideau, D., Tavernier, C., Jaouen, H., Blase, X.: Onsite matrix elements of the tight-binding hamiltonian of a strained crystal: Application to silicon, germanium, and their alloys. Phys. Rev. B **79**(24), 245201 (2009)
11. Rideau, D., Feraille, M., Ciampolini, L., Minondo, M., Tavernier, C., Jaouen, H., Ghetti, A.: Strained Si, Ge, and $Si_{1-x}Ge_x$ alloys modeled with a first-principles-optimized full-zone $\mathbf{k}\cdot\mathbf{p}$ method. Phys. Rev. B **74**(19), 195208 (2006)
12. Rieger, M.M., Vogl, P.: Electronic-band parameters in strained $Si_{1-x}Ge_x$ alloys on $Si_{1-y}Ge_y$ substrates. Phys. Rev. B **48**(19), 14,276–14,287 (1993)
13. Shiri, D., Kong, Y., Buin, A., Anantram, M.P.: Strain induced change of bandgap and effective mass in silicon nanowires. Appl. Phys. Lett. **93**(7), 073114 (2008)
14. Sverdlov, V., Baumgartner, O., Kosina, H., Selberherr, S., Schanovsky, F., Esseni, D.: The linear combination of bulk bands-method for electron and hole subband calculations in strained silicon films and surface layers. In: International Workshop on Computational Electroncis, pp. 49–52 (2009)
15. Sverdlov, V., Ungersboeck, E., Kosina, H., Selberherr, S.: Effects of shear strain on the conduction band in silicon: An efficient two-band $\mathbf{k}\cdot\mathbf{p}$ theory. In: Proc. European Solid-State Device Research Conf., pp. 386–389 (2007)
16. Sverdlov, V., Ungersboeck, E., Kosina, H., Selberherr, S.: Current transport models for nanoscale semiconductor devices. Mater. Sci. Eng. R **58**(6–7), 228–270 (2008)
17. Takagi, S.I., Hoyt, J.L., Welser, J.J., Gibbons, J.F.: Comparative study of phonon-limited mobility of two-dimensional electrons in strained and unstrained Si metal-oxide-semiconductor field-effect transistors. J. Appl. Phys. **80**(3), 1567–1577 (1996)
18. Uchida, K., Kinoshita, A., Saitoh, M.: Carrier transport in (110) nMOSFETs: Subband structure, non-parabolicity, mobility characteristics, and uniaxial stress engineering. In: Intl. Electron Devices Meeting, pp. 1019–1021 (2006)
19. Uchida, K., Krishnamohan, T., Saraswat, K.C., Nishi, Y.: Physical mechanisms of electron mobility enhancement in uniaxial stressed MOSFETs and impact of uniaxial stress engineering in ballistic regime. In: Intl. Electron Devices Meeting, pp. 129–132 (2005)
20. Ungersboeck, E., Dhar, S., Karlowatz, G., Sverdlov, V., Kosina, H., Selberherr, S.: The effect of general strain on band structure and electron mobility of silicon. IEEE Trans. Electron Devices **54**(9), 2183–2190 (2007)
21. VASP: Vienna Ab-initio Simulation Program. Kresse, G., Hafner, J.: Phys. Rev. B **47**, 558 (1993); *ibid.* B **49**, 14251 (1994); Kresse, G., Fertmueller, J.: Phys. Rev. B **54**, 11169 (1996); Computs. Mat. Sci. **6**, 15 (1996)
22. Yu, P., Cardona, M.: Fundamentals of Semiconductors. Springer, Berlin (2003)

Chapter 10
Electron Subbands in Silicon in the Effective Mass Approximation

10.1 Arbitrary Substrate Orientation

In MOSFETs the charge is transferred in a surface inversion layer close to the interface between silicon and an oxide. The surface layer is created by applying an appropriate voltage to the gate electrode. In this case a potential well at the oxide interface in silicon is formed. The charge carriers are confined in this potential well. Charge carriers are free to move along the interface $\mathbf{r} = (x, y)$ and the wave function along the interface can be taken in a plane wave form. In the orthogonal direction, however, the carrier motion is confined, so the total wave function is in the form:

$$\Psi(\mathbf{r}, z) = \exp(i\mathbf{kr})\psi_{n\mathbf{k}}(z). \tag{10.1}$$

The width of the well is comparable to the de-Broglie wave length of charge carriers in the quantization direction. The subband energies E are found from the Schrödinger equation for the envelope wavefunctions $\psi_{n\mathbf{k}}(z)$:

$$[T - eU(z)] \exp(i\mathbf{kr})\psi_{n\mathbf{k}}(z) = E_{n\mathbf{k}} \exp(i\mathbf{kr})\psi_{n\mathbf{k}}(z), \tag{10.2}$$

where $U(z)$ is the potential energy and T is the kinetic energy operator. In case it is presented by the corresponding $\mathbf{k}\cdot\mathbf{p}$ expressions for the valence or for the conduction band the potential energy term $U(z)$ is substituted with $U(z)\mathbb{1}$ where $\mathbb{1}$ is the unit matrix. The wave function is then a vector with six component in the former or a spinor with two components in the latter case.

The conduction band in silicon consists of the three pairs of equivalent valleys oriented along the crystallographic axes x, y, z. To calculate the subband structure for any substrate orientation, one has to transform the Hamiltonian by means a unitary transformation from the crystallographic system x_c, y_c, z_c to the interface coordinate system in which the potential energy is in the form $U(z)$. The unitary transformation matrix from the crystallographic coordinate system to the interface coordinate system is given by

V. Sverdlov, *Strain-Induced Effects in Advanced MOSFETs*, Computational Microelectronics, DOI 10.1007/978-3-7091-0382-1_10, © Springer-Verlag/Wien 2011

$$\mathbf{U} = \begin{pmatrix} \cos\phi\cos\theta & -\sin\phi & \cos\phi\sin\theta \\ \sin\phi\cos\theta & \cos\phi & \sin\phi\sin\theta \\ -\sin\theta & 0 & \cos\theta \end{pmatrix}, \tag{10.3}$$

and involves a rotation of ϕ about the z_c axis followed by a subsequent rotation of θ about the new y axis. The direction of the quantization axis is given by

$$\mathbf{e}_z = (\cos\phi\sin\theta, \sin\phi\sin\theta, \cos\theta). \tag{10.4}$$

If a single-band effective mass approximation is applied [1,2], the kinetic energy is in the form (9.1). The momentum operator and the reciprocal effective mass tensor $\hat{\nu} = \hat{m}^{-1}$ in the interface coordinate system can be written as:

$$p_j = \sum_k U_{jk} p_k^{(c)}, \tag{10.5}$$

$$\nu_{ij} = \sum_k U_{ik} U_{jk} \nu_{kk}^{(c)}. \tag{10.6}$$

Here, U_{jk} are the elements of a unitary matrix, $\nu_{kk}^{(c)} = 1/m_k$, and $m_k = m_t, m_l$ are the transversal or longitudinal effective masses in silicon. The kinetic energy is then in the form:

$$T = \frac{1}{2} \sum_{i,j} \nu_{ij} p_i p_j, \qquad i,j = x,y,z. \tag{10.7}$$

where $p_j = -i\hbar(\partial/\partial x_j)$ denotes the momentum operator, and ν_{ij} is the reciprocal effective mass tensor.

The function $\psi_{n\mathbf{k}}(z)$ satisfies the equation:

$$\frac{\hbar^2}{2}\nu_{33}\frac{\mathrm{d}^2\xi}{\mathrm{d}z^2} + i\hbar^2(\nu_{13}k_x + \nu_{23}k_y)\frac{\mathrm{d}\xi}{\mathrm{d}z} + (\grave{E} - U(z))\psi_{n\mathbf{k}}(z) = 0, \tag{10.8}$$

where

$$\grave{E} = E - \frac{\hbar^2}{2}(\nu_{11}k_1^2 + 2\nu_{12}k_1k_2 + \nu_{22}k_2^2). \tag{10.9}$$

Following Stern and Howard [3], the first derivative in (10.8) can be eliminated using the substitution

$$\psi(z) = \zeta(z)\exp\left(-\frac{iz}{\nu_{33}}(\nu_{13}k_1 + \nu_{23}k_2)\right). \tag{10.10}$$

The differential equation for $\zeta(z)$ takes the form

$$\frac{\mathrm{d}^2\zeta_i(z)}{\mathrm{d}z^2} + \frac{2m_\perp}{\hbar^2}(E_i - U(z))\zeta_i(z) = 0, \quad \text{using} \quad m_\perp = 1/\nu_{33}. \tag{10.11}$$

The subband energies E_i and the functions $\zeta_i(z)$ do not depend on the wave vector $\mathbf{k}$ for the boundary condition $\psi_{n\mathbf{k}}(0) = 0$ at the silicon/oxide interface [3]. This boundary condition can be applied, when the potential barrier at the interface is high enough so that the penetration of the wave function into the oxide can be neglected. The eigenfunctions $\zeta(z)$ satisfy the boundary conditions $\lim_{z\to\infty} \zeta_i(z) = \lim_{z\to 0} \zeta_i(z) = 0$ and are assumed to be normalized:

$$\int_0^\infty dz \zeta_i(z) = 1. \tag{10.12}$$

The energy spectrum is given by [3]

$$\begin{aligned} E(k_1, k_2) = E_i + \frac{\hbar^2}{2}\Bigg[&\left(\nu_{11} - \frac{\nu_{13}^2}{\nu_{33}}\right) k_1^2 + 2\left(\nu_{12} - \frac{\nu_{13}\nu_{23}}{\nu_{33}}\right) k_1 k_2 \\ &+ \left(\nu_{22} - \frac{\nu_{23}^2}{\nu_{33}}\right) k_2^2 \Bigg]. \end{aligned} \tag{10.13}$$

The energy levels E_i for a given value of $m_\perp$ generate a set of subband minima called subband ladder. The value of the quantization mass depends on the substrate orientation. Since the conduction band of silicon consists of three pairs of valleys, in the effective mass approximation there exist three ladders of subbands for an arbitrary substrate orientation. For a particular substrate orientation an additional degeneracy between the subband ladders is achieved. According to (10.11) the ladder with the largest quantization mass $m_\perp$ has the lowest energy. Following a convention, the subband ladders are called unprimed, primed, and double primed in compliance with the energy of the lowest subband in a ladder beginning with the lowest energy [1].

The effective mass approximation eases the computational effort considerably, however, as it will be demonstrated in the next section, a more general approach based on the $\mathbf{k}\cdot\mathbf{p}$ Hamiltonian is required to describe the electron subband structure in stressed silicon-on-insulator MOSFETs with ultra-thin silicon body, and the hole subbands in the valence band. The potential energy $U(z)$ is determined by the electrostatic self-consistent potential $\Phi(z)$ defined by all charges in the inversion layer including electrons. Thus, in order to find the subbands in the inversion layer the Schrödinger equation (10.2) must be solved together with the Poisson equation determining the shape of the electrostatic potential $\Psi(z)$:

$$\nabla^2 \Phi = -4\pi \frac{e}{\kappa_{sc}} [N_d(z) + p(z) - n(z)]. \tag{10.14}$$

Here, $N_d(z)$ is the doping profile in the semiconductor, and $p(z)$ and $n(z)$ denote the hole and electron concentration, respectively. The boundary conditions for the Poisson equation in a surface layer are: $\lim_{z\to\infty} \Phi(z) = 0$ far away from the interface in the bulk and

$$\kappa_{ox}\frac{\mathrm{d}\Phi}{\mathrm{d}z}\bigg|_{z=0^-} = \kappa_{sc}\frac{\mathrm{d}\Phi}{\mathrm{d}z}\bigg|_{z=0^+}, \tag{10.15}$$

at the silicon-oxide interface. In (10.15), (10.14) κ_{ox} and κ_{sc} are the dielectric permittivity of the oxide and semiconductor.

The dispersion (10.13)

$$E(k_1, k_2) = E_i + \frac{\hbar^2}{2}\mathbf{k}^T\hat{\mathbf{M}}^{-1}\mathbf{k}, \tag{10.16}$$

represents an ellipse with the principal axes rotated with respect to the axes x and y of the crystallographic coordinate system. The corresponding effective masses $m_{\parallel,1}$ and $m_{\parallel,2}$ associated to motion parallel to the interface are obtained by diagonalizing the matrix

$$\hat{\mathbf{M}}^{-1} = \begin{pmatrix} \nu_{11} - \nu_{13}^2/\nu_{33} & \nu_{12} - \nu_{13}\nu_{23}/\nu_{33} \\ \nu_{12} - \nu_{13}\nu_{23}/\nu_{33} & \nu_{22} - \nu_{23}^2/\nu_{33} \end{pmatrix}. \tag{10.17}$$

The eigenvalues $m_{\parallel,1}^{-1}$ and $m_{\parallel,2}^{-1}$ are calculated by solving the secular equation

$$\det\left(\hat{\mathbf{M}}^{-1} - m_{\parallel}^{-1} I\right) = 0, \tag{10.18}$$

where I denotes the two-dimensional unit matrix.

10.2 Substrate Orientation (001)

The interface coordinate system coincides with the crystallographic system for (001) substrate orientation. In the principal crystallographic system the inverse effective mass tensors describing the energy dispersion of the three valley pairs are

$$\nu'^{(1)} = \begin{pmatrix} \frac{1}{m_l} & 0 & 0 \\ 0 & \frac{1}{m_t} & 0 \\ 0 & 0 & \frac{1}{m_t} \end{pmatrix}, \quad \nu'^{(2)} = \begin{pmatrix} \frac{1}{m_t} & 0 & 0 \\ 0 & \frac{1}{m_l} & 0 \\ 0 & 0 & \frac{1}{m_t} \end{pmatrix}, \quad \text{and} \quad \nu'^{(3)} = \begin{pmatrix} \frac{1}{m_t} & 0 & 0 \\ 0 & \frac{1}{m_t} & 0 \\ 0 & 0 & \frac{1}{m_l} \end{pmatrix}. \tag{10.19}$$

After the kinetic energy operator is determined, the subband structure is found by solving (10.2). Because the potential energy term includes the potential created by carriers confined in the surface layer, (10.2) must be solved simultaneously with the Poisson equation in a self-consistent manner. Subband structure of (001) surface layers in the effective mass approximation is described in details in [1].

10.3 Substrate Orientation (110)

The principal crystallographic system does not coincide with an interface system and a coordinate transformation is necessary to determine the quantization and transport masses for (110) oriented substrate. From the unit vector

$$\mathbf{e}_n^{(110)} = \frac{1}{\sqrt{2}}(1,1,0) = (\cos\phi\sin\theta, \sin\phi\sin\theta, \cos\theta) \tag{10.20}$$

defining the substrate orientation, the angles $\theta = 90°$ and $\phi = 45°$ of the coordinate transformation can be identified. The transformation matrix is written in the form

$$\mathbf{U} = \begin{pmatrix} 0 & -1/\sqrt{2} & 1/\sqrt{2} \\ 0 & 1/\sqrt{2} & 1/\sqrt{2} \\ -1 & 0 & 0 \end{pmatrix}. \tag{10.21}$$

The reciprocal effective mass tensors for the three valley pairs are found from (10.6)

$$\nu^{(1,2)} = \begin{pmatrix} \frac{1}{m_t} & 0 & 0 \\ 0 & \frac{m_t+m_l}{2m_t m_l} & \frac{m_t+m_l}{2m_t m_l} \\ 0 & \frac{m_t+m_l}{2m_t m_l} & \frac{m_t+m_l}{2m_t m_l} \end{pmatrix}, \nu^{(3)} = \begin{pmatrix} \frac{1}{m_l} & 0 & 0 \\ 0 & \frac{1}{m_t} & 0 \\ 0 & 0 & \frac{1}{m_t} \end{pmatrix}. \tag{10.22}$$

The quantization masses $m_{\perp}^{(1,2)} = 1/\nu_{33}^{(1,2)} = 2m_l m_t/(m_t + m_l)$ for the two valley pairs labeled 1 and 2 are equal. The quantization mass of the remaining valley is $m_{\perp}^{(3)} = m_t$. Because $2m_l m_t/(m_t + m_l) > m_t$, the two pairs of valleys with the larger quantization mass $m_{\perp}^{(1,2)}$ form the lowest fourfold degenerate unprimed subband ladder. The remaining valleys with $m_{\perp}^{(3)} = m_t$ belong to the primed twofold degenerate subband ladder. For the transport masses the eigenvalue problem (10.18) for $\mathbf{M}_{\nu}^{-1}$ has to be solved, which for (110) substrate yields

$$\mathbf{M}^{-1}{}_{(1,2)} = \begin{pmatrix} \frac{1}{m_t} & 0 \\ 0 & \frac{2}{m_t+m_l} \end{pmatrix}, \qquad \mathbf{M}_{(3)}^{-1} = \begin{pmatrix} \frac{1}{m_l} & 0 \\ 0 & \frac{1}{m_t} \end{pmatrix}. \tag{10.23}$$

Therefore, the transport masses for the unprimed subband ladder are

$$m_{\parallel,1}^{(1,2)} = m_t, \quad m_{\parallel,2}^{(1,2)} = (m_l + m_t)/2, \tag{10.24}$$

while for the primed ladder

$$m_{\parallel,1}^{(3)} = m_l, \quad m_{\parallel,2}^{(3)} = m_t, \tag{10.25}$$

are obtained. Opposite to the subbands at (001) substrate, where the unprimed ladder is twofold degenerate and the primed ladder is fourfold degenerate, at a (110) interface the unprimed ladder is fourfold and the primed ladder is twofold degenerate.

10.4 Substrate Orientation (111)

The substrate orientation vector $\mathbf{e}_n^{(111)}$ for a (111) substrate is $\frac{1}{\sqrt{3}}(1, 1, 1)$. The axes of the principal crystallographic systems have to be rotated by the angles $\cos\theta = 1/\sqrt{3}$ ($\theta \approx 54.74°$) and $\phi = 45°$. The inverse effective mass tensors for the three valley pairs are

$$\nu^{(1)} = \begin{pmatrix} \frac{m_t+5m_l}{6m_t m_l} & \frac{m_l-m_t}{2\sqrt{3}m_t m_l} & \frac{m_t-m_l}{3\sqrt{2}m_t m_l} \\ \frac{m_l-m_t}{2\sqrt{3}m_t m_l} & \frac{m_t+m_l}{2m_t m_l} & \frac{m_l-m_t}{\sqrt{6}m_t m_l} \\ \frac{m_t-m_l}{3\sqrt{2}m_t m_l} & \frac{m_l-m_t}{\sqrt{6}m_t m_l} & \frac{m_t+2m_l}{3m_t m_l} \end{pmatrix}, \tag{10.26}$$

$$\nu^{(2)} = \begin{pmatrix} \frac{m_t+5m_l}{6m_t m_l} & \frac{m_t-m_l}{2\sqrt{3}m_t m_l} & \frac{m_t-m_l}{3\sqrt{2}m_t m_l} \\ \frac{m_t-m_l}{2\sqrt{3}m_t m_l} & \frac{m_t+m_l}{2m_t m_l} & \frac{m_t-m_l}{\sqrt{6}m_t m_l} \\ \frac{m_t-m_l}{3\sqrt{2}m_t m_l} & \frac{m_t-m_l}{\sqrt{6}m_t m_l} & \frac{m_t+2m_l}{3m_t m_l} \end{pmatrix}, \tag{10.27}$$

$$\nu^{(3)} = \begin{pmatrix} \frac{2m_t+m_l}{3m_t m_l} & 0 & \frac{\sqrt{2}(m_l-m_t)}{3m_t m_l} \\ 0 & \frac{1}{m_t} & 0 \\ \frac{\sqrt{2}(m_l-m_t)}{3m_t m_l} & 0 & \frac{m_t+2m_l}{3m_t m_l} \end{pmatrix}. \tag{10.28}$$

Thus, all valleys have the same quantization mass $m_3^{(1,2,3)} = \frac{3m_t m_l}{m_t+2m_l}$ and therefore belong to the same sixfold degenerate subband ladder. The inverse transport masses are obtained as the eigenvalues of $\mathbf{M}_\nu^{-1}$

$$\mathbf{M}^{-1}_{(1,2)} = \begin{pmatrix} \frac{3(m_l+m_t)}{2m_t(2m_l+m_t)} & \frac{\sqrt{3}(m_l-m_t)}{2m_t(2m_l+m_t)} \\ \frac{\sqrt{3}(m_l-m_t)}{2m_t(2m_l+m_t)} & \frac{m_l+5m_t}{4m_l m_t+2m_t^2} \end{pmatrix}, \quad \mathbf{M}^{-1}_{(3)} = \begin{pmatrix} \frac{3}{2m_l+m_t} & 0 \\ 0 & \frac{1}{m_t} \end{pmatrix}, \tag{10.29}$$

from where one obtains

$$m_{\parallel,1}^{(1,2,3)} = m_t, \quad \text{and} \quad m_{\parallel,2}^{(1,2,3)} = \frac{2m_l + m_t}{3}, \tag{10.30}$$

for all three valley pairs. The subband ladder is sixfold degenerate.

Table 10.1 Principal effective masses of the silicon subband ladders for three substrate orientations. $m_{\|,1}$, and $m_{\|,2}$ denote the transport masses, while $m_{\perp}$ stands for the quantization mass

Surface orientation	Degeneracy	$m_{\|,1}$	$m_{\|,2}$	$m_{\perp}$	Ladder
(001)	2	m_t	m_t	m_l	Unprimed
	4	m_l	m_t	m_t	Primed
(110)	4	m_t	$\frac{m_t + m_l}{2}$	$\frac{2m_t m_l}{m_t + m_l}$	Unprimed
	2	m_l	m_t	m_t	Primed
(111)	6	m_t	$\frac{2m_l + m_t}{3}$	$\frac{3m_t m_l}{m_t + 2m_l}$	Unprimed

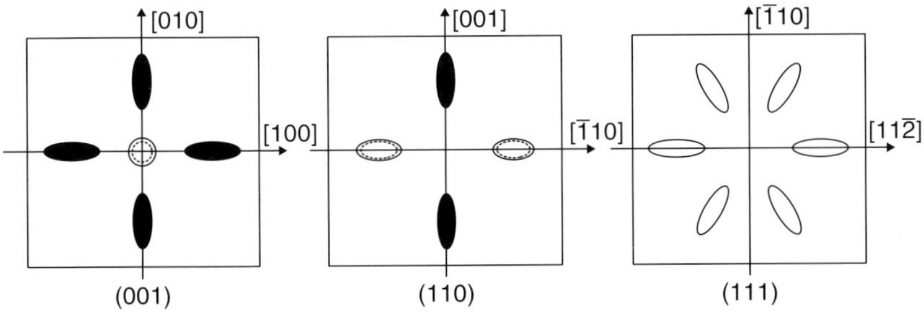

Fig. 10.1 Schematic subband structure at (001) (*left panel*), (110) (*middle panel*), and (111) (*right panel*) interfaces. At (001) and (110) surfaces the subbands belong to the twofold and fourfold degenerate ladders, while at a (111) interface there exists one ladder with degeneracy equal six

In Table 10.1 the principal effective masses of silicon for the three substrate orientations are summarized [3]. In relaxed silicon the six conduction band valleys form a set of two subband ladders for substrate orientation (001) and (110), whereas for (111) oriented substrate only one subband ladder with sixfold degeneracy is formed, as it is schematically shown in Fig. 10.1

References

1. Ando, T., Fowler, A.B., Stern, F.: Electronic properties of two-dimensional systems. Rev. Mod. Phys. **54**(2), 437–672 (1982)
2. Bastard, G.: Superlattice band structure in the envelope-function approximation. Phys. Rev. B **24**(10), 5693–5697 (1981)
3. Stern, F., Howard, W.E.: Properties of semiconductor surface inversion layers in the electric quantum limit. Phys. Rev. **163**(3), 816–835 (1967)

Chapter 11
Electron Subbands in Thin Silicon Films

11.1 Numerical Methods for Subband Structure Calculations

Strain and hybrid orientation techniques are among the most important concepts to increase the performance of modern MOSFETs. The reason for the mobility enhancement lies in the band structure modification caused by stress. Multi-gate FinFETs and ultra-thin silicon body-based Silicon-On-Insulator (SOI) FETs are considered as perfect candidates for the 22 nm technology node and beyond. Modification of the subband structure of inversion channels is the reason for improved transport characteristics of strained devices. Strong size quantization leads to a formation of quasi-two-dimensional subbands in carrier systems within thin silicon films. The electron subband structure is usually approximated by six equivalent minima located close to the X-points in the Brillouin zone. The dispersion of the conduction band valley is usually described by a parabolic approximation with the transversal masses m_t and the longitudinal mass m_l. Isotropic non-parabolicity takes into account deviations in the density of states at higher energies. In ultra-thin body (UTB) FETs, however, the band non-parabolicity affects the subband energies substantially, and it was recently indicated that anisotropic, direction-dependent non-parabolicity could explain a peculiar mobility behavior at high carrier concentrations in a FET with (110) UTB orientation [32]. Therefore, a more refined description of the conduction band dispersion beyond the usual single-band non-parabolic approximation is needed. Another reason to challenge this standard approximation is its inability to address properly the band structure modification under stress, where a more general description is needed [33]. Indeed, shear strain also modifies substantially both the longitudinal [27, 34] and transversal [13, 27, 33, 34] effective masses. Any dependence of the effective masses on stress is neglected within the single-band description of the conduction band and can only be introduced phenomenologically. In order to describe the dependence of the effective mass on stress a single-band description is not sufficient, and coupling to other bands has to be taken into account.

Several options are available. The **k·p** theory is a well established method to describe the band structure analytically [16]. Recently, a 30-bands **k·p** theory was introduced [20]. Although universal, it cannot provide an explicit analytical solution

V. Sverdlov, *Strain-Induced Effects in Advanced MOSFETs*, Computational Microelectronics, DOI 10.1007/978-3-7091-0382-1_11, © Springer-Verlag/Wien 2011

for the energy dispersion. An efficient two-band **k·p** theory [4,13,27,34] reproduces the band structure at low energies quite well as shown in Chap. 9.

We use the empirical non-local pseudopotential method (EPM) for numerical band structure calculations. The parameters of the EPM are adjusted in order to reproduce the measurable quantities of semiconductors, energy gap and effective masses, as well as results of the first principle density-functional calculations. The method includes spin-orbit coupling. In our calculations of the silicon band structure we used the parameters from [21]. As follows from Chap. 5, the EPM results are the most accurate, when compared to the first principle density-functional band structure calculations obtained with the VASP [35].

Accurate modeling of the subband structure is mandatory in modern TCAD tools in order to be able to describe MOSFET performance enhancement. The "Linear Combination of Bulk Bands" (LCBB) method [8] approximates the wave function in a size-quantized system as a superposition of the bulk Bloch functions found from a pseudopotential method. The coefficients in the linear combination depend on the confinement potential. In its original version the LCBB method did not include stress and was not able to describe the valence band due to lack of spin-orbit interaction. With a generalization to include spin-orbit interaction and strain the LCBB method can be used to prove that less computationally demanding methods based on the **k·p** Hamiltonian are accurate enough thus providing a viable alternative to electron spectrum calculations in confined systems based on a tight-binding approach. The two-band **k·p** Hamiltonian (9.1)–(9.3) describes the bulk structure of the conduction band accurately up to energies of 0.5 eV [28]. It includes a shear strain component which is neglected in the parabolic approximation [4, 13, 28]. Shear strain is responsible for effective mass modification and is therefore an important source of the electron mobility enhancement in ultra-thin silicon films [32, 34]. We also demonstrate by comparison with LCBB calculations that the six-band **k·p** Hamiltonian accurately describes the hole subband structure.

11.2 "Linear Combination of Bulk Bands" Method

In the LCBB method the solution

$$\psi E(\mathbf{r}) = \sum_{n,k_z} A_n(\mathbf{k}, k_z)\,|n, \mathbf{k}, k_z\rangle \tag{11.1}$$

of the Schrödinger equation $(H_0 + V(z))\,\psi_E(\mathbf{r}) = E\psi_E(\mathbf{r})$ with the confinement potential $V(z)$ is obtained using the complete set of the Bloch functions $|n, \mathbf{k}, k_z\rangle$ $(\mathbf{k} = (k_x, k_y))$ of the Hamiltonian $H_0 = \frac{\mathbf{p}^2}{2m_0} + U(\mathbf{r})$ of bulk silicon [8]:

$$\left(\frac{\mathbf{p}^2}{2m_0} + U(\mathbf{r})\right)|n, \mathbf{k}, k_z\rangle = E_{FB}(\mathbf{k}, k_z)\,|n, \mathbf{k}, k_z\rangle\,.$$

Therefore, the method is suitable to get the dispersion relations for any confining potential in silicon. In this work we model the thin silicon film of the thickness t by approximating $V(z)$ with a square well potential although the generalization to an arbitrary potential profile is straightforward. The hight of the potential barrier corresponds to the potential energy barrier at the semiconductor-oxide interface.

The advantage of the LCBB method is that it allows to use accurate band structure and Bloch functions obtained by a suitable numerical method without any additional approximation. The Schrödinger equation written in the basis of (11.1) takes the form [8]:

$$\begin{aligned} &\textstyle\sum_{n,k_z} \langle n, \mathbf{k}, k_z|\, V(z)\, |n, \mathbf{k}, k_z\rangle\, A_n(\mathbf{k}, k_z) \\ &= (E_n(\mathbf{k}) - E_{FB}(\mathbf{k}, k_z))\, A_n(\mathbf{k}, k_z), \end{aligned} \tag{11.2}$$

where $E_n(\mathbf{k})$ is the n-the subband energy at $\mathbf{k}$. It is important to stress that in the summation of (11.2) k_z must vary in the whole periodicity interval of the reciprocal lattice vector space from $[-2\pi/a_0, 2\pi/a_0]$ along the k_z direction [8]. The matrix element of the confining potential can be expressed via the overlap integrals of the periodic parts of the Bloch functions $\langle u_{n,\mathbf{k},k_z} \left| u_{n',\mathbf{k},k_z'} \right\rangle$ as

$$\begin{aligned} &\langle n, \mathbf{k}, k_z|\, V(z) \left| n', \mathbf{k}, k_z' \right\rangle \\ &= \textstyle\sum_{(\mathbf{g},g_z)} \langle u_{n,\mathbf{k}-\mathbf{g},k_z-g_z} \left| u_{n',\mathbf{k},k_z'} \right\rangle \\ &\times V(k_z' - k_z + g_z)\delta_{\mathbf{k},\mathbf{k}'+\mathbf{g}}, \end{aligned} \tag{11.3}$$

where $(\mathbf{g}, g_z)$ is the reciprocal lattice vector,

$$V(k_z) = \int \frac{dz}{L_z} V(z) \exp(i k_z z)$$

is the Fourier harmonic of the confining potential and $\delta_{\mathbf{k},\mathbf{k}'+\mathbf{g}}$ is the Kronecker's delta [8]. The Fourier transform $V(q)$ decays rapidly for $q > 2\sqrt{g_z^2 + g^2}$. This makes it sufficient to take only a few terms in the summation over the reciprocal lattice vectors in (11.3) to obtain the desired accuracy of in the subband energies $E_M(\mathbf{k})$.

Important ingredients of the LCBB method are the overlap integrals (11.3) of the periodic parts of the Bloch functions. Overlap integrals of the Bloch functions from the first and second conduction band computed with pseudopotentials from [34] are in excellent agreement with previously published results [8] as demonstrated in Fig. 11.1. The advantage of the pseudopotential method [34] is that it provides the band structure in strained silicon. The method also includes the spin-orbit coupling which is important to describe the valence band.

We first investigate the subband structure of a (001) silicon film. The dispersion relations for the unprimed subbands in a 5.4 nm thick relaxed Si film in [110] direction are shown in Fig. 11.2. The subband calculations based on the full-band

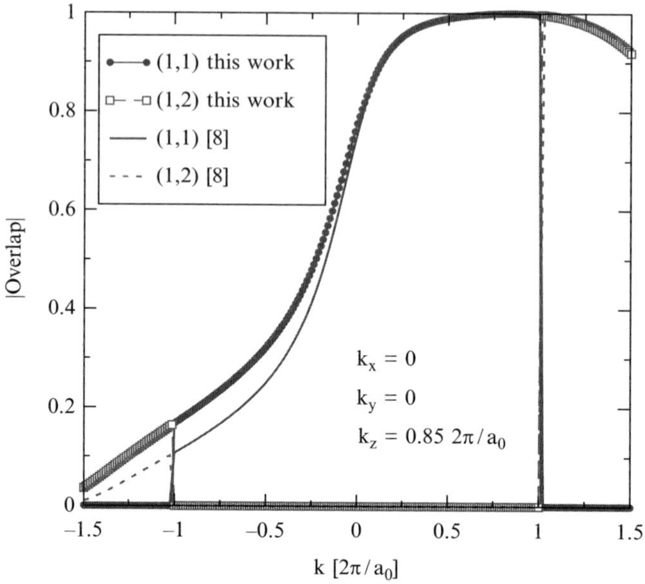

Fig. 11.1 Comparison of the overlap integrals obtained with the two sets of pseudopotentials from [8] and [34]. (i, j) means the overlap integral between the Bloch functions of the conduction bands i and j

Fig. 11.2 Dispersion relations of the unprimed subbands for the relaxed (001) 5.4 nm thick Si film. The subband degeneracy is lifted for finite momentum in [110] direction

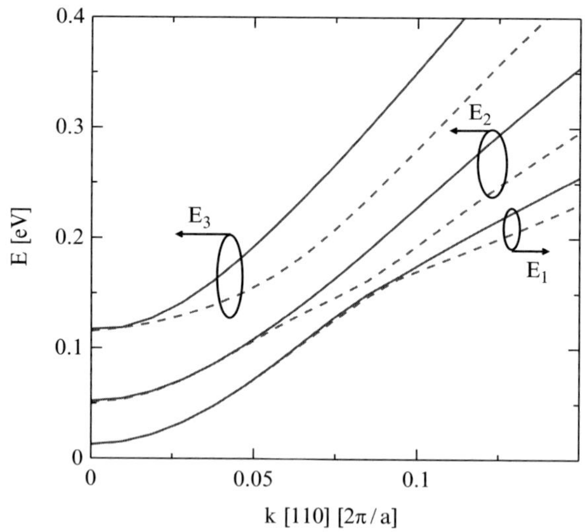

consideration demonstrate that the two unprimed subbands with the same quantum number, which are completely equivalent in the effective mass approximation, develop quite a difference in energies for finite momentum in [110] direction. This result suggests that even the effective masses of the two unprimed subbands with

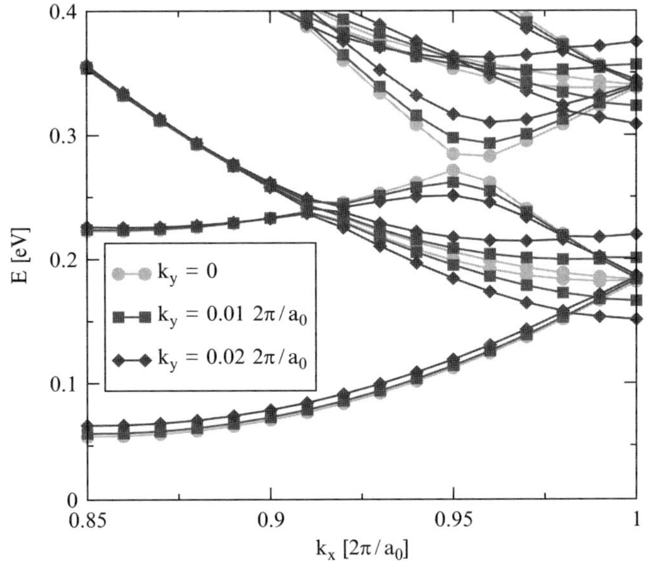

Fig. 11.3 Primed subband dispersion relations for a (001) unstressed 5.4 nm thick Si film in [001] direction. The third valley [8] splits for finite k_y. It indicates that the third valley may originate from the two [100] primed subbands

the same quantum number may be different, an observation that has never been mentioned before.

Dispersion relations of primed subbands in a 5.4 nm (001) silicon film are shown in Fig. 11.3 at $k_y = 0$ and also for non-zero values of k_y. For $k_y = 0$ we can clearly see the third valley with the minimum at the X-point [7, 8]. However, for $k_y \neq 0$ it clearly splits in two branches, with the energy of the lower branch decreasing while increasing k_y and moving away from the X-point in [100] direction. For $k_y = 0.15(2\pi/a_0)$ and $k_x = 2\pi/a_0$ the energy of the lower branch becomes equal to the energy of the unprimed subband at the minimum $k_x = 0.85(2\pi/a_0)$, $k_y = 0$. A similar behavior is observed, when the value of k_y is inverted to $k_y \to -k_y$. This behavior indicates that the third valley originates from a pair of primed subbands with minima along [100] direction. Thus, one can conclude that the feature interpreted in [8] as the "third" valley is the cross section of the primed subband dispersions with the minima in [100] direction by the $k_y = 0$ plane.

Next we consider an example of a (001) film stressed along [110] direction. Stress in [110] direction generates the shear strain component which causes a profound modification of the bulk dispersion [13, 27, 34]. Due to the shear strain component the degeneracy between the unprimed subbands with the same quantum number is lifted even at $k_x = k_y = 0$ resulting in a large strain-induced valley splitting (Fig. 11.4). Substantial differences in the dispersion along [110] (solid lines) and $[1\bar{1}0]$ (dashed lines) directions indicate strong modification of the effective masses due to strain and reduced thickness [27]. Finally, we have carried the subband

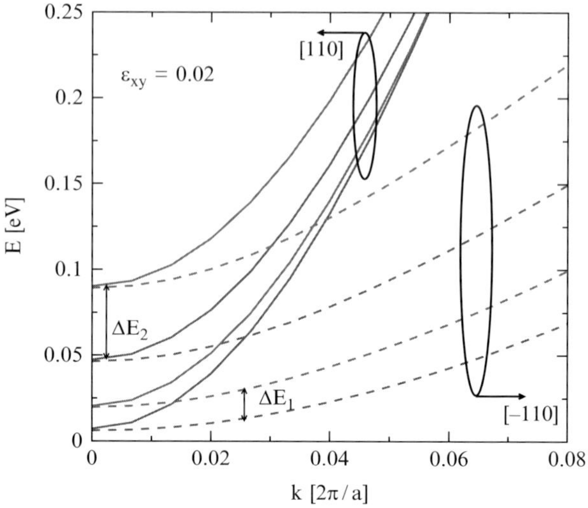

Fig. 11.4 Strain-induced splitting between unprimed subbands is observed in a (001) 5.4 nm thick Si film. Strain dependence of the effective masses results in the difference between the subband dispersions in [110] and [1$\bar{1}$0] directions

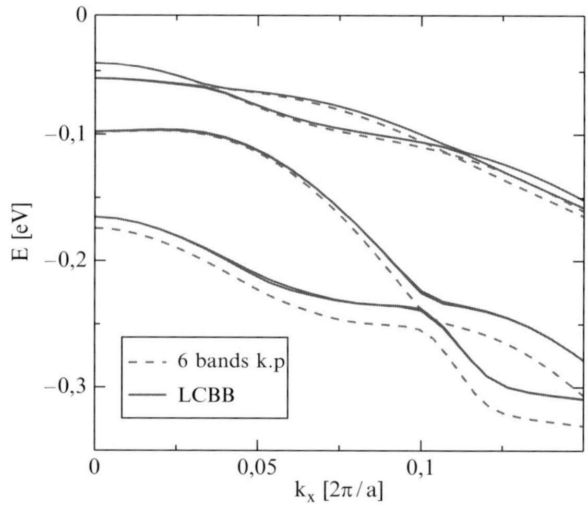

Fig. 11.5 Dispersion relation of hole subbands in a (001) relaxed Si film of 5.4 nm thickness obtained with the LCBB and the six-bands **k·p** method. Good agreement is achieved thanks to the inclusion of spin-orbit interaction into the EPM full band calculations

calculations for the valence band. The non-local pseudopotentials [21, 34] include spin-orbit interaction and thus the generalized LCBB method accurately describes the hole subbands. A comparison of subbands obtained with the LCBB method and the 6-bands **k·p** method is shown in Fig. 11.5. Spin-orbit interaction is included in LCBB leading to the formation of split-off subbands.

Now we investigate in details the peculiarities of the electron subband structure by using the two-band **k·p** method. The main interest will be to understand the unprimed subband splitting induced by strain.

11.3 Unprimed Subbands in (001) Films: Analytical Consideration

In this section we analyze the structure of unprimed subbands in thin (001) oriented silicon films based on the two-band **k·p** Hamiltonian (9.1). We perform an analytical analysis of the subband structure of the UTB silicon films under uniaxial strain. For simplicity the assumption of a square well potential with infinite potential walls is used. This assumption is only valid as long as the ground subband energy is much higher than the amplitude of the potential profile variation within the film. As soon as this assumption breaks down, a generalized numerical treatment of the **k·p** Schrödinger and Poisson equations is required. It is shown [3] that this numerical procedure can be implemented: the electron subband structure is obtained self-consistently thus completing the analysis of uniaxially strained UTB silicon films and inversion layers.

The two-band **k·p** Hamiltonian of a [001] valley in the vicinity of the X-point of the Brillouin zone in Si must be in the form (9.1) [4]:

$$H = \left(\frac{\hbar^2 k_z^2}{2m_l} + \frac{\hbar^2 (k_x^2 + k_y^2)}{2m_t} + V(z) \right) \mathbb{1} + \left(2D\varepsilon_{xy} - \frac{\hbar^2 k_x k_y}{M} \right) \sigma_z + \frac{\hbar^2 k_z k_0}{m_l} \sigma_x \tag{11.4}$$

where $\sigma_{x,z}$ are the Pauli matrices, $\mathbb{1}$ is the 2×2 unit matrix, $k_0 = 0.15 \times 2\pi/a_0$ is the position of the valley minimum relative to the X-point in relaxed Si, k_i with $i \in \{x, y, z\}$ is the wave vector, ε_{xy} denotes the shear strain component in physics notations, $M^{-1} = m_t^{-1} - m_0^{-1}$, and $D = 14\,\text{eV}$ is the shear strain deformation potential [4, 13, 28, 34].

The confining potential $V(z)$ along the [001] direction modulates the conduction band profile. As long as the ground subband energy is much higher than the amplitude of the potential profile within the film, the confining potential in an ultra-thin silicon film can be approximated as square well potential with infinite walls (Fig. 11.6). This approximation is applied to single- as well as to double-gate structures.

For a potential with infinite potential walls the wave function must be set to zero at the boundaries. This simplification allows an analytical analysis of the unprimed subband structure in (001) thin silicon films.

Let us briefly outline the procedure to obtain the subband dispersion relations.

In the two-band model the wave function is a spinor with two components:

$$\psi = \begin{pmatrix} a(\mathbf{k}) \\ b(\mathbf{k}) \end{pmatrix} e^{i\mathbf{k}\cdot\mathbf{r}}, \tag{11.5}$$

where $a(\mathbf{k})$ and $b(\mathbf{k})$ are constants which depend on the wave vector $\mathbf{k}$. Substitution of (11.5) into (11.4) and as assumption that the potential in the thin film is zero leads to the following eigenvalue problem:

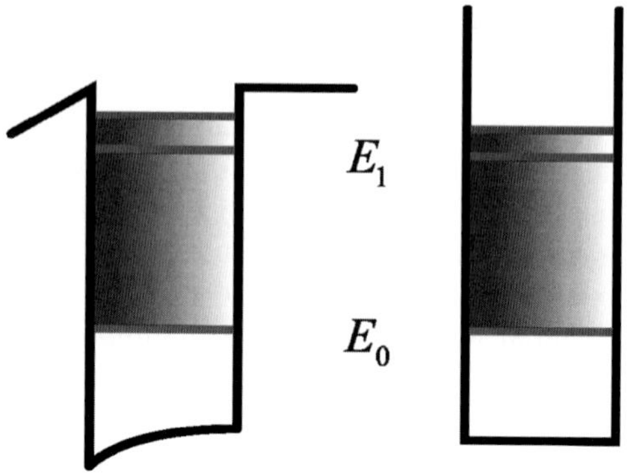

Fig. 11.6 Potential in an ultra-thin SOI film of a single-gate MOSFET (*left*) and a corresponding model square well potential with infinite walls

$$(H - E)\begin{pmatrix} a(\mathbf{k}) \\ b(\mathbf{k}) \end{pmatrix} e^{i\mathbf{k}\cdot\mathbf{r}} = 0. \tag{11.6}$$

Taking the determinant of (11.6) and setting it to zero results in the energy dispersion relation of the system:

$$E(\mathbf{k}) = \frac{\hbar^2 k_z^2}{2m_l} + \frac{\hbar^2(k_x^2 + k_y^2)}{2m_t} \pm \sqrt{\delta^2 + \left(\frac{\hbar^2 k_z k_0}{m_l}\right)^2} \tag{11.7}$$

with

$$\delta = \left(D\varepsilon_{xy} - \frac{\hbar^2 k_x k_y}{\mathrm{M}}\right). \tag{11.8}$$

This is the same dispersion as (9.4), however, only discrete k_z values are allowed in a thin film. To find these values and the subband quantization energies at fixed k_x and k_y, the off-diagonal term $\hbar^2 k_x k_y / M$ leading to the conduction band non-parabolicity is included into the definition of δ in (11.8).

For each energy E there exist four solutions of k_z (Fig. 11.7). Figure 11.7 shows that the dispersion $E(\mathbf{k})$ is even with respect to the X-point at $k = 0$. Therefore, from the four roots k_z only two are independent. The missing two roots are obtained by alternating the signs of the values of the two independent roots k_1 and k_2. The values of the roots are not always real. For non-zero shear strain the gap between the two conduction bands is opened at the X-point (Fig. 8.2). For energies within the gap the k_2 and $-k_2$ values are imaginary.

The wave function at the energy E is a superposition of the four eigenvectors with k_z being the solutions of the equation $E = E(k_z)$:

$$\psi(z)= \begin{pmatrix} a(k_1) \\ b(k_1) \end{pmatrix} e^{ik_1\cdot z} + \begin{pmatrix} a(-k_1) \\ b(-k_1) \end{pmatrix} e^{-ik_1\cdot z} + \begin{pmatrix} a(k_2) \\ b(k_2) \end{pmatrix} e^{ik_2\cdot z} + \begin{pmatrix} a(-k_2) \\ b(-k_2) \end{pmatrix} e^{-ik_2\cdot z}. \tag{11.9}$$

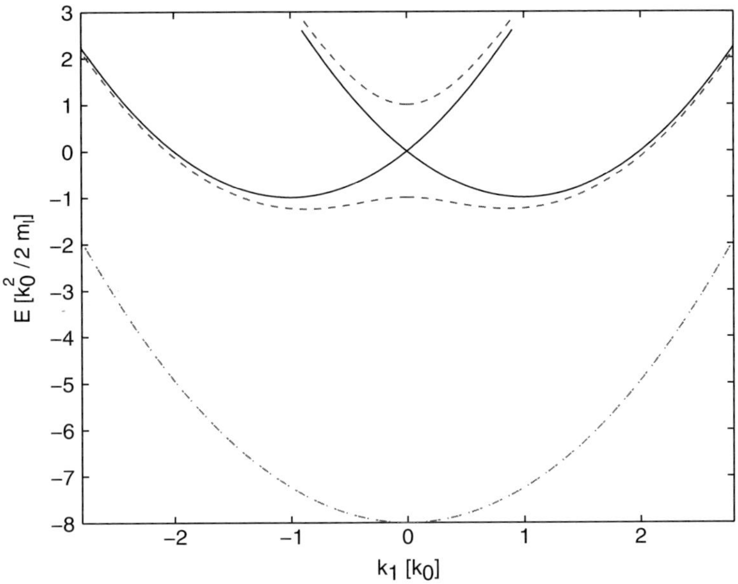

Fig. 11.7 Conduction band profile close to the X-point for $\eta = 0$ (*solid lines*), $\eta = 0.5$ (*dashed lines*), and $\eta = 4$ (*dashed-dotted line*)

The subband quantization energies are obtained by setting to zero both components of the spinor at both film interfaces. This results in a system of four linear homogeneous equations for the coefficients in the linear combination. To find the dispersion equation we introduce the ratio $c(k_z)$ between $b(k_z)$ and $a(k_z)$. $c(k_z)$ is an odd function with respect to k_z. Here we assume that k_x and k_y are constants, so only k_z is preserved as an argument of the functions:

$$c(k_z) = \frac{b(k_z)}{a(k_z)} = -\frac{\frac{\hbar^2 k_z k_0}{m_l}}{\frac{\hbar^2 k_z^2}{2m_l} + \frac{\hbar^2(k_x^2+k_y^2)}{2m_l} + \delta - E(k_z)}. \tag{11.10}$$

The boundary conditions

$$\psi(z = \pm t/2) = 0 \tag{11.11}$$

for a symmetric confining potential demands that the solution satisfies $a(-k_z) = \pm a(k_z)$. This leads to the wave function in the form

$$\begin{aligned}\psi(z) = \;& a(k_1)\left(\begin{pmatrix}1\\ c(k_1)\end{pmatrix} e^{ik_1\cdot z} \pm \begin{pmatrix}1\\ -c(k_1)\end{pmatrix} e^{-ik_1\cdot z}\right)\\ &+ a(k_2)\left(\begin{pmatrix}1\\ c(k_2)\end{pmatrix} e^{ik_2\cdot z} \pm \begin{pmatrix}1\\ -c(k_2)\end{pmatrix} e^{-ik_2\cdot z}\right),\end{aligned} \tag{11.12}$$

After some simplifications, the two independent systems of equations are obtained:

$$a(k_1)\cos(k_1 t/2) + a(k_2)\cos(k_2 t/2) = 0, \tag{11.13}$$

$$a(k_1)c(k_1)\sin(k_1 t/2) + a(k_2)c(k_2)\sin(k_2 t/2) = 0, \tag{11.14}$$

$$a(k_1)\sin(k_1 t/2) + a(k_2)\sin(k_2 t/2) = 0, \tag{11.15}$$

$$a(k_1)c(k_1)\cos(k_1 t/2) + a(k_2)c(k_2)\cos(k_2 t/2) = 0. \tag{11.16}$$

Expressing $a(k_1)$ with (11.13) and (11.15) and putting them into (11.14) and (11.16) leads to these two conditions:

$$\tan(k_1 t/2) = \frac{c(k_2)}{c(k_1)} \quad \tan(k_2 t/2), \tag{11.17}$$

$$\cot(k_1 t/2) = \frac{c(k_2)}{c(k_1)} \quad \cot(k_2 t/2). \tag{11.18}$$

In case when the solution k_2 is imaginary, the corresponding trigonometric functions are substituted with the hyperbolic ones:

$$\tan(k_1 t/2) = \frac{ic(k_2)}{c(k_1)} \quad \tanh(k_2 t/2), \tag{11.19}$$

$$\cot(k_1 t/2) = -\frac{ic(k_2)}{c(k_1)} \quad \coth(k_2 t/2). \tag{11.20}$$

After introducing the dimensionless variables defined as

$$X_{1,2} = \frac{k_{1,2}}{k_0}, \Delta = \frac{2\hbar^2 k_0^2}{m_l}, \varepsilon = \frac{E}{E_0}, \eta = \frac{2\delta}{\Delta}, \tag{11.21}$$

and a few calculation steps presented for convenience in the Appendix (11.9.1), the equations (11.18) can be written in the form:

$$\tan(X_1 k_0 t/2) = \frac{X_2}{X_1} \frac{\eta \pm \sqrt{\eta^2 + X_1^2}}{\eta \pm \sqrt{\eta^2 + X_2^2}} \tan(X_2 k_0 t/2), \tag{11.22}$$

$$\cot(X_1 k_0 t/2) = \frac{X_2}{X_1} \frac{\eta \pm \sqrt{\eta^2 + X_1^2}}{\eta \pm \sqrt{\eta^2 + X_2^2}} \cot(X_2 k_0 t/2). \tag{11.23}$$

For the numerical solution it is convenient to get rid of the trigonometric functions with singularities and to rewrite the equations as

$$\sin(X_1 k_0 t/2)\cos(X_2 k_0 t/2) = \frac{X_2}{X_1}\frac{\eta \pm \sqrt{\eta^2+X_1^2}}{\eta \pm \sqrt{\eta^2+X_2^2}} \cdot \sin(X_2 k_0 t/2)\cos(X_1 k_0 t/2), \tag{11.24}$$

$$\cos(X_1 k_0 t/2)\sin(X_2 k_0 t/2) = \frac{X_2}{X_1}\frac{\eta \pm \sqrt{\eta^2+X_1^2}}{\eta \pm \sqrt{\eta^2+X_2^2}} \cdot \cos(X_2 k_0 t/2)\sin(X_1 k_0 t/2). \tag{11.25}$$

The two variables X_1 and X_2 still coexist in these equations. Therefore, to find either of them, an extra relation to re-express X_1 as a function of X_2 or vice versa is needed. As demonstrated in Appendix(11.9.1), these relations are:

$$X_1^2 = X_2^2 + 4 + 4\sqrt{X_2^2 + \eta^2}, \tag{11.26}$$

$$X_2^2 = X_1^2 + 4 - 4\sqrt{X_1^2 + \eta^2}. \tag{11.27}$$

The procedure for obtaining the energies of unprimed subbands in (001) thin films is as follows. First, one has to re-express one of the X's (say X_2) via another (X_1) with help of (11.27) (or (11.26)) and solve the (11.24) and (11.25) X as a function of the parameter η. Because of the trigonometric functions present in (11.24) and (11.25) there will be two infinite ladders of solutions as functions of η corresponding to the unprimed subband ladders. The subband energy is computed by substituting the solution k_1 into (11.7).

11.3.1 Dispersion Relations from an Auxiliary Tight-Binding Model

The dispersion relations (11.22) and (11.22) were obtained from the two-band **k·p** theory (11.4). The **k·p** theories are usually applied, when the characteristic scale of the variation of envelope function is much less than the lattice constant. In contrast, an atomistic tight-binding approach allows to obtain the whole Bloch function and is free from the limitations of the **k·p** method. Following Boykin [5], we introduce an auxiliary tight-binding model defined on a lattice of sites each containing two localized orbitals $\alpha(z)$ and $\beta(z)$. The Bloch functions $u(z)$ and $v(z)$ at the X-points corresponding to the two bands considered within the two-band **k·p** theory are expressed via the orbitals as

$$u(z) = \sum_n \alpha(z - na) \tag{11.28}$$

and

$$\nu(z) = \sum_n i\beta(z - na), \tag{11.29}$$

where a is the lattice constant, and n is an integer. The Bloch function for an arbitrary k_z counted with respect to the X-point is written in the form [16]:

$$\Psi(z, k_z) = \sum_n \exp(ik_z na)(a(k_z)u(z - na) + ib(k_z)\nu(z - na)). \tag{11.30}$$

The coefficients $a(k_z)$ and $b(k_z)$ are determined by the two-band **k·p** model as:

$$\begin{pmatrix} \frac{\hbar^2 k_z^2}{2m_l} + \frac{\hbar^2(k_x^2+k_y^2)}{2m_t} - \delta - E & \frac{\hbar^2 k_z k_0}{m_l} \\ \frac{\hbar^2 k_z k_0}{m_l} & \frac{\hbar^2 k_z^2}{2m_l} + \frac{\hbar^2(k_x^2+k_y^2)}{2m_t} + \delta - E \end{pmatrix} \begin{pmatrix} a(k_z) \\ b(k_z) \end{pmatrix} = 0. \tag{11.31}$$

As shown in Fig. 11.7, for a particular energy E there exist two pairs of solutions of $k_z = \pm k_1$ and $K_z = \pm k_2$. The wave function for a finite array of N sites corresponding to the energy E is:

$$\begin{aligned}\Psi_\pm(z, E) = \sum_{i=1,2}\sum_{n=-N/2}^{N/2} & C_i(\exp(ik_i na)(a(k_i)u(z - na) + ib(k_i)\nu(z - na)) \\ & + \exp(-ik_i na)(a(k_i)u(z - na) - ib(k_i)\nu(z - na))),\end{aligned} \tag{11.32}$$

where C_i is a constant. Taking into account strong on-site localization of the orbitals α and β and their mutual orthonormality, the allowed values of k_1 or k_2 in the well with infinite potential walls are found from the condition that the wave function is zero at the film interfaces $z = t/2$. This results in two sets of equations (k_2 is assumed real) corresponding to the plus sign in (11.32) [5]:

$$a(k_1)\cos(k_1 t/2) + a(k_2)\cos(k_2 t/2) = 0, \tag{11.33}$$

$$b(k_1)\sin(k_1 t/2) + b(k_2)\sin(k_2 t/2) = 0. \tag{11.34}$$

For the plus sign one obtains

$$a(k_1)\sin(k_1 t/2) + a(k_2)\sin(k_2 t/2) = 0, \tag{11.35}$$

$$b(k_1)\cos(k_1 t/2) + b(k_2)\cos(k_2 t/2) = 0. \tag{11.36}$$

Equations (11.33), (11.34) result in the dispersion equations, correspondingly:

$$\cot(k_1 t/2) = \frac{a(k_2)b(k_1)}{a(k_1)b(k_2)} \quad \cot(k_2 t/2) \tag{11.37}$$

$$\tan(k_1 t/2) = \frac{a(k_2)b(k_1)}{a(k_1)b(k_2)} \quad \tan(k_2 t/2) \tag{11.38}$$

Interestingly, the dispersion equations (11.38) are the same as (11.18). The only condition used to obtain (11.38), was the smallness of the localization orbital radius as compared to the film thickness. Therefore, this condition determines the region of validity of the dispersion equations (11.18) and (11.38).

Let us now reformulate (11.24) and (11.25) for the sum and the difference of X_1 and X_2. First we introduce the transformation rules for y_n and $\bar{y}_n$ as,

$$y_n = \frac{X_1 - X_2}{2} \quad \text{and} \quad \bar{y}_n = \frac{X_1 + X_2}{2} \tag{11.39}$$

or

$$(y_n + \bar{y}_n)^2 = X_1^2 \quad \text{and} \quad (y_n - \bar{y}_n)^2 = X_2^2. \tag{11.40}$$

We only show the derivation for (11.24), due to the similarity with (11.25). Using the above given transformation and rewriting (11.24) to separate y_n and $\bar{y}_n$ leads to the following expression:

$$\sin(y_n k_0 t) + \sin(\bar{y}_n k_0 t) = \frac{c(X_2)}{c(X_1)} \cdot (-\sin(y_n k_0 t) + \sin(\bar{y}_n k_0 t)) \tag{11.41}$$

Further simplification steps result in:

$$\sin(y_n k_0 t) = \frac{c(X_2) - c(X_1)}{c(X_2) + c(X_1)} \sin(\bar{y}_n k_0 t) \tag{11.42}$$

Now, as it is shown in Appendix(11.9.2), one can re-express $\bar{y}_n$ as function of y_n

$$\bar{y}_n^2 = \frac{1 - y_n^2 - \eta^2}{1 - y_n^2}. \tag{11.43}$$

The derivation of the fraction containing $c(k_1)$ and $c(k_2)$ can also be found in Appendix (11.9.2). The resulting dispersion equation is finally in the form:

$$\sin(y_n k_0 t) = \pm \frac{\eta\, y_n \sin\left(\sqrt{\frac{1-y_n^2-\eta^2}{1-y_n^2}} k_0 t\right)}{\sqrt{(1 - y_n^2)(1 - \eta^2 - y_n^2)}} \tag{11.44}$$

This equation can be easily analyzed allowing to understand the role of shear strain and non-parabolicity on the structure of unprimed subbands in thin (001) silicon films.

11.4 Strain-Induced Valley Splitting

11.4.1 Small Strain Values

For zero stress the ratio on the right hand side of (11.44) is equal to zero, and the standard quantization condition $q_n = \pi n / k_0 t$ is recovered. This condition is obtained from either of the two equations, therefore, the subbands are twofold degenerate. For non-zero shear strain and/or $k_x \neq 0$ and $k_y \neq 0$ the parameter $\eta \neq 0$. Due to the plus/minus sign in the right-hand side of (11.44), the equation splits into two non-equivalent branches for $\eta \neq 0$. (11.44) is nonlinear and can be solved only numerically. However, for small η the solution can be sought in the form $y_n = q_n \pm \zeta$, where ζ is small. Substituting $y_n = q_n$ into the right-hand side of (11.44) and solving the equation by perturbations with respect to ζ, we obtain for the difference between the energies of the two unprimed subbands with the same quantum number n:

$$\Delta E_n = 2\left(\frac{\pi n}{k_0 t}\right)^2 D\varepsilon_{xy} k_0 t \, \frac{\sin(k_0 t)}{|1 - q_n^2|}. \tag{11.45}$$

The two ladders of unprimed subband are conceived by the two [001] valleys, which are degenerate in the bulk. Interestingly, the degeneracy is lifted in thin films, when the dispersion in [001] direction is non-parabolic. This is why the energy splitting (11.45) is often referred to as valley splitting.

The valley splitting is shown to be linear in strain for small shear strain values and to depend strongly on the film thickness [29]. In accordance with earlier publications [2,5,8], the valley splitting is inversely proportional to the third power of k_0 and the third power of the film thickness t. Due to the cosine term in the right-hand side the value of the valley splitting oscillates with film thickness, in accordance with [5,8]. In contrast to previous works, the subband splitting is proportional to the gap δ at the X-point, (and not at the Γ-point [2, 18, 19]) which depends strongly on shear strain. Thus the application of uniaxial [110] stress to [001] ultra-thin Si film generates a valley splitting proportional to strain. This effect can be used to generate controllably the valley splitting larger than the spin Zeeman energy, a fact which has not been yet appreciated in spin-related applications of silicon.

11.4.2 High Values of Shear Strain

For high strain values the dispersion (11.8) of the lowest conduction band become parabolic again (shown in Fig. 11.7) and the quantization levels in a square well potential with a parabolic band must be recovered in this limit. We note that in the limit $\delta \gg E_0$ $X_2 = 2\sqrt{-\eta}$ and equations (11.22) and (11.23) take the form [30]:

$$\tan(X_1 k_0 t/2) \approx \frac{b(X_1)}{a(X_1)} \frac{1}{\sqrt{\eta}}, \tag{11.46}$$

$$\cot(X_1 k_0 t/2) \approx \frac{b(X_1)}{a(X_1)} \frac{1}{\sqrt{\eta}}. \tag{11.47}$$

For large η (11.46) has the solution $X_1 = \pi(2n-1)/k_0 t$ while (11.47) gives $X_1 = 2\pi n/k_0 t$ which results in the well-known quantization result $X_1 = \pi m/k_0 t, m = 1, 2, 3, \cdots$ for subbands in an infinite potential square well with a single parabolic band. For the difference in energy ΔE_n between the two subbands with the same quantum number n (which are degenerate at $\eta = 0$) we get $\Delta E_n = E_1(4n-1)$ in the limit of large η.

11.4.3 Numerical Solutions

The numerical solutions were obtained in two ways. First, (11.22) and (11.23) are solved to obtain for the subband structures. The value $X_2 = \sqrt{X_1^2 + 4 - 4\sqrt{\eta^2 + X_1^2}}$ becomes imaginary at high strain values. A special care must be taken to choose the correct branch of $\sqrt{X_2^2 + \eta^2}$ in (11.22)/(11.23) or alternatively (11.24)/(11.25). The sign of $\sqrt{X_2^2 + \eta^2}$ must be alternated after the expression becomes zero, as it is displayed in Fig. 11.8.

In order to be sure that the results are correct and to bypass the limitations of (11.22) and (11.23) which were obtained for a square well potential with infinite potential walls, a direct discretization of the differential operators in the Hamiltonian (11.4) on a mesh, with the subsequent numerical solution of the corresponding system has been performed [3]. The later method allows to include the finiteness of the potential barrier at the interface of a film, to consider arbitrary film orientation, and to solve the Schrödinger equation self-consistently with the Poisson equation. The equation determined by the Hamiltonian (11.4) is elliptic, so no spurious solutions have been observed after discretization, and the solution of the eigenvalue problem is straightforward. Numerical results for (001) silicon films shown below obtained by both methods were equivalent.

Figures 11.9 and 11.10 show the energies of the subbands as a function of shear strain for two different film thicknesses. Shear strain opens the gap between the two conduction bands at the X-point making the dispersions non-parabolic [28], which makes the (11.22) and (11.23) non-equivalent. This removes the subband degeneracy and introduces the valley splitting. Figures 11.11 and 11.12 show the energy difference between two unprimed subbands ΔE_n as a function of strain for the same quantum number n. At unrealistically large strain values when the valley minimum is located at the X-point, and the dispersion, according to (11.7) becomes parabolic again, the well-known quantization conditions $X_1 = \pi m/k_0 t, m = 1, 2, 3, \cdots$ are perfectly observed in Figs. 11.11 and 11.12.

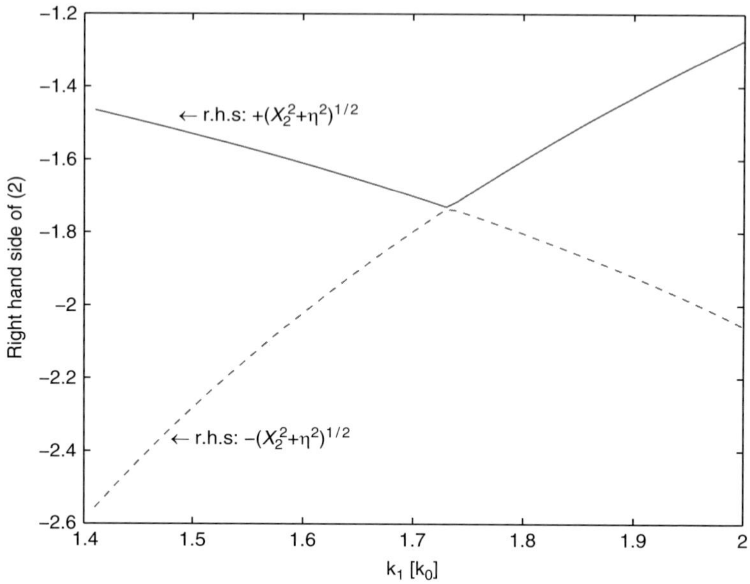

Fig. 11.8 The right hand side of (11.22) and (11.23) plotted close to the point $\sqrt{\eta^2 + X_2^2} = 0$. It is clearly seen that the sign of the square root must be alternated at this point

Fig. 11.9 Subband quantization energies E_n (normalized to the ground subband energy) for a film thickness of 3.3 nm. The valley splitting appears for non-zero shear strain η

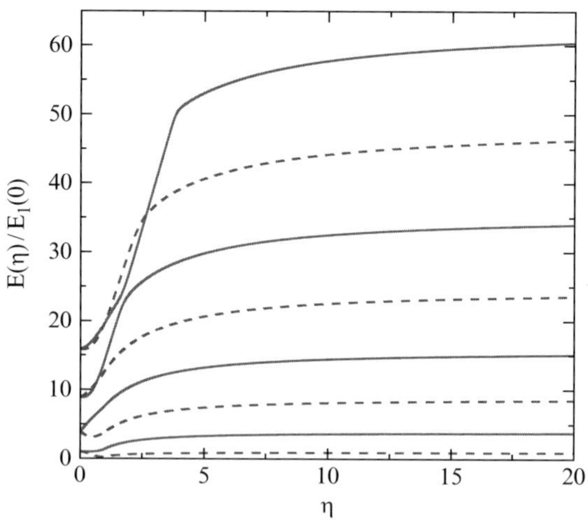

Figure 11.13 shows the valley splitting for several film thicknesses as a function of shear strain. In ultra-thin films already at moderate stress levels the splitting energy becomes larger than $k_\mathrm{B}T$. It is seen again that for small strain values the splitting is linear in shear strain. For large strain the quantization relations in an infinite square well potential with a single parabolic band are recovered resulting in the largest subband splitting. Uniaxial stress is currently used to enhance performance

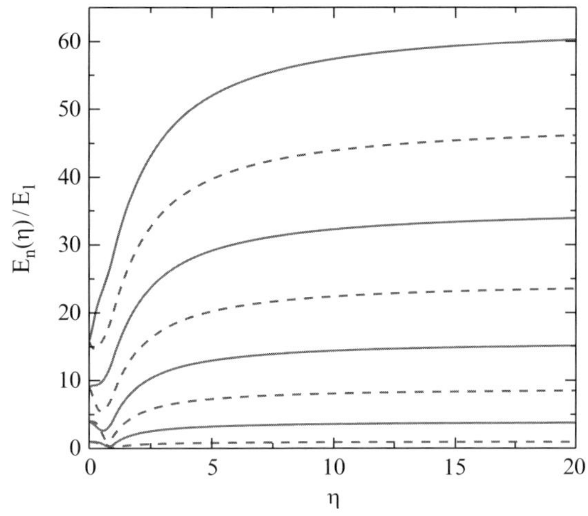

Fig. 11.10 The same as in Fig. 11.9 for a film thickness 6.5 nm. The valley splitting depends strongly on the film thickness. The valley splitting is maximal at high strain values

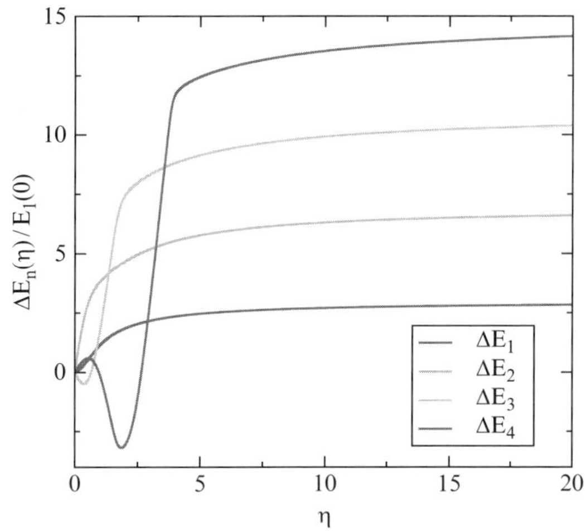

Fig. 11.11 Difference of the subband quantization energies ΔE_n (normalized to the ground subband energy) from (11.24) and (11.25) for a film thickness of 3.3 nm. The valley splitting appears for non-zero shear strain η

of modern MOSFETs, where it is introduced in a controllable way. Therefore, the valley splitting can be controlled by adjusting strain and thickness t.

11.5 Effective Mass of the Unprimed Subbands

In Fig. 11.14 the dependences of the transversal masses on shear strain in bulk silicon (9.9), (9.10) (lines) and of the effective masses for the two lowest subbands with $n = 1$ (symbols) along [110] and $[1\bar{1}0]$ in a film of a thickness $t = 10.86$ nm are compared. The transversal masses of the two ground subbands are in good agreement with (9.9), (9.10). However, when the film thickness decreases a substantial discrepancy between (9.9), (9.10) and the numerical solutions in films appears,

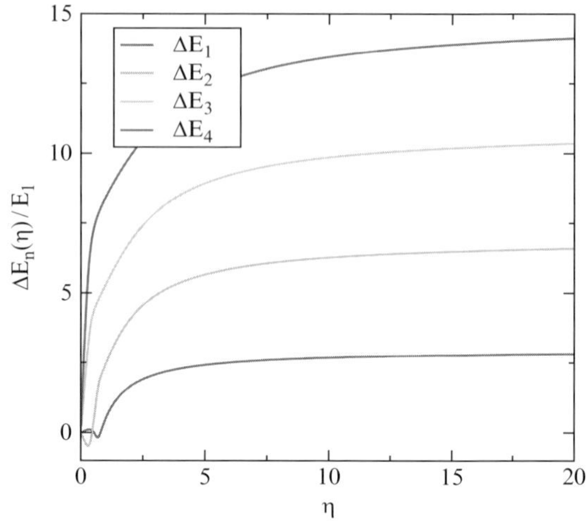

Fig. 11.12 $\Delta E_n(\eta)$ for a film thickness of 6.5 nm. The splitting depends strongly on the film thickness

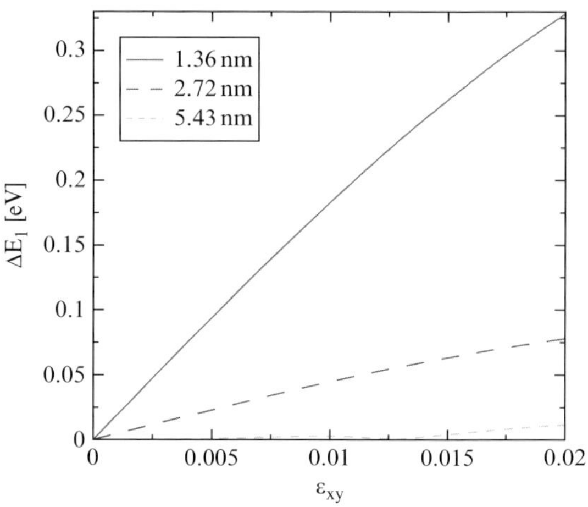

Fig. 11.13 Splitting induced by shear strain for several film thicknesses is shown. As can be seen for ultra-thin films the splitting is already larger than $k_B T$ for moderate stress levels

especially at high values of strain, as shown in Fig. 11.15. Figure 11.16 shows that the curvature effective masses along [110] and $[1\bar{1}0]$ become different for decreasing film thickness even in relaxed films. The dependence of the effective masses of the two ground subbands without strain on film thickness is shown in Fig. 11.17. It does not of course mean that the subbands are parabolic, and the effective mass along [110] or $[1\bar{1}0]$ is computed as the inverse of the corresponding second derivative. The non-parabolic subband dispersions are demonstrated in Figs. 11.18 and 11.19. For a given k_x and k_y the subband with a lower energy has the equi-energy surfaces in the form of the unification of the two ellipses with the effective masses m_1 and m_2, while a subband with a higher energy has the intersection of the two ellipses as equi-energy surfaces.

The difference between the bulk transversal masses (9.9), (9.10) and the subband masses in thin films is due to the rapidly growing value of the right-hand side in

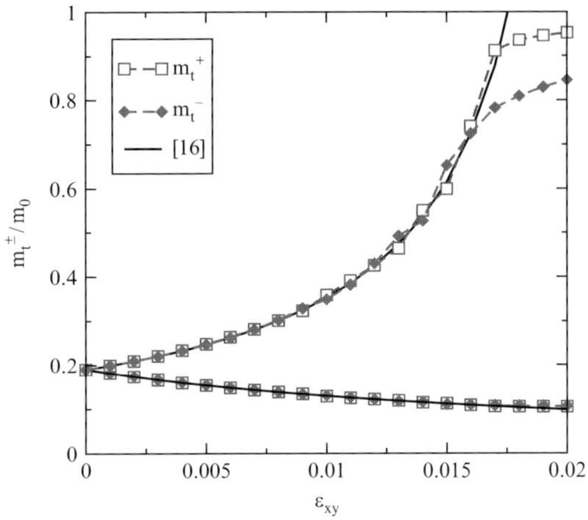

Fig. 11.14 The dependences of the effective masses on shear strain for the two lowest subbands in a film of a thickness $t = 10.86\,\mathrm{nm}$ (*symbols*). The dependences of the transversal masses on shear strain in bulk silicon (9.9), (9.10) are also shown (*lines*)

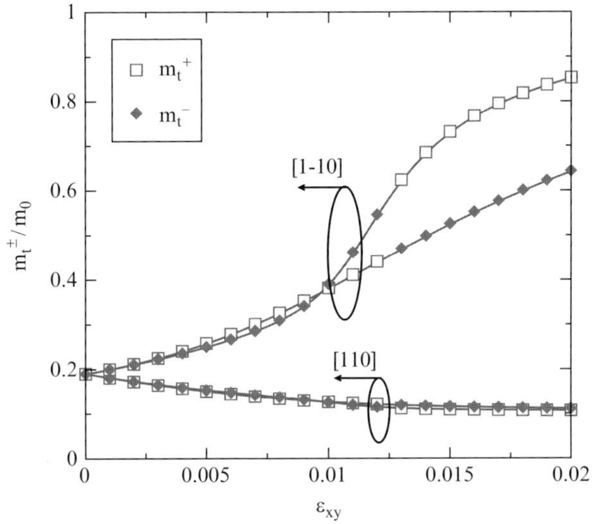

Fig. 11.15 The dependences of the effective masses on shear strain for the two lowest subbands in a film of a thickness $t = 5.43\,\mathrm{nm}$

(11.44) with the film thickness decreased. The right-hand side is exactly zero in relaxed films, when either k_x or k_y is equal to zero. This is why the minimum of two ground subbands in relaxed films is at the same position $k_x = k_y = 0$. As soon as shear strain is non-zero or both $k_x \neq 0$ and $k_y \neq 0$, the energies of the two ground subbands becomes different.

In order to justify the dependences of the subband effective masses or, rather, of the second derivatives along [110] and $\left[1\bar{1}0\right]$ on the film thickness let us solve (11.44) by perturbation assuming η to be small. In the zeroth order the right-hand side can be ignored, and one finds $y_n^0 = q_n = \frac{\pi n}{k_0 t}$.

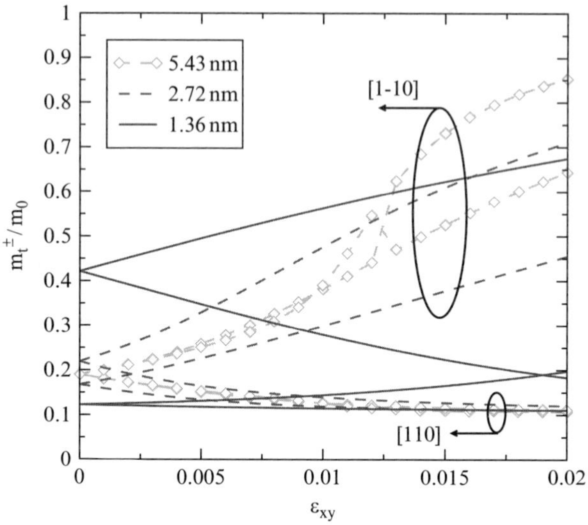

Fig. 11.16 The dependences of the effective masses on shear strain for the two lowest subbands for several film thicknesses

Fig. 11.17 Ultra thin films exhibit different effective masses for the two ground subbands even without stress. The thinner the film the more pronounced is the difference in effective masses

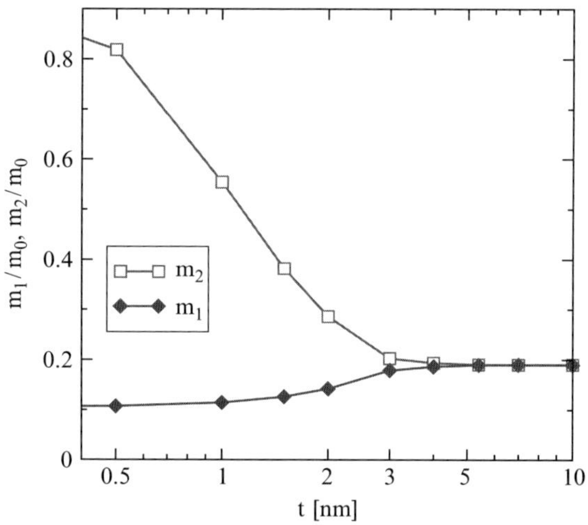

Substituting $y_n^0 = q_n = \frac{\pi n}{k_0 t}$ into the right-hand side of (11.44) and solving for small correction to y_n, the dispersion relation for the unprimed subbands n can be obtained as

$$E_n^{\pm} = \frac{\hbar^2}{2m_l}\left(\frac{\pi n}{t}\right)^2 + \frac{\left(k_x^2 + k_y^2\right)}{2m_t} \pm \left(\frac{\pi n}{k_0 t}\right)^2 \frac{\left|D\varepsilon_{xy} - \frac{\hbar^2 k_x k_y}{M}\right|}{k_0 t \left|1 - \left(\frac{\pi n}{k_0 t}\right)^2\right|} \sin(k_0 t). \tag{11.48}$$

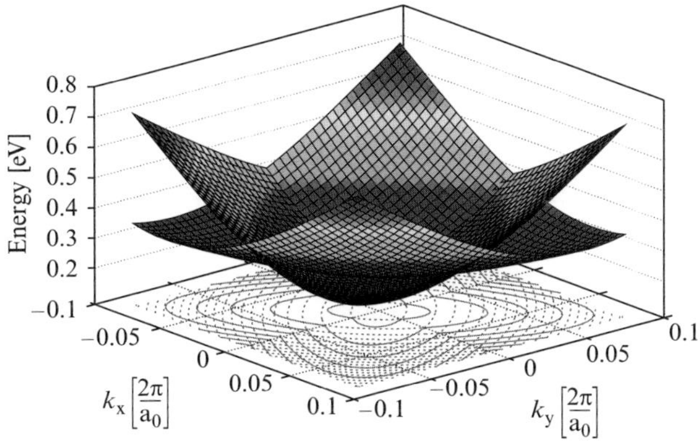

Fig. 11.18 For a film thickness of 1.36 nm the dispersion of the two ground subbands is shown. Without strain the subbands are degenerate at the minimum

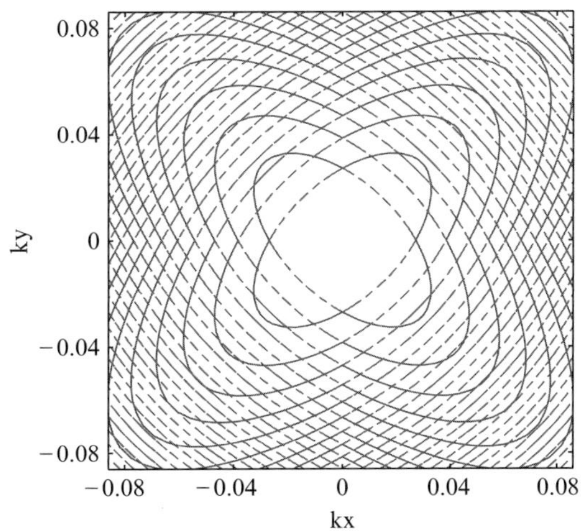

Fig. 11.19 The two ground subbands are shown as contour plots. The subband dispersion of one of them is described by the intersection of the two ellipses, while the other subband dispersion is given by the unification of the two ellipses

(11.48) confirms that the subband degeneracy is preserved only in the absence of shear strain and when either $k_x = 0$ or $k_y = 0$. (11.48) shows that the unprimed subbands for thin films are not equivalent anymore, even in relaxed films. Deriving the expression for the effective masses out of (11.48) reveals two independent effective masses in [110] direction for the unprimed subbands with the same quantum number n without strain:

$$m_{(1,2)} = \left(\frac{1}{m_t} \pm \frac{1}{M} \left(\frac{\pi n}{k_0 t} \right)^2 \frac{\sin(k_0 t)}{k_0 t \left| 1 - \left(\frac{\pi n}{k_0 t} \right)^2 \right|} \right)^{-1} . \qquad (11.49)$$

This behavior is in agreement with the effective masses found numerically for the two ground subbands in relaxed thin films shown in Fig. 11.17.

11.6 Valley Splitting in Magnetic Field and Point Contacts

In Sect. 11.4 it was demonstrated that shear strain can be effectively used to increase the splitting between the valleys beyond the Zeeman spin splitting. This allows effective separation between the valley and spin quantum numbers [12], opening silicon to spintronic applications. Experimentally, the valley splitting is studied in silicon layers epitaxially grown on (001) SiGe substrates. Because of the lattice mismatch, the silicon layer experiences biaxial tension. Tensile biaxial stress partly removes the valley degeneracy as it was described in Chap. 9. Experiments show that the degeneracy between the two remaining valleys in silicon layers is also removed.

Existing experimental results for silicon quantum wells can be separated in two groups by the methods used to obtained the data.

- Measurements of the valley splitting in an external magnetic field in the silicon quantum wells grown on SiGe substrates. Due to a mismatch in the lattice constants between silicon and SiGe a biaxial tensile stress is created in a (100) silicon quantum well. Due to size quantization in the well, the sixfold degeneracy between the bulk conduction band valleys is enhanced. The remaining degeneracy between the two lowest unprimed subbands is lifted by applying a perpendicular magnetic field [12, 14, 15]. The corresponding splitting deduced from the Shubnikov-de-Haas oscillations using a wide Hall bar geometry shows a strikingly linear dependence on the magnetic field [12]. This linear dependence previously observed at high fields was recently confirmed to exist down to values of the magnetic field as small as 0.3 T [12]. The valley or, more correctly, the lowest subband splitting is in the range of a few μeV for small fields [12, 14, 15]. This value obtained from experiments is much smaller than the one predicted theoretically [5].
- The second set of experiments on valley splitting is based on the conductivity measurements through a constriction or a point contact. The constriction is created by applying a negative bias to the gates patterned on top of the two-dimensional electron gas in a quantum well. The potential well gets depleted under the gates, which leads to the formation of a point-like contact between the two-dimensional electron gas reservoirs. The conductivity of a point contact is quantized in units of $e^2/\hbar$ at low temperature. The conductivity steps correspond to transversal channels and, therefore, can be labeled by valley, subband, and spin quantum numbers. By carefully analyzing these steps it was first demonstrated in [31] that the valley splitting can be a few meV without external magnetic field. This value is larger than typical spin Zeeman splitting. Recently, a similar study was reported in [12], where a valley splitting of several meV was confirmed. It was also observed that the valley splitting can be controlled by additional confinement in nanostructures. The additional confinement is achieved

by either changing the gate voltage or by an external magnetic field, showing that the valley splitting can be controlled through both physical and magnetic confinement.

The valley splitting has been known from the mid sixties [10] and has been the subject of investigations since. A comprehensive overview of the earlier theoretical models of the electric break-through at the Γ-point by Ohkawa and Uemura [18, 19] and the surface scattering theory by Sham and Nakayama [22] is given in the famous review of Ando *et al.* [2]. In the Ohkawa-Uemura theory the intervalley coupling is induced by the conduction band bending at the Γ-point. In the Sham-Nakayama theory the coupling is due to possible electron transition from one valley to another as a result of electron scattering at the abrupt interface. These two models are considered to be not completely independent and are related due to similar expressions for the predicted valley splitting [2]. Recently, the theory of intervalley coupling has been extended [11] to explain experimental data [12]. Inspired by a first principles theory [9], intervalley coupling was introduced *phenomenologically* at the heterostructure interface. The strength of the intervalley interaction parameter can be calibrated to reproduce results from a tight-binding model [5].

The most striking feature found experimentally [12] is the difference in scales between the two values for the valley splitting obtained from the two experimental setups described in the previous section. In order to make the theoretical predictions qualitatively consistent with the experimental data, a quantum well with an orientation slightly tilted from the (001) orientation was considered in [11]. The quantum well developes atomic steps at the interfaces associated with the global tilt [11]. The surface disorder is due to a slight misalignment of the wafer from (001) direction. For a strong field the magnetic length becomes smaller than the typical thickness variation length which results in large valley splitting. According to [12], this is also the reason why in a laterally confined electron system of quantum point contacts the valley splitting is large. At week magnetic fields the magnetic length extends to many steps of thickness variations. An electron is then effectively moving in a quantum well slightly tilted from the (001) orientation. where the valley splitting is suppressed [12]. In order to recover an experimentally observed linear dependence of the valley splitting on the magnetic field strength a particular disorder model for step profiles must be adopted [11].

An advantage of the two-band **k·p** model (11.4) is that it provides the value of the valley splitting in a confined system and allows to calculate intervalley coupling without introducing any additional fitting parameters. As it is shown below, a linear dependence of the valley splitting on magnetic field can be obtained from the two-band **k·p** model alone. The model also explains a larger value of the valley splitting in a [110] oriented point contact.

11.6.1 Valley Splitting in Magnetic Fields

In relaxed (001) silicon films the Landau levels in an orthogonal magnetic field B are found from (11.48) by using the Bohr-Sommerfeld quantization condition:

$$E_m^{(1,2)} = \hbar\omega_c\left(m + \frac{1}{2}\right)\frac{\pi}{4\arctan(\sqrt{m_{(1,2)}/m_{(2,1)}})}, \tag{11.50}$$

where the masses are determined by (11.49), and

$$\omega_c = \frac{eB}{\sqrt{m_1 m_2}c}$$

is the cyclotron frequency, e is the electron charge, and c is the speed of light. According to (11.50), the difference $|E_m^{(1)} - E_m^{(2)}|$ is linear regarding the magnetic field. In Shubnikov-de-Haas experiments there will now occur two sets of resistance oscillations with slightly different periods the in inverse magnetic field. Because of the small difference between the masses the difference in the periods will also be small. However, at the Fermi level the quantum number m, which is proportional to the ratio of the Fermi energy to the cyclotron frequency, is typically very large and may lead to a splitting of several hundreds μeV. The difference in the periods can be interpreted as an appearance of an additional energy shift between the equivalent unprimed valleys. Most importantly, the shift is linear in the magnetic field. The linear dependence of splitting between the valleys on the magnetic field will be also observed even in the presence of a small intrinsic constant valley splitting as long as this splitting is much smaller than the Fermi energy. This splitting is possible due to a remaining shear strain and/or conduction band non-parabolicity which is not accounted for in the two-band **k·p** theory (11.4) and is usually several tens of μeV, thus much smaller than the Fermi energy. For a 10 nm thick silicon film grown on SiGe it follows from (11.50) that the valley splitting can be several tens of μeV in a magnetic field of 1 T, which is consistent with the experimental observations [12].

11.6.2 Valley Splitting in a Point Contact

We consider a point contact in [110] direction realized by confining an electron system of a thin silicon film laterally by depleting the area under additional gates. Without strain the low-energy effective Hamiltonian in the point contact can be written as:

$$H_{(1,2)} = \frac{\hbar^2 k_x'^2}{2m_{(1,2)}} + \frac{\hbar^2 k_y'^2}{2m_{(2,1)}} + \frac{1}{2}\kappa x'^2 + V_b \tag{11.51}$$

where the primed variables are along the [110] and $[1\bar{1}0]$ axes, the effective masses are determined by (11.49), κ is the spring constant of the point contact confinement potential in $[1\bar{1}0]$ direction, and V_b is a gate voltage dependent conduction band shift in the point contact [36]. The dispersion relation of propagating modes within the point contact is written

$$H_{(1,2)} = \frac{\hbar^2 k_y'^2}{2m_{(2,1)}} + \hbar\omega_{(1,2)}\left(p + \frac{1}{2}\right) + V_b \qquad (11.52)$$

where $\omega_{(1,2)} = \kappa/m_{(1,2)}$. Since the energy minima of the two propagating modes with the same p are separated, they are resolved in the conductance experiment through the point contact as two distinct steps. The valley splitting is $\Delta E_p = \hbar p|\omega_1 - \omega_2|$. The difference in the effective masses (11.49) and, correspondingly, the valley splitting can be greatly enhanced by reducing the effective thickness t of the quasi-two-dimensional electron gas, which is usually the case in a gated electron system, when the inversion layer is formed.

In a [100] oriented point contact without strain the effective Hamiltonian is

$$H^{\pm} = \frac{(k_x^2 + k_y^2)}{2m_t} \pm \left(\frac{\pi n}{k_0 t}\right)^2 \frac{\left|\frac{\hbar^2 k_x k_y}{M}\right|}{k_0 t \left|1 - \left(\frac{\pi n}{k_0 t}\right)^2\right|} \sin(k_0 t) + \frac{1}{2}\kappa x^2. \qquad (11.53)$$

Due to symmetry with respect to k_y the subband minima in a point contact are always degenerate. For this reason the valley splitting in [100] oriented point contacts is greatly reduced as compared to point contacts in [110] direction.

11.7 Primed Subbands in Ultra-Thin (001) Silicon Films

The shear strain component in the [110] direction does not affect the primed valleys along [100] and [010] direction, except for a small irrelevant minimum displacement from the [100] or [010] axes [1] due to the term (8.40) in the Hamiltonian. Interestingly, recent Linear Combination of Bulk Bands calculations [23] and calculations based on the density-functional theory [17] uncover the dependence of the transversal effective mass m_t of the primed subbands on the silicon film thickness t. Here we analyze the dependence of the primed subbands effective mass via the two-band **k·p** Hamiltonian (11.4). In the case of the primed subbands in (001) silicon films the quantization is done along the k_x (or k_y) direction. In the Hamiltonian

$$H = \begin{pmatrix} \frac{\hbar^2 k_z^2}{2m_l} + \frac{\hbar^2 k_x^2}{2m_t} + \frac{\hbar^2 k_y^2}{2m_t} - \frac{\hbar^2}{M} k_x k_y + U(x) & \frac{\hbar^2 k_z k_0}{m_l} \\ \frac{\hbar^2 k_z k_0}{m_l} & \frac{\hbar^2 k_z^2}{2m_l} + \frac{\hbar^2 k_x^2}{2m_t} + \frac{\hbar^2 k_y^2}{2m_t} + \frac{\hbar^2}{M} k_x k_y + U(x) \end{pmatrix}. \qquad (11.54)$$

The quantization is along the x-axis, so the substitution $k_x = -i\hbar d/dx$ must be performed, while the other two components k_y and k_z are set to be constants.

The Hamiltonian (11.54) has a structure similar to (11.4): it contains the first and the second derivatives. The substantial difference is that the first derivative $k_y/M(-i\hbar d/dx)$ is multiplied by the variable k_y, contrary to the corresponding first-order term in (11.4) $k_0/m_l(-i\hbar d/dz)$, where $k_0 = 0.152\pi/a_0$ is a constant. Let us approximate the confining potential $U(x)$ by a square well potential with infinite walls. Because the structure of the Hamiltonians (11.54) and (11.4) and thus the dispersions are similar (in fact, the dispersion is described by the same relation (9.4)), one can claim that for any energy E there will be two pairs of the solutions for k_x which we will also call here $\pm k_1$ and $\pm k_2$. Repeating the calculations presented in Appendix 11.9 the relation between the two pairs of roots can be written in the form:

$$\frac{k_2^2(k_1)}{k_0^2} = \frac{k_1^2}{k_0^2} + 4\frac{k_y^2}{k_0^2}\frac{m_t^2}{M^2} - 4\sqrt{\frac{k_1^2}{k_0^2}\frac{k_y^2}{k_0^2}\frac{m_t^2}{M^2} + \frac{k_z^2}{k_0^2}\frac{m_t^2}{m_l^2}}. \tag{11.55}$$

The dispersion equations are in the form:

$$\sin\left(\frac{k_1 t}{2}\right)\cos\left(\frac{k_2(k_1)t}{2}\right) - \cos\left(\frac{k_1 t}{2}\right)\sin\left(\frac{k_2(k_1)t}{2}\right) f_{\pm}(k_1) = 0, \tag{11.56}$$

where the functions $f_{pm}(k_1)$ are defined as:

$$f_{\pm}(k_1) = \begin{cases} \dfrac{\frac{k_z}{k_0}\frac{m_t}{m_l} \mp \sqrt{\frac{k_1^2}{k_0^2}\frac{k_y^2}{k_0^2}\frac{m_t^2}{M^2} + \frac{k_z^2}{k_0^2}\frac{m_t^2}{m_l^2}}}{\frac{k_1}{k_0}} \dfrac{\frac{k_2(k_1)}{k_0}}{\frac{k_z}{k_0}\frac{m_t}{m_l} \mp \sqrt{\frac{k_2(k_1)^2}{k_0^2}\frac{k_y^2}{k_0^2}\frac{m_t^2}{M^2} + \frac{k_z^2}{k_0^2}\frac{m_t^2}{m_l^2}}}, & \frac{k_1^2}{k_0^2}\frac{k_y^2}{k_0^2}\frac{m_t^2}{M^2} < 4\frac{k_y^4}{k_0^4}\frac{m_t^4}{M^4} - \frac{k_z^2}{k_0^2}\frac{m_t^2}{m_l^2} \\ \dfrac{\frac{k_z}{k_0}\frac{m_t}{m_l} \mp \sqrt{\frac{k_1^2}{k_0^2}\frac{k_y^2}{k_0^2}\frac{m_t^2}{M^2} + \frac{k_z^2}{k_0^2}\frac{m_t^2}{m_l^2}}}{\frac{k_1}{k_0}} \dfrac{\frac{k_2(k_1)}{k_0}}{\frac{k_z}{k_0}\frac{m_t}{m_l} \pm \sqrt{\frac{k_2(k_1)^2}{k_0^2}\frac{k_y^2}{k_0^2}\frac{m_t^2}{M^2} + \frac{k_z^2}{k_0^2}\frac{m_t^2}{m_l^2}}}, & \frac{k_1^2}{k_0^2}\frac{k_y^2}{k_0^2}\frac{m_t^2}{M^2} > 4\frac{k_y^4}{k_0^4}\frac{m_t^4}{M^4} - \frac{k_z^2}{k_0^2}\frac{m_t^2}{m_l^2} \end{cases} \tag{11.57}$$

After the ladder of solutions $k_1^n(k_y, k_z)$ is found, the energy of the primed subbands is computed as

$$E_n(k_y, k_z) = \frac{k_y^2}{2m_t} + \frac{k_z^2}{2m_l} + \frac{k_1^{(n)2}(k_y, k_z)}{2m_t} - \sqrt{\frac{k_z^2 k_0^2}{m_l^2} + \frac{k_y^2 k_1^{(n)2}}{M}} \tag{11.58}$$

11.7.1 *Effective Mass of Primed Subbands*

We solve the dispersion (11.55)–(11.58) numerically. Alternatively, like in the case of unprimed subbands, we discretize the differential operators $k_x = -id/dx$ in

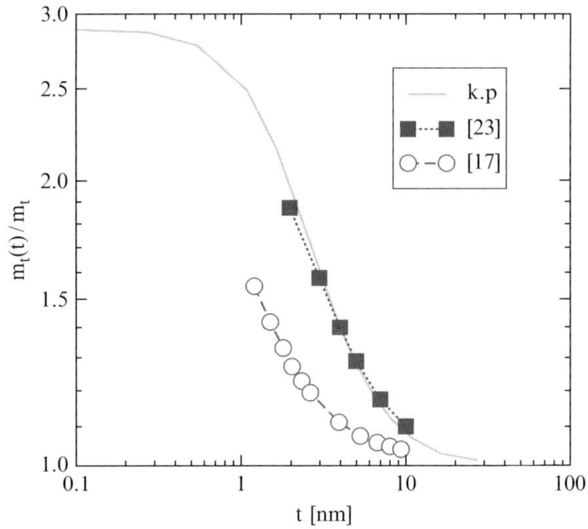

Fig. 11.20 Comparison between the effective masses for the lowest ground subband obtained with fullband LCBB calculations [23] (*filled symbols*), calculations from [17] (*open symbols*) and the two band **k·p** model (*solid line*). Excellent agreement with the full band calculations are achieved

the Hamiltonian (11.48) on a mesh in the x direction and solve the eigenvalue problem numerically. Both approaches yield the same results. The dependence of the transversal masses of primed subbands in (001) relaxed silicon film is shown in Fig. 11.20. The predictions of the two-band **k·p** theory are in excellent agreement with the results of the Linear Combination of the Bulk Bands method with a potential barrier of 3 eV at the film interface [23] and are also consistent with the DFT calculations from [17].

11.8 Substrate Orientations Different from (001)

There is growing experimental evidence [32, 33] that the parabolic approximation is not sufficient to accurately describe the subband parameters (effective masses and subband energies) in ultra-thin films under uniaxial stress. For instance, the effective masses of the unprimed subbands in a (001) film depend on shear strain [33] and silicon film thickness [26]. These effects are ignored in the parabolic band approximation but well described by the two-band **k·p** model (9.1)–(9.3) of the conduction band as it was demonstrated above.

We are now going to demonstrate that the two-band **k·p** model (9.1)–(9.3) allows an accurate description of the dependences of the subband energies and effective masses in thin silicon films with orientations different from (001).

11.8.1 Rotation of the Hamiltonian

To describe subbands in a film with an arbitrary crystal orientation a proper rotation of the Hamiltonian is required. Assuming $\mathbf{e_n}$ to be the quantization direction in the crystallographic coordinate system, let us define a unitary transformation $\mathbf{U_c}$ which transforms the quantization direction into $\mathbf{e_z} = (0, 0, 1)^{\mathrm{T}}$

$$\mathbf{e_z} = \mathbf{U_c^{-1}} \mathbf{e_n}. \tag{11.59}$$

Rotation of the wave vector $\mathbf{k} \mapsto \mathbf{U_c k}$ with $\mathbf{k} = (k_x, k_y, k_z)^{\mathrm{T}}$ is required. When the direction of the quantization direction $\mathbf{e_n}$ in the crystallographic system is determined by the spherical angles φ and ϑ, the transformation from the crystal coordinate system to the coordinate system is written as

$$\mathbf{U_c} = \begin{bmatrix} \cos(\varphi)\cos(\vartheta) & -\sin(\varphi) & \cos(\varphi)\sin(\vartheta) \\ \sin(\varphi)\cos(\vartheta) & \cos(\varphi) & \cos(\varphi)\sin(\vartheta) \\ -\sin(\vartheta) & 0 & \cos(\vartheta) \end{bmatrix}.$$

Furthermore, the different valley types along $\nu = \mathrm{x, y, z}$ axes need to be considered. Using the transformations

$$\mathbf{U_{v,x}} = \begin{bmatrix} 0 & 0 & -1 \\ 0 & 1 & 0 \\ 1 & 0 & 0 \end{bmatrix}, \quad \mathbf{U_{v,y}} = \begin{bmatrix} 0 & 0 & -1 \\ 1 & 0 & 0 \\ 0 & 1 & 0 \end{bmatrix}, \quad \mathbf{U_{v,z}} = \mathbf{I}$$

the total transformation matrix for valley ν is given by [3]

$$\mathbf{U}_\nu = \mathbf{U}_{\mathbf{v},\nu} \mathbf{U_c}. \tag{11.60}$$

To take strain into account for each valley type, the strain tensor is transformed by

$$\varepsilon \mapsto \mathbf{U}^{\mathrm{T}}_{\mathbf{v},\nu} \varepsilon \mathbf{U}_{\mathbf{v},\nu}. \tag{11.61}$$

Therefore, the diagonal blocks of the two-band $\mathbf{k}\cdot\mathbf{p}$ Hamiltonian (11.4) for arbitrary substrate orientation and valley type ν are rewritten as

$$\mathrm{H}_{\mp,\nu} = E_c(z) + \Xi_{\mathrm{u}} \varepsilon_{\mathrm{zz}} + \frac{\hbar^2}{2} \mathbf{k}^{\mathrm{T}} \mathbf{U}^{\mathrm{T}}_\nu \begin{bmatrix} \frac{1}{m_t} & 0 & 0 \\ 0 & \frac{1}{m_t} & 0 \\ 0 & 0 & \frac{1}{m_l} \end{bmatrix} \mathbf{U}_\nu \mathbf{k}$$
$$\mp \begin{bmatrix} 0 & 0 & \frac{\hbar^2 k_0}{m_l} \end{bmatrix} \mathbf{U}_\nu \mathbf{k}.$$

The off-diagonal coupling components are transformed

$$\mathrm{H}_{\mathrm{off},\nu} = D\varepsilon_{\mathrm{xy}} - \mathbf{k}^{\mathrm{T}}\mathbf{U}_{\nu}^{\mathrm{T}} \begin{bmatrix} 0 & \frac{\hbar^2}{M} & 0 \\ \frac{\hbar^2}{M} & 0 & 0 \\ 0 & 0 & 0 \end{bmatrix} \mathbf{U}_{\nu}\mathbf{k}.$$

To find subbands, the replacement $k_z \rightarrow -\mathrm{i}\,\partial_z$ is introduced. The discretization is realized with a finite difference scheme with hard wall boundary conditions. The resulting eigenvalue problem gives rise to discrete energies describing the subband structure.

11.8.2 Thin (110) Oriented Silicon Films

To describe the subbands in a film with (110) crystal orientation we appropriately rotate the Hamiltonian for each pair of the valleys and resolve the subband structure numerically. The zero boundary conditions for the wave functions at the interfaces are applied.

Figure 11.21 demonstrates the thickness dependence of the effective mass of the ground unprimed subbands of $(\bar{1}10)$ films in the [010] direction. Results are in agreement with those obtained by the pseudo-potential method [23]. The effective mass increase indicates that transport properties in ultra-thin $(\bar{1}10)$ films in [010] direction are degraded with decreasing thickness t.

Figure 11.22 shows the subband energy dependence of the film thickness for the lowest primed and unprimed subbands. Surprisingly, in ultra-thin $(\bar{1}10)$ films the twofold degenerate subband becomes lower in energy than the fourfold degenerate one. This is a direct manifestation of the increased non-parabolicity [3] of the

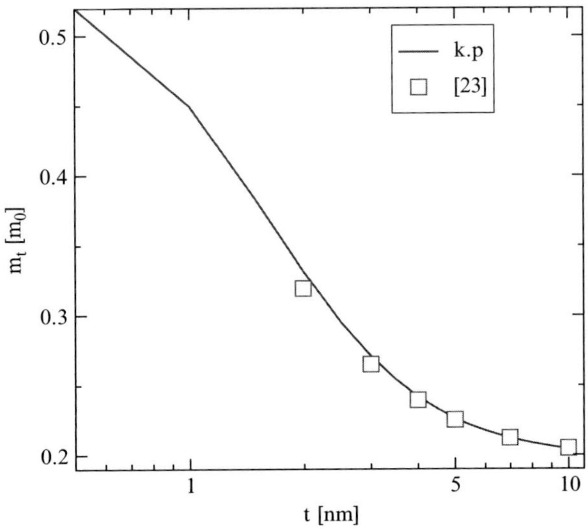

Fig. 11.21 Effective mass in [010] direction dependence on thickness for the fourfold degenerate ground subband in a $(\bar{1}10)$ film. The results obtained with help of **k·p** model are compared against the results of pseudo-potential calculations [23]

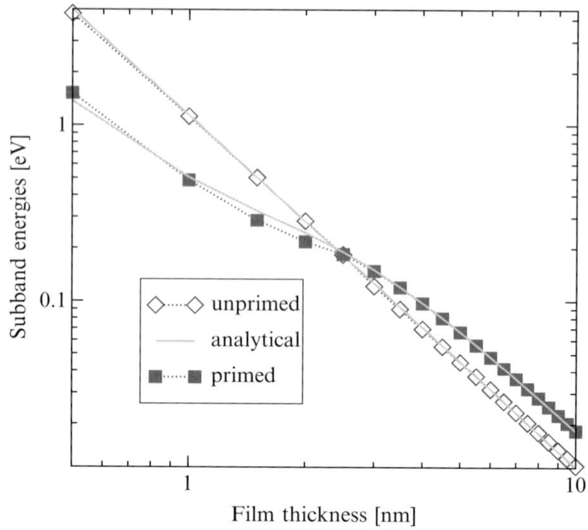

Fig. 11.22 Dependence of the primed and unprimed subband energies on the film thickness t in $(\bar{1}10)$ films. Analytical results (11.62) and (11.63) are shown by lines

[001] valleys in the [110] quantization direction [32]. The twofold degenerate subbands are characterized by the lighter mass m_t in [110] direction. Thus, the [110] channel direction becomes beneficial in FinFETs with ultra-thin bodies. Analytical results for the twofold degenerate subbands also shown in Fig. 11.22 are obtained by substituting k_x by its quantized values $k_x = \pi n/t$ in the dispersion

$$E_n(k_y, k_z) = \frac{\hbar^2}{2m_l} k_z^2 + \frac{\hbar^2}{2m_t}\left[\frac{\pi^2 n^2}{t^2} + k_y^2\right] - \frac{\hbar^2}{m_t}\sqrt{k_0^2 k_z^2 + \left[\frac{m_l D \varepsilon_{xy}}{\hbar^2} + \left(k_y^2 + \frac{\pi^2 n^2}{t^2}\right)\frac{m_l}{M}\right]^2}, \qquad (11.62)$$

where, as before, $k_0 = 0.152\pi/a_o$, ε_{xy} denotes the shear strain component in the crystallographic system, and $D = 14\,\text{eV}$ is the shear strain deformation potential. The analytical expression for the fourfold degenerate subbands are obtained with

$$E_n(k_y, k_z) = \frac{\hbar^2 \pi^2 n^2 (m_l + m_t)}{4 m_l m_t t^2}, \qquad (11.63)$$

with the corresponding quantization mass [24]. Figure 11.23 demonstrates that (11.62) describes accurately not only the minimum position but also the dispersion in [110] (k_y) direction of the twofold degenerate unprimed subbands in a $t = 3\,\text{nm}$ thin relaxed silicon film.

Tensile uniaxial stress along the channel enhances electron transport in [110] direction. In a (110) thin film the uniaxial stress favors to altering the relative energy order of primed and unprimed subband ladders. Due to an increase of the quantization mass m_x in the $[\bar{1}10]$ direction for increased tensile strain and higher

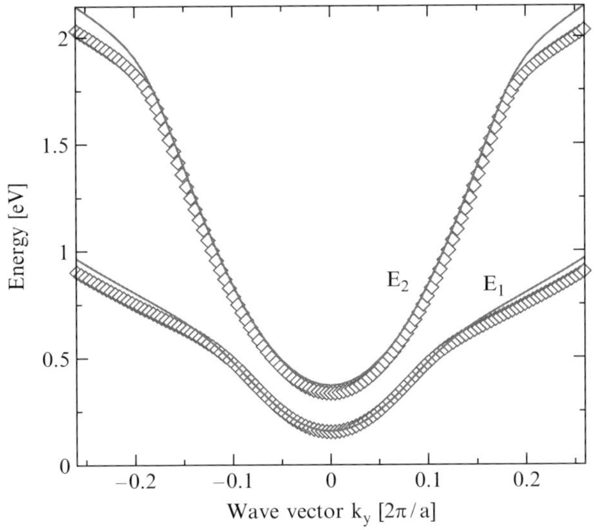

Fig. 11.23 Twofold degenerate subband dispersion in the [110] direction for a $(\bar{1}10)$ film with $t = 3\,\text{nm}$. Lines are obtained with (11.62)

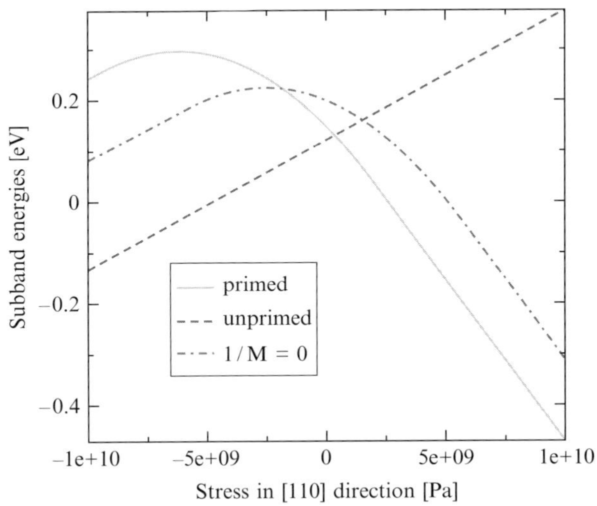

Fig. 11.24 Crossing between the first primed and unprimed subbands in a $t = 3\,\text{nm}$ $(\bar{1}10)$ film as a function of uniaxial stress in [110] direction

non-parabolicity at reduced thicknesses this transition happens at low stress values as compared to the case when these effects are ignored by setting $1/M = 0$, as shown in Fig. 11.24. The effective mass in the transport direction is, however, reduced with increased stress for decreased film thickness t as demonstrated in Fig. 11.25, which results in higher mobility. Thus, we conclude that FinFETs

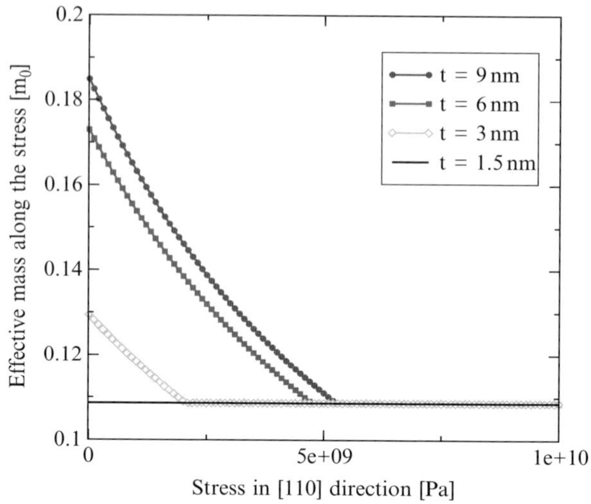

Fig. 11.25 Dependence of the transport effective mass in the first twofold degenerate subbands on tensile stress along the [110] direction, for several thicknesses of $(\bar{1}10)$ films

with ultra-thin body tensely stressed along the [110] channel have superior transport characteristics over [010] FinFETs because of the subband ladders inversion and a lighter effective mass of the lowest subband [25]. This behavior was recently observed experimentally [6].

11.9 Appendix

11.9.1 Re-Expressing X_1 as a Function of X_2

Following dimensionless units are introduced:

$$X = \frac{k_z}{k_0} \quad E_0 = \frac{\hbar^2 k_0^2}{m_l}$$
$$\varepsilon = \frac{E}{E_0} \quad \eta = \frac{\delta}{E_0} \tag{11.64}$$

The energy dispersion (11.7) takes the form:

$$\varepsilon(X) = \frac{X^2}{2} \pm \sqrt{\eta^2 + X^2} + \frac{m_l(k_x^2 + k_y^2)}{2m_t k_0^2}. \tag{11.65}$$

Setting the determinant of the dimensionless Hamiltonian to zero gives X as a function of energy ε:

$$\left(\frac{X^2}{2} - X - \varepsilon\right)\left(\frac{X^2}{2} + X - \varepsilon\right) - \eta^2 = 0$$

$$\text{or} \quad \left(\frac{X^2}{2} - \varepsilon\right)^2 - X^2 - \eta^2 = 0. \tag{11.66}$$

Introducing

$$\xi = X^2, \tag{11.67}$$

allows to re-express the fourth order equation (11.66) as a second order equation

$$\frac{\xi^2}{4} - \xi\varepsilon + \varepsilon^2 - \xi - \eta^2 = 0 \tag{11.68}$$

The solution are

$$\xi = 2(1+\varepsilon) \pm \sqrt{4(1+\varepsilon)^2 - 4(\varepsilon^2 - \eta^2)} \text{ or}$$

$$\xi = 2(1+\varepsilon) \pm 2\sqrt{1 + 2\varepsilon + \eta^2}. \tag{11.69}$$

The solutions can also be written as

$$\xi = \left(1 \pm \sqrt{1 + 2\varepsilon + \eta^2}\right)^2 - \eta^2 \tag{11.70}$$

Thus, for X values one gets the following expressions:

$$X_1^2 = \left(1 + \sqrt{1 + 2\varepsilon + \eta^2}\right)^2 - \eta^2 \tag{11.71}$$

$$X_2^2 = \left(1 - \sqrt{1 + 2\varepsilon + \eta^2}\right)^2 - \eta^2. \tag{11.72}$$

Using the following identities

$$\frac{X_1^2 + X_2^2}{2} = 2(1+\varepsilon), \tag{11.73}$$

$$\frac{X_1^2 - X_2^2}{2} = 2\sqrt{1 + 2\varepsilon + \eta^2} \tag{11.74}$$

results in the desired expressions $X_1(X_2)$ and $X_2(X_1)$:

$$X_1^2 = \left(1 + \frac{X_1^2 - X_2^2}{4}\right)^2 - \eta^2, \tag{11.75}$$

$$X_2^2 = \left(1 - \frac{X_1^2 - X_2^2}{4}\right)^2 - \eta^2 \text{ or} \tag{11.76}$$

$$X_1^2 = X_2^2 + 4 + 4\sqrt{X_2^2 + \eta^2}, \tag{11.77}$$

$$X_2^2 = X_1^2 + 4 - 4\sqrt{X_1^2 + \eta^2}. \tag{11.78}$$

The corresponding expression for the function $c(X)$ in dimensionless units is

$$c(X) = -\frac{X}{\eta \pm \sqrt{\eta^2 + X^2}}. \tag{11.79}$$

11.9.2 Expressing the Dispersion Equations in Terms of $X_1 \pm X_2$

Let us define the following variables:

$$y_n = \frac{X_1 + X_2}{2}, \tag{11.80}$$

$$\bar{y}_n = \frac{X_1 - X_2}{2}. \tag{11.81}$$

By using (11.75), (11.76) one obtains:

$$(\bar{y}_n \pm y_n)^2 + \eta^2 = \left(1 \pm \frac{X_1^2 - X_2^2}{4}\right)^2, \tag{11.82}$$

which is rewritten as

$$\bar{y}_n^2 + y_n^2 + \eta^2 = 1 + \frac{X_1^2 - X_2^2}{4}, \tag{11.83}$$

$$\bar{y}_n\, y_n = \frac{X_1^2 - X_2^2}{4}, \tag{11.84}$$

It then follows that

$$\bar{y}_n^2 + y_n^2 + \eta^2 = 1 + \bar{y}_n^2\, y_n^2, \tag{11.85}$$

$$\bar{y}_n^2\left(1 - y_n^2\right) = 1 - y_n^2 - \eta^2, \tag{11.86}$$

$$\bar{y}_n^2 = \frac{1 - y_n^2 - \eta^2}{1 - y_n^2}. \tag{11.87}$$

Thus, $\bar{y}_n$ is a function of y_n. Similar, y_n is a function of $\bar{y}_n$:

$$y_n^2 = \frac{1 - \bar{y}_n^2 - \eta^2}{1 - \bar{y}_n^2}. \tag{11.88}$$

Let us express the fraction $F \equiv \frac{c(X_2)-c(X_1)}{c(X_2)+c(X_1)}$ in (11.42) in dimensionless units:

$$F \equiv \frac{X_2\left(\eta \pm \sqrt{\eta^2+X_1^2}\right) - X_1\left(\eta \pm \sqrt{\eta^2+X_2^2}\right)}{X_2\left(\eta \pm \sqrt{\eta^2+X_1^2}\right) + X_1\left(\eta \pm \sqrt{\eta^2+X_2^2}\right)}. \tag{11.89}$$

Substituting X_1 with (11.75) and X_2 with (11.76) and using the identity $\sqrt{X_{1,2}^2+\eta^2}=|1\pm\frac{X_1^2-X_2^2}{4}|=|1\pm\bar{y}_n y_n|$ one rewrites F as a function of $\bar{y}_n$ and y_n:

$$F \equiv \frac{(\bar{y}_n - y_n)\,(\eta \pm (1+\bar{y}_n y_n)) - (\bar{y}_n + y_n)\,(\eta \pm (1-\bar{y}_n y_n))}{(\bar{y}_n - y_n)\,(\eta \pm (1+\bar{y}_n y_n)) + (\bar{y}_n + y_n)\,(\eta \pm (1-\bar{y}_n y_n))}, \tag{11.90}$$

Further simplifications result in:

$$F \equiv -\frac{y_n\left(\eta \pm 1 \mp \bar{y}_n^2\right)}{\bar{y}_n\left(\eta \pm 1 \mp y_n^2\right)}. \tag{11.91}$$

Substituting (11.87) into (11.91) leads to

$$F \equiv \frac{-y_n \eta\left(1 \pm \frac{\eta}{1-y_n^2}\right)}{\left(\eta \pm 1 \mp y_n^2\right)\sqrt{\frac{1-y_n^2-\eta^2}{1-y_n^2}}}, \tag{11.92}$$

which gives the corresponding term in (11.44):

$$I \equiv \mp \frac{y_n\,\eta}{\sqrt{(1-y_n^2)(1-y_n^2-\eta^2)}}. \tag{11.93}$$

References

1. Rideau, D., Feraille, M., Michaillat, M., Niquet, Y.M., Tavernier, C., Jaouen, H.: On the validity of the effective mass approximation and the Luttinger **k·p** model in fully depleted SOI MOSFETs. Solid State Electron. **53**(4), 452–461 (2009)
2. Ando, T., Fowler, A.B., Stern, F.: Electronic properties of two-dimensional systems. Rev. Mod. Phys. **54**(2), 437–672 (1982)
3. Baumgartner, O., Karner, M., Sverdlov, V., Kosina, H.: Electron subband structure in strained silicon UTB films from the Hensel-Hasegawa-Nakayama model: Part 2 efficient self-consistent numerical solution of the **k·p** Schrödinger equation. Solid State Electron. **54**(2), 143–148 (2010)
4. Bir, G.L., Pikus, G.E.: Symmetry and strain-induced effects in semiconductors. Willey, New York - Toronto (1974)
5. Boykin, T.B., Klimeck, G., Oyafuso, F.: Valence band effective-mass expressions in the $sp^3d^5s^*$ empirical tight-binding model applied to a Si and Ge parametrization. Phys. Rev. B **69**(11), 115201 (2004)

6. Chen, J., Saraya, T., Hiramoto, T.: Experimental study on uniaxially stressed gate-all-around silicon nanowires nMOSFETs on (110) silicon-on-insulator. In: Semiconductor Device Research Symposium, 2009. ISDRS '09. International, pp. 1–2 (2009)
7. Esseni, D., Palestri, P.: Fullbandbulk quantization analysis reveals a third valley in (001) silicon invesrion layers. IEEE Electron Device Lett. **24**(5), 353–355 (2005)
8. Esseni, D., Palestri, P.: Linear combination of bulk bands method for investigating the low-dimensional electron gas in nanostructured devices. Phys. Rev. B **72**(16), 165342 (2005)
9. Foreman, B.A.: First-principles envelope-function theory for lattice-matched semiconductor heterostructures. Phys. Rev. B **72**(16), 165345 (2005)
10. Fowler, A.B., Fang, F.F., Howard, W.E., Stiles, P.J.: Magneto-oscillatory conductance in silicon surfaces. Phys. Rev. Lett. **16**(20), 901–903 (1966)
11. Friesen, M., Chutia, S., Tahan, C., Coppersmith, S.N.: Valley splitting theory of SiGe/Si/SiGe quantum wells. Phys. Rev. B **75**(11), 115318 (2007)
12. Goswami, S., Slinker, K.A., Friesen, M., McGuire, L.M., Truitt, J.L., Tahan, C., Klein, L.J., Chu, J.O., Mooney, P.M., van der Weide, D.W., Joynt, R., Coppersmith, S.N., Eriksson, M.A., Orellana, P.: Controllable valley splitting in silicon quantum devices. Nat. Phys. **3**(8), 41–45 (2007)
13. Hensel, J.C., Hasegawa, H., Nakayama, M.: Cyclotron resonance in uniaxially stressed silicon. II. Nature of the covalent bond. Phys. Rev. **138**(1A), A225–A238 (1965)
14. Khrapai, V.S., Shashkin, A.A., Dolgopolov, V.T.: Strong enhancement of the valley splitting in a two-dimensional electron system in silicon. Phys. Rev. B **67**(11), 113305 (2003)
15. Lai, K., Pan, W., Tsui, D.C., Lyon, S., Mühlberger, M., Schäffler, F.: Two-flux composite fermion series of the fractional quantum hall states in strained Si. Phys. Rev. Lett. **93**(15), 156805 (2004)
16. Luttinger, J.M., Kohn, W.: Motion of electrons and holes in perturbed periodic fields. Phys. Rev. **97**(4), 869–883 (1955)
17. Martinez, A., Kalna, K., Sushko, P., Shluger, A., Barker, J., Asenov, A.: Impact of body-thickness-dependent band structure on scaling of double-gate MOSFETs: A DFT/NEGF study. Nanotechnology, IEEE Transactions **8**(2), 159–166 (2009)
18. Ohkawa, F.J., Uemura, Y.: Theory of valley splitting in an n-channel (100) inversion layer of Si: I. Formulation by extended zone effective mass theory. J. Physical Soc. Japan **43**(3), 907–916 (1977)
19. Ohkawa, F.J., Uemura, Y.: Theory of valley splitting in an n-channel (100) inversion layer of Si: II. Electric break through. J. Physical Soc. Japan **43**(3), 917–924 (1977)
20. Rideau, D., Feraille, M., Ciampolini, L., Minondo, M., Tavernier, C., Jaouen, H., Ghetti, A.: Strained Si, Ge, and $Si_{1-x}Ge_x$ alloys modeled with a first-principles-optimized full-zone $\mathbf{k}{\cdot}\mathbf{p}$ method. Phys. Rev. B **74**(19), 195208 (2006)
21. Rieger, M.M., Vogl, P.: Electronic-band parameters in strained $Si_{1-x}Ge_x$ alloys on $Si_{1-y}Ge_y$ substrates. Phys. Rev. B **48**(19), 14,276–14,287 (1993)
22. Sham, L.J., Nakayama, M.: Effective-mass approximation in the presence of an interface. Phys. Rev. B **20**(2), 734–747 (1979)
23. van der Steen, J.L., Esseni, D., Palestri, P., Selmi, L., Hueting, R.: Validity of the parabolic effective mass approximation in silicon and germanium n-MOSFETs with different crystal orientations. IEEE Trans. Electron Devices **54**(8), 1843–1851 (2007)
24. Stern, F., Howard, W.E.: Properties of semiconductor surface inversion layers in the electric quantum limit. Phys. Rev. **163**(3), 816–835 (1967)
25. Sverdlov, V., Baumgartner, O., Selberherr, S.: Subband parameters in strained (110) silicon films from the Hensel-Hasegawa-Nakayama model of the conduction band. In: Semiconductor Device Research Symposium, 2009. ISDRS '09. International, pp. 1–2 (2009)
26. Sverdlov, V., Baumgartner, O., Windbacher, T., Schanovsky, F., Selberherr, S.: Thickness dependence of the effective masses in a strained thin silicon film. In: Proceeings of International Conference on Simulation of Semiconductor Processes and Devices, pp. 1–4 (2009)
27. Sverdlov, V., Karlowatz, G., Dhar, S., Kosina, H., Selberherr, S.: Two-band $\mathbf{k}{\cdot}\mathbf{p}$ model for the conduction band in silicon: Impact of strain and confinement on band structure and mobility. Solid State Electron. **52**, 1563–1568 (2008)

28. Sverdlov, V., Ungersboeck, E., Kosina, H., Selberherr, S.: Effects of shear strain on the conduction band in silicon: An efficient two-band **k·p** theory. In: Proceedings of European Solid-State Device Research Conference, pp. 386–389 (2007)
29. Sverdlov, V., Ungersboeck, E., Kosina, H., Selberherr, S.: Current transport models for nanoscale semiconductor devices. Mater. Sci. Eng. R **58**(6–7), 228–270 (2008)
30. Sverdlov, V.A., Selberherr, S.: Electron subband structure and controlled valley splitting in silicon thin-body SOI FETs: Two-band **k·p** theory and beyond. Solid State Electron. **52**(12), 1861–1866 (2008)
31. Takashina, K., Ono, Y., Fujiwara, A., Takahashi, Y., Hirayama, Y.: Valley polarization in Si(100) at zero magnetic field. Phys. Rev. Lett. **96**(23), 236801 (2006) DOI 10.1103/PhysRevLett.96.236801
32. Uchida, K., Kinoshita, A., Saitoh, M.: Carrier transport in (110) nMOSFETs: Subband structure, non-parabolicity, mobility characteristics, and uniaxial stress engineering. In: International Electron Devices Meeting, pp. 1019–1021 (2006)
33. Uchida, K., Krishnamohan, T., Saraswat, K.C., Nishi, Y.: Physical mechanisms of electron mobility enhancement in uniaxial stressed MOSFETs and impact of uniaxial stress engineering in ballistic regime. In: International Electron Devices Meeting, pp. 129–132 (2005)
34. Ungersboeck, E., Dhar, S., Karlowatz, G., Sverdlov, V., Kosina, H., Selberherr, S.: The effect of general strain on band structure and electron mobility of silicon. IEEE Trans. Electron Devices **54**(9), 2183–2190 (2007)
35. VASP: Vienna Ab-initio Simulation Program. Kresse, G., Hafner, J.: Phys. Rev. B **47**(558), (1993); ibid. B **49**(14251), (1994); Kresse, G., Fertmueller, J.: Phys. Rev. B **54**(11169), (1996); Computs. Mat. Sci. **6**(15), (1996)
36. van Wees, B.J., van Houten, H., Beenakker, C.W.J., Williamson, J.G., Kouwenhoven, L.P., van der Marel, D., Foxon, C.T.: Quantized conductance of point contacts in a two-dimensional electron gas. Phys. Rev. Lett. **60**(9), 848–850 (1988)

Chapter 12
Demands of Transport Modeling in Advanced MOSFETs

12.1 TCAD Tools: Technological Motivation and General Outlook

Integrated circuits (IC) play a key role in modern digital information society. Superior computational performance is achieved by making transistor faster and assembling more and more elements on a chip, This is achieved by scaling the MOSFET size down. In the past decade the minimum feature size of transistor has been successfully reduced which allowed to double the number of transistors on a chip every second year. This trend is expected to continue in the next decade, as predicted and institutionalized by the International Technology Roadmap for Semiconductors [1] and supported by demonstration of MOSFETs with the gate length as short as 6 nm [28].

The success of microelectronics technology is partly enabled and supported by sophisticated Technology Computer-Aided Design (TCAD) tools which are used to assist in IC development and engineering at practically all stages from process definition to circuit optimization. At the current moment, the TCAD tools allows to reduce research and development costs by approximately 35–40% [1]. Most TCAD tools are based on semi-classical macroscopic transport models. From an engineering point of view, semi-classical models, like the drift-diffusion transport model, have enjoyed an amazing success due to their relative simplicity, numerical robustness, and the ability to perform two- and three-dimensional simulations on large unstructured meshes [150]. However, with device size dramatically reduced and new technology elements and materials introduced, the TCAD tools based on a standard semi-classical transport description begin to show shortcomings.

From the viewpoint of transport modeling, the problem is twofold. First, with the downscaling the driving field and its gradient increase dramatically in the short channel. As a result the carrier distribution along the channel can no longer be described even by the shifted and heated Maxwellian distribution. In order to properly account for hot-carrier and non-local effects, the drift-diffusion and even the energy transport model have to be improved to incorporate the substantial modifications in the distribution function.

V. Sverdlov, *Strain-Induced Effects in Advanced MOSFETs*, Computational Microelectronics, DOI 10.1007/978-3-7091-0382-1_12, © Springer-Verlag/Wien 2011

The second reason for semi-classical modeling tools to gradually loose their validity lies in the particle-wave duality of carriers. The carrier motion can be described with the classical Newton law only, when the characteristic size of the device is much larger than the corresponding de-Broglie electron wave length. When the device dimensions are getting comparable to the carrier wave length, the carriers can no longer be treated as classical point-like particles, and effects originating from the quantum-mechanical nature of propagation begin to determine transport in ultra-scaled devices.

A typical example of quantum effects in MOSFETs is the quantization of carrier motion in the potential well of the inversion layer at the silicon/dielectric interface. The size of the channel is usually in the order of a few nanometers, which is also the size of the electron de-Broglie wave length. In this confining potential only levels with certain, quantized energies are allowed. This results in the formation of subbands, as it was shown in Sect. 11 and, therefore, transport can no longer be accurately described by the classical equations in three dimensions, and a new description based on two-dimensional subbands must be adopted.

Another quantum-mechanical effect in modern MOSFETs is the tunneling of charge carriers through classically forbidden regions. Tunneling leads to increasing leakage through thin gate dielectrics. This gate leakage is responsible for an increased power consumption in modern MOSFETs. In 65 nm MOSFETs the gate dielectrics are already so thin that the leakage current leads to critically high values of generated power. The conventional thinning down of gate dielectrics becomes no longer possible, and new technological solutions must be introduced to continue scaling and keeping the heat generation within reasonable limits. The use of alternative gate oxide materials with higher permittivity (high-k dielectrics) and metal gates [119, 122, 132] Natarajan, S. helps significantly reducing the gate leakage current and allowed to successfully continuing scaling to the 45 nm and the 32 nm nodes. Strain introduced at 90 nm [176, 177] as an important MOSFET drive current and performance booster is perfectly compatible with the new high-k/metal gate technology [122, 132].

Several options for future technology nodes are currently under extensive investigation. One of them is multi-gate FinFET device architectures with improved channel control and reduced short-channel effects can eventually be employed [129]. Another option is to replace silicon in the channel with an advanced material characterized by substantially higher mobility [142, 143]. Both options are compatible with high-k/metal gate technology and strain-induced current boost.

Tunneling is not only affecting the gate leakage currents. In ultra-scaled MOSFETs with gate lengths of 10 nm and smaller quantum-mechanical tunneling under the barrier in the transport direction from source to drain becomes important [170]. When carrier scattering is neglected, a coherent transport approach based on the solution of the Schrödinger equation provides an accurate description for quantum-mechanical transport. This method is fast and efficient and can be implemented into commercial TCAD tools. However, since all devices operate at room temperature, carrier scattering in silicon-based FETs is still important, and transport is significantly affected by scattering even in ultra-scaled FETs [134]. Recent studies indicate that the crossover from diffusive to ballistic transport in nano-wire transistors may occur at a much shorter distance than previously anticipated [54]. An adequate transport model for ultra-scaled MOSFETs must therefore account for quantum-mechanical coherent and dissipative scattering effects simultaneously.

Modern TCAD tools have to be flexible enough to address challenges due to upcoming technological changes resulting from the use of new materials and structures. They have to describe properly transport in silicon and new advanced materials depending on strain and must be prepared to adequately describe the new quantum-mechanical phenomena which will determine transport in ultra-scaled CMOS and post-CMOS devices.

12.1.1 Brief History of TCAD Transport Modeling

The first fully numerical transport description was already suggested in 1964 by Gummel [64] for the one-dimensional bipolar transistor. The approach was further developed and applied to pn junctions [24] and to impact ionization avalanche transit-time diodes by Scharfetter and Gummel [148]. The first application of a solution of the two-dimensional Poisson equation to metal-oxide-semiconductor (MOS) structures was performed by Loeb [108] and Schroeder and Muller [149]. The first simultaneous solutions of coupled continuity and Poisson equations applied to junction field effect transistors [84] and to bipolar transistors [159] go back to 1969.

Since these pioneering works on transport modeling many different approaches have been applied to practically all important devices, and the number of papers in the field has grown exponentially. Today the modeling of transport in modern ICs has matured into a well established field with active research, intensive software development, and vast commercial applications. Many textbooks, monographs, and reviews devoted to theoretical and computational aspects of transport modeling in ICs have been published. Probably the first monograph covering practically all aspects from modeling and discretization to applications is [150]. Not pretending to cover all the literature we mention only the most recent monographs relevant to the review. Various transport models were summarized in [113], while new approaches to transport are well described in [22].

As the costs of development and maintenance of modern sophisticated TCAD software have significantly increased, only few large semiconductor companies can afford to support their own TCAD development team. There is a fairly large number of commercial TCAD software products available on the market [158, 174] which serve most of the industrial demands. Numerous TCAD tools developed at universities have the advantage that they are freely distributed software [2, 70, 179]. These tools regain their popularity due to the concept of a complex approach to simulations based on different levels of precision and complexity. It makes these tools valuable not only for pure educational and research purposes. Semiconductor manufacturing companies which need more refined simulations of complex phenomena often consult with researchers from universities to verify their data obtained experimentally.

12.1.2 Transport Modeling: Formulation of the Problem

In order to analyze a semiconductor device under general operating conditions, a mathematical model has to be formulated first. Regardless of the complexity of carrier dynamics inside the semiconductor, two equations are an important part of any model and always have to be included. The Poisson equation relates the density $\rho(\mathbf{r}, t)$ of charged carriers in the device to the electrostatic potential $\phi(\mathbf{r}, t)$:

$$\mathrm{div}(\kappa\, \mathrm{grad}\, \phi(\mathbf{r}, t)) = -\rho(\mathbf{r}, t), \tag{12.1}$$

where κ is the dielectric permittivity. The Poisson equation must be supplemented by the proper boundary conditions, which usually are of the Dirichlet or Neumann type. The Dirichlet boundary condition fixes the potential at the domain boundary and is conveniently applied to model the interface between an external electrical contact and the device. The Neumann boundary condition sets the normal component of the electric field at the interface $E_n = -(\mathbf{n}, \mathrm{grad}\, \phi(\mathbf{r}))$ to zero in order to isolate the simulation domain from the dielectric environment.

The second equation is the continuity equation, relating the current density $\mathbf{j}(\mathbf{r}, t)$ to the time derivative of the charge density:

$$\frac{\partial \rho(\mathbf{r}, t)}{\partial t} = \mathrm{div}\, \mathbf{j}(\mathbf{r}, t). \tag{12.2}$$

In semiconductors two types of particles carrying positive (holes) and negative (electrons) charges are present. By introducing the electron $n(\mathbf{r}, t)$ and hole $p(\mathbf{r}, t)$ concentrations, the continuity equation can be written for each carrier type as:

$$\begin{aligned} -q\frac{\partial n(\mathbf{r}, t)}{\partial t} &= \mathrm{div}\, \mathbf{j}_e(\mathbf{r}, t) - qR(\mathbf{r}, t), \\ +q\frac{\partial p(\mathbf{r}, t)}{\partial t} &= \mathrm{div}\, \mathbf{j}_p(\mathbf{r}, t) + qR(\mathbf{r}, t), \end{aligned} \tag{12.3}$$

where $R(\mathbf{r}, t)$ is the electron-hole generation rate.

In order to form a closed set of equations and complete formulation of the mathematical model, the system of (12.1) and (12.3) must be supplemented by the material properties relating the current density $\mathbf{j}(\mathbf{r}, t)$ to the electron $n(\mathbf{r}, t)$ and hole $p(\mathbf{r}, t)$ concentrations and the electric field $\mathbf{E}(\mathbf{r}, t) = -\mathrm{grad}\, \phi(\mathbf{r}, t)$. Derivation of these relations is a challenging task which requires an accurate consideration of physical mechanisms responsible for carrier motion inside the device.

Although we do not have the intention to completely cover this enormously large field of research here, we present several important examples and outline some difficulties and challenges in transport description in modern MOSFETs with strain. We begin with the semi-classical description of carrier dynamics inside the device, which is justified when the characteristic device size is much larger than the corresponding electron wave length. In ultra-scaled devices quantum effects start playing an important role. Different types of quantum potential and density gradient corrections can be introduced. A more refined description treats the carriers motion quantum-mechanical in the confinement direction resulting in the subband description of carriers in surface layers and single- and double-gate structures. Since the channel length in MOSFETs is large compared to the scattering length, a semi-classical transport description can be applied the current direction. However, transport descriptions based on a fully quantum-mechanical approach dealing with the dissipative quantum transport will be needed in ultra-scaled devices. Beginning with the description of ballistic coherent transport, the methods based on the Non-Equilibrium Green's Function (NEGF), on the density matrix and the Wigner function to include scattering are briefly discussed. For quantum devices it is sometimes easier to compute the total current without resorting to the continuity equations (12.3), however, for any transport model, classical or quantum, (12.3) must always be satisfied. A particular emphasis will be put on inclusion of strain effects into the transport simulations.

12.2 Semi-Classical Transport

The system of an ensemble of classical particles is conveniently described by the single-particle distribution function $f(\mathbf{r}, \mathbf{k}, t)$ in phase space, formed by position $\mathbf{r}$ and momentum $(\hbar\mathbf{k})$. It can be shown [82] that in case of slow varying perturbations in space and time the distribution function satisfies the Boltzmann equation [150]:

$$\frac{\partial f}{\partial t} + \mathbf{v} \cdot \nabla_r f + \frac{s_\nu \mathrm{q}}{\hbar} \mathbf{E} \cdot \nabla_{\mathbf{k}} f = \left(\frac{\partial f}{\partial t} \right)_{\text{coll}}, \tag{12.4}$$

Here, $\mathbf{v} = \nabla_{\mathbf{k}} E(k)$ is the carrier velocity, m is the effective mass. To avoid the confusion with the electric field $\mathbf{E}$, the dispersion is denoted as $\mathcal{E} = E(k) = \hbar^2 k^2/(2m)$ in this section. The sign function s_ν distinguishes between negatively charged electrons $s_n = -1$ and positive holes, $s_p = 1$. The right-hand side in (12.4) represents the collision operator due to phonons, impurities, interfaces, and other scattering sources.

For realistic structures, a direct numerical solution of this equation by discretization of the phase space is computationally expensive. This is why the TCAD tools do not usually solve the Boltzmann equation and are based on simplified transport models. Approximate solutions can be obtains by using the method of moments of the distribution function. Defining the moments of the distribution function $f(\mathbf{r}, \mathbf{k}, t)$, one consecutively obtained to the drift-diffusion model [64], the energy-transport models [164], or the six-moments transport model [58]. Transport models based on the moments of the Boltzmann equation are well accepted in TCAD.

Transport models uses several material characteristics as input parameters. For the drift-diffusion model the mobility represents the most relevant parameter [150] relating the drift velocity to the driving force. Mobility dependence on material, structure, and physical effects, like the remote Coulomb scattering or soft-phonon scattering at the interface of high-k materials, must be quantified. Demands on mobility modeling also arise from currently investigated effects to increase transistor on-current. Such effects are for instance mobility enhancement in strained silicon channels, possibly in combination with a favorable substrate orientation and channel orientation.

TCAD tools are using simplified transport models, because solving the Boltzmann equation with a Monte Carlo technique is computationally very expensive.

Although computationally expensive to obtain the device current-voltage characteristics, a solution of the Boltzmann equation by Monte Carlo methods is convenient for calibration of the transport models based on the moments of the distribution function. In particular, a Monte Carlo solution of the Boltzmann equation is necessary to obtain the accurate closure relations, expressing higher moments via the moments of lower order. The solution of the Boltzmann equation with the Monte Carlo technique is also useful, because it allows to introduce and control scattering mechanisms at the microscopic level and, most importantly, to incorporate the peculiarities of the semiconductor band structure modification with strain.

Although the drift-diffusion and energy-transport models have been used for a long time, calibration of the model parameters to incorporate strain is needed. With down-scaling of device feature sizes the drift-diffusion and even energy-transport models are gradually losing their validity due to large driving electric fields changing rapidly within the devices giving rise to hot-carrier and non-local effects. Thus, an extension of the models to the computationally more sophisticated transport model is necessary. The derivation of the most advanced semi-classical transport model, the six-moments model, is sketched next.

12.2.1 From Drift-Diffusion to Higher Moments Equations

The drift-diffusion model has been very successful due to its efficiency and numerical robustness. These properties make feasible two- and three-dimensional numerical studies of transport in fairly complex device structures. The robustness comes from the fact that in this approach the current density is given by a flow with the gradient of the quasi-Fermi level as the driving force. One should keep in mind several shortcomings of this model when it is applied to miniaturized devices: the hot-carrier effects described by the dependence of mobility on the driving field are difficult to include correctly and the non-local effects such as velocity overshoot are completely neglected. Higher-order transport models like the hydrodynamic transport [11] and the energy transport [164] models are designed to overcome some of these shortcomings of the drift-diffusion model. The energy-transport model additionally takes into account the carrier energy balance. There are, however, several problems with the energy-transport model to be used in TCAD applications. One of them is that it typically tends to overestimate the non-local effects and thus the on-current of a device. It has been demonstrated that, with the heated Maxwellian assumption [58] implicit in the model, the high energy tail of the carrier distribution is considerably overestimated. This may result in unacceptable errors in the estimation of the hot carrier induced gate tunneling current [52]. Another example is poor description of transport in partially-depleted silicon-on-insulator (SOI) devices. Because of hot carrier diffusion into the floating body of the device overestimated the energy-transport model may even fail completely in predicting the device characteristics [61–63], and the application of transport models including higher order moments is required.

Recently the six-moments transport model has been introduced [58, 60, 61, 95]. This model includes additional information on the shape of the distribution function as compared to the energy transport model, while it is still computationally far more efficient than the solution of the Boltzmann equation by Monte Carlo methods. The derivation is based on equations for statistical averages defined as

$$\Phi = \frac{1}{4\pi^3} \int \Phi(\mathbf{k})\, f(\mathbf{r}, \mathbf{k}, t)\, \mathrm{d}^3 k, \tag{12.5}$$

where $\Phi(\mathbf{k})$ is a weight function in $\mathbf{k}$-space. For simplicity we assume that the energy band is isotropic and parabolic, although generalization to a non-parabolic energy band is possible [56]. In order to derive the six-moments model, the following weight functions are chosen:

$$\begin{aligned} &\Phi_0 = 1, \quad \Phi_2 = \mathscr{E}, \quad \Phi_4 = \mathscr{E}^2, \\ &\Phi_1 = \hbar\, \mathbf{k}, \; \Phi_3 = \mathbf{v}\, \mathscr{E}, \; \Phi_5 = \mathbf{v}\, \mathscr{E}^2. \end{aligned} \tag{12.6}$$

Taking the moment of the Boltzmann equation gives the following general moment equation,

$$\frac{\partial \langle \Phi_j \rangle}{\partial t} + \nabla_{\mathbf{r}} \cdot \langle \mathbf{v} \otimes \Phi_j \rangle - s_\nu\, \mathrm{q}\, \mathbf{E} \cdot \langle \nabla_{\mathbf{p}} \otimes \Phi_j \rangle = \int \mathrm{d}^3 k\, \Phi_j \left(\frac{\partial f}{\partial t} \right)_{\mathrm{coll}}, \tag{12.7}$$

where $\otimes$ denotes the tensor product. In order to obtain a closed set of equations for moments several approximations have to be introduced. One is concerned with the moments of the scattering integral, which are frequently approximated using a macroscopic relaxation time expression:

$$\int \mathrm{d}^3 k \Phi_j \left(\frac{\partial f}{\partial t} \right)_{\mathrm{coll}} \cong -\frac{\langle \Phi \rangle - \langle \Phi \rangle_0}{\tau_\Phi}. \tag{12.8}$$

The distribution function can be separated into a symmetric part and an antisymmetric part. In the diffusion approximation it is assumed that the antisymmetric part is small compared to the symmetric part. It implies that the displacement in k-space of the distribution function and correspondingly the drift velocity is small. As a consequence, the symmetric part of the distribution function depends only on the absolute value of **k**, and the distribution function can be written as:

$$f(\mathbf{k}) = f_S(|\mathbf{k}|) + f_A(\mathbf{k}). \tag{12.9}$$

With this assumption all tensor-valued averages become scalar quantities. It can be argued that at technical frequencies the time derivatives of the fluxes can be neglected. As a result one obtains the following balance equations

$$\frac{\partial\langle 1\rangle}{\partial t} + \nabla\cdot\langle\mathbf{v}\rangle = 0, \tag{12.10}$$

$$\frac{\partial\langle\mathscr{E}\rangle}{\partial t} + \nabla\cdot\langle\mathbf{v}\,\mathscr{E}\rangle - s_\nu\,\mathrm{q}\,\mathbf{E}\cdot\langle\mathbf{v}\rangle = -\frac{\langle\mathscr{E}\rangle - \langle\mathscr{E}\rangle_0}{\tau_{\mathscr{E}}}, \tag{12.11}$$

$$\frac{\partial\langle\mathscr{E}^2\rangle}{\partial t} + \nabla\cdot\langle\mathbf{v}\,\mathscr{E}^2\rangle - s_\nu\,2\,\mathrm{q}\,\mathbf{E}\cdot\langle\mathbf{u}\,\mathscr{E}\rangle = -\frac{\langle\mathscr{E}^2\rangle - \langle\mathscr{E}^2\rangle_0}{\tau_{\mathscr{E}^2}}, \tag{12.12}$$

and the following flux equations

$$\frac{2}{3}\,\nabla\langle\mathscr{E}\rangle - s_\nu\,\mathrm{q}\,\mathbf{E}\,\langle 1\rangle = -m_\nu\,\frac{\langle\mathbf{v}\rangle}{\tau_m}, \tag{12.13}$$

$$\frac{2}{3}\,\nabla\langle\mathscr{E}^2\rangle - s_\nu\,\frac{5}{3}\,\mathrm{q}\,\mathbf{E}\,\langle\mathscr{E}\rangle = -m_\nu\,\frac{\langle\mathbf{v}\,\mathscr{E}\rangle}{\tau_S}, \tag{12.14}$$

$$\frac{2}{3}\,\nabla\langle\mathscr{E}^3\rangle - s_\nu\,\frac{7}{3}\,\mathrm{q}\,\mathbf{E}\,\langle\mathscr{E}^2\rangle = -m_\nu\,\frac{\langle\mathbf{v}\,\mathscr{E}^2\rangle}{\tau_K}. \tag{12.15}$$

In order to obtain the six-moments model we introduce the following quantities, in analogy to the energy transport models:

$$\langle 1\rangle = \nu,\quad \langle\mathscr{E}\rangle = \frac{3}{2}\,\mathrm{k_B}\,\nu\,T_\nu,\quad \langle\mathscr{E}^2\rangle = \frac{15}{4}\,\mathrm{k_B^2}\,\nu\,T_\nu\,\Theta_\nu,\quad \langle\mathscr{E}^3\rangle = \frac{105}{8}\,\mathrm{k_B^3}\,\nu\,M_6$$

$$\langle\mathbf{v}\rangle = \frac{\mathbf{J}_\nu}{s_\nu\,\mathrm{q}},\quad \langle\mathbf{v}\,\mathscr{E}\rangle = \mathbf{S}_\nu,\quad \langle\mathbf{v}\,\mathscr{E}^2\rangle = \mathbf{K}_\nu.$$

Here, ν is the carrier concentration, T_ν denotes the carrier temperature, $\mathbf{J}_\nu$ is the electrical current density, and $\mathbf{S}_\nu$ stands for the energy flux density. As compared to the energy-transport models, for the six-moments model, the new variables are a second order temperature Θ_ν, the moment of sixth order M_6, and a flux $\mathbf{K}_\nu$ related to the kurtosis of the distribution function. With generation-recombination terms added, the balance equations for the new variables become

$$\nabla\cdot\mathbf{J}_\nu = -s_\nu\,\mathrm{q}\left(\frac{\partial\nu}{\partial t} + R_\nu\right), \tag{12.16}$$

$$\nabla \cdot \mathbf{S}_\nu = -C_4 \frac{\partial(\nu\, T_\nu)}{\partial t} + \mathbf{E} \cdot \mathbf{J}_\nu - C_4\, \nu \frac{T_\nu - T_L}{\tau_{\mathscr{E}}} + G_{\mathscr{E}\nu}, \tag{12.17}$$

$$\nabla \cdot \mathbf{K}_\nu = -C_5 \frac{\partial(\nu\, T_\nu\, \Theta_\nu)}{\partial t} + 2\, s_\nu\, \mathrm{q}\, \mathbf{E} \cdot \mathbf{S}_\nu - C_5\, \nu \frac{T_\nu\, \Theta_\nu - T_L^2}{\tau_\Theta} + G_{\Theta_\nu}, \tag{12.18}$$

$$C_4 = \frac{3}{2}\,\mathrm{k_B}, \qquad C_5 = \frac{15}{4}\,\mathrm{k_B^2}. \tag{12.19}$$

We note that generation-recombination terms may depend on both electron and hole distribution functions in an integral, non-local manner [150], which makes the task of solving the corresponding equations extremely difficult. Therefore, generation-recombination terms have to be modeled carefully using knowledge from the solid-state physics of semiconductors and may represent a significant challenge [59].

The system of equations for the densities must be completed with the flux equations, which are written as

$$\mathbf{J}_\nu = -C_1 \left(\nabla(\nu\, T_\nu) - s_\nu \frac{\mathrm{q}}{\mathrm{k_B}}\, \mathbf{E}\, \nu \right), \qquad C_1 = s_\nu\, \mathrm{k_B}\, \mu_\nu, \tag{12.20}$$

$$\mathbf{S}_\nu = -C_2 \left(\nabla(\nu\, T_\nu\, \Theta_\nu) - s_\nu \frac{\mathrm{q}}{\mathrm{k_B}}\, \mathbf{E}\, \nu\, T_\nu \right), \qquad C_2 = \frac{5}{2} \frac{\mathrm{k_B^2}}{\mathrm{q}} \frac{\tau_S}{\tau_m}\, \mu_\nu, \tag{12.21}$$

$$\mathbf{K}_\nu = -C_3 \left(\nabla(\nu\, M_6) - s_\nu \frac{\mathrm{q}}{\mathrm{k_B}}\, \mathbf{E}\, \nu\, T_\nu\, \Theta_\nu \right), \qquad C_3 = \frac{35}{4} \frac{\mathrm{k_B^3}}{\mathrm{q}} \frac{\tau_K}{\tau_m}\, \mu_\nu, \tag{12.22}$$

where the mobility

$$\mu_\nu = \mathrm{q}\tau_m / m_\nu \tag{12.23}$$

is introduced.

The drift-diffusion transport model consists of the continuity equation (12.16) and the current relation (12.20). The latter is decoupled from the higher order equation by introducing a closure assumption for the second order moment, $T_\nu = T_L$. The physical meaning of this assumption is that the carrier gas is in equilibrium with the lattice.

The energy-transport model additionally takes into account the carrier energy balance equation (12.17) and the energy flux equation (12.21). To close the system of equations, an assumption on the fourth order moment has to be introduced. The assumption of a heated Maxwellian distribution for the symmetric part of the distribution function gives the closure relation $\Theta_\nu = T_\nu$.

Going one step further in the model hierarchy one obtains a transport model of sixth order. A balance equation for the average squared energy (12.18) and the related flux equation (12.22) are added. To close the equation system, the moment of sixth order M_6 has to be approximated using the lower order moments. For a Maxwellian distribution function and parabolic energy bands one would obtain $M_6 = T_\nu^3$. However, as the six-moments model does not require the heated Maxwellian approximation by treating the kurtosis of the distribution function as an unknown, a more general closure relation is desirable. Sonoda *et al.* [162] proposed a similar six-moments transport model, however, with a very restrictive closure relation. In [57] an empirical closure relation has been proposed taking into account also the second order temperature Θ_ν.

$$M_6 = T_\nu^3 \left(\frac{\Theta_\nu}{T_\nu}\right)^c . \tag{12.24}$$

From Monte Carlo simulations, which are an accurate reference, the value of $c = 2.7$ has been estimated [55]. Compared to the energy-transport models, the six-moments model requires two additional relaxation times, namely the relaxation time of the second order temperature τ_Θ, and the kurtosis flux relaxation time τ_K. Since analytical models for these new parameters are not available, tabulated values obtained from bulk Monte Carlo simulations [55] can be used.

12.2.2 Model Verification

In order to investigate the validity of macroscopic transport models we have carried out extensive Monte Carlo simulations of transport through a silicon n-doped - intrinsic - n-doped ($n - i - n$) structure. The length of the two heavily doped contacts ($N_D = 10^{20}$ cm^{-3}) is kept constant, while the length of the intrinsic channel ($N_A = 10^{16}$ cm^{-3}) varies. The band structure of silicon is approximated by six valleys with parabolic dispersion relation. Electron scattering with acoustic and optical phonons as well as with ionized impurities is taken into account. The steady state Boltzmann equation can also be conveniently solved by expanding the angular dependence of the distribution function $f(\mathbf{r}, \mathbf{k})$ on $\mathbf{k}$ using a complete set of spherical harmonics $Y_{lm}(\theta, \phi)$:

$$f(\mathbf{r}, \mathbf{k}) = \sum_{lm} f_{lm}(\mathbf{r}, k) Y_{lm}(\theta, \phi), \tag{12.25}$$

where the θ and ϕ are the polar angles between the electric field $\mathbf{E}$ and $\mathbf{k}$. In the low-field limit one can truncate the expansion (12.25) after the terms with $l = 1$, which in case of parabolic isotropic bands and randomizing elastic scattering results in a drift-diffusion transport model with low-field mobility. As is shown in [67], in case of elastic scattering this approximation gives good results for silicon where the valleys are not isotropic. For general scattering processes and realistic band structures as well as at higher driving fields more terms in the expansion (12.25) are needed [135].

The average velocity determines carrier transport in the structure. The average velocity as a moment of the distribution function can be computed along the device and is thus a more sensitive measure of validity for a particular transport model than any integral characteristic of a device like the total current. A typical result for the

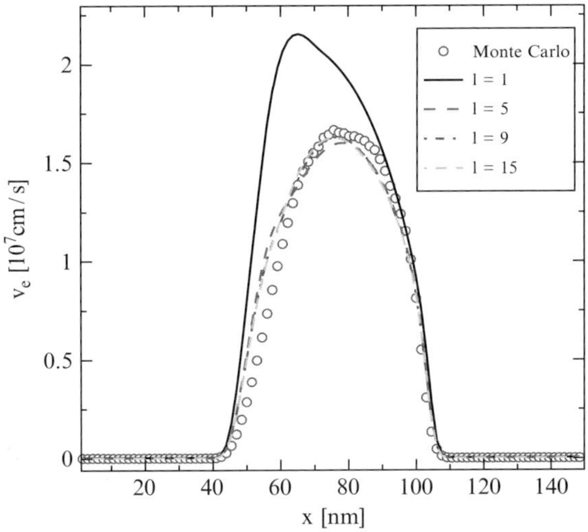

Fig. 12.1 Average velocity along the $n-i-n$ structure computed by solving the Boltzmann transport equation with the Monte Carlo method and using the spherical harmonic expansion method, with different numbers of terms included in the series (12.25). It is demonstrated that even in a short-channel structure the series expansion with terms only up to $l = 9$ included provides excellent results as compared to the more time consuming Monte Carlo data

average velocity of carriers along an $n-i-n$ structure with an intrinsic region length of 40 nm computed with the Monte Carlo method is shown in Fig. 12.1. The solution of the Boltzmann equation by spherical harmonic expansion, with different number of terms in the series (12.25) is also displayed in Fig. 12.1. In such a short device the series (12.25) truncated at $l = 1$ gives only a poor approximation for the velocity profile. At the same time it is demonstrated that the spherical harmonic expansion (12.25) including terms up to $l = 9$ gives a perfect result. The inclusion of terms with $l > 9$ into (12.25) does not change the velocity profile confirming the rapid convergence of the series (12.25). Therefore, the spherical harmonic expansion method gives excellent results with only a few terms included into (12.25) even for short-channel devices. For this reason the method is much less time consuming than the Monte Carlo solution of the Boltzmann equation. However, Monte Carlo data are needed in order to validate the applicability of the spherical expansion method to describe transport in short-channel devices.

Current-voltage characteristics computed with the Monte Carlo (identical to those computed with spherical harmonic expansion method) and using the macroscopic transport models based on the moments of the distribution function are shown in Figs. 12.2 and 12.3, for several channel lengths. It is demonstrated that for long devices (1 μm) the drift-diffusion, the energy transport model, and the six-moments model give almost equivalent results which are in perfect agreement with the results of the spherical harmonic expansion method. For a device with $L_{ch} = 250$ nm the drift-diffusion model underestimates the current. Since the carrier temperature is constant, the drift-diffusion model does not account for any non-local effects and

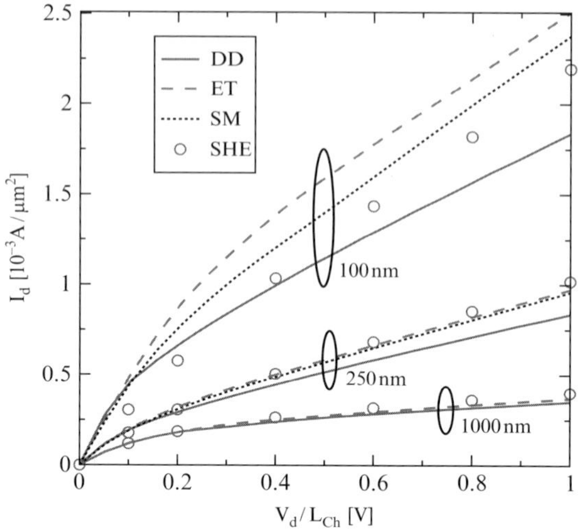

Fig. 12.2 Current-voltage characteristics for devices with different channel length computed with the spherical harmonic expansion method and using macroscopic transport models based on moments of the distribution function. Here DD stands for the drift-diffusion model, ET for energy-transport model, SM for the six-moments model, and SHE for spherical harmonic expansion

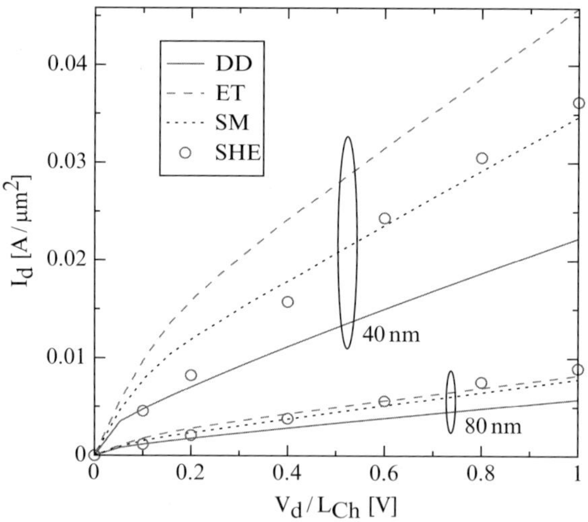

Fig. 12.3 The same as in Fig. 12.2 for shorter devices. The six-moments model gives results closest to the results of Monte Carlo and the spherical Harmonic expansion methods

cannot capture the non-local transport inside short-channel devices. This causes the accuracy of the drift-diffusion model to decrease for gate lengths shorter than 250 nm, where the restriction of constant carrier temperature must be relaxed. Due to the temperature gradient, heat flow and thermal diffusion appear. The drift-diffusion transport model must be augmented with the energy flow.

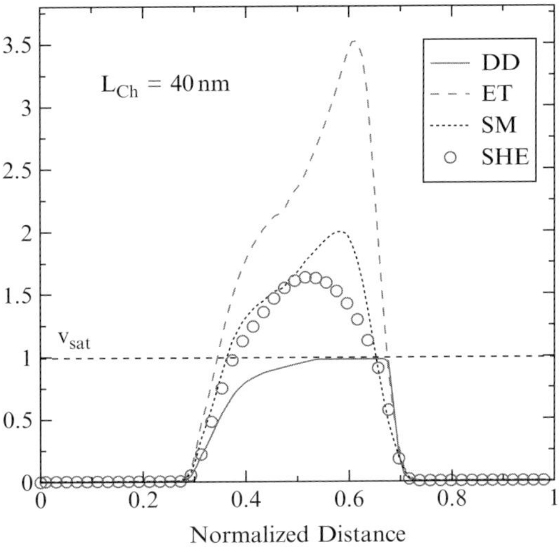

Fig. 12.4 Average velocity along a device with 40 nm channel length computed with the macroscopic transport models and with the spherical harmonic expansion method. While the drift-diffusion model underestimates the velocity and current, the energy-transport model overestimates the velocity in short-channel devices. The six-moments model gives best results as compared to the velocity obtained by the spherical harmonic expansion method

The energy-transport model additionally akes into account the energy flux equation and the carrier energy balance equation. The model, however, overestimates the drive current as it was mentioned before. Figure 12.4 illustrates the average velocity profile in a device with a 40 nm long channel. The drift-diffusion model underestimates the average velocity while the energy-transport model overestimates it. In order to reduce this spurious velocity overshoot effect the next moments should be included for devices with L_{ch} shorter than 100 nm.

This is accomplished by introducing a transport model of sixth order. A balance equation for the average squared energy and the related flux equation are added. To close the equation system, the moment of sixth order has to be approximated using lower order moments. From Monte Carlo simulations serving as an accurate reference, an empirical closure relation has been derived (12.24). The inclusion of higher moments improves the transport model significantly. The current-voltage characteristics are reproduced fairly well even in devices as short as 40 nm as demonstrated in Fig. 12.3, because of the more accurate results for the velocity (Fig. 12.4). The relative error of the current computed by the macroscopic transport models presented in Fig. 12.5 as function of the channel length clearly demonstrates the applicability of the six-moments model to describe the output characteristics of devices with a channel as short as 40 nm.

Since the energy-transport model takes into account the energy flux equation in addition to the carrier energy balance equation, it requires modeling of two mobilities for the current density and the energy flux for each carrier type, one relaxation

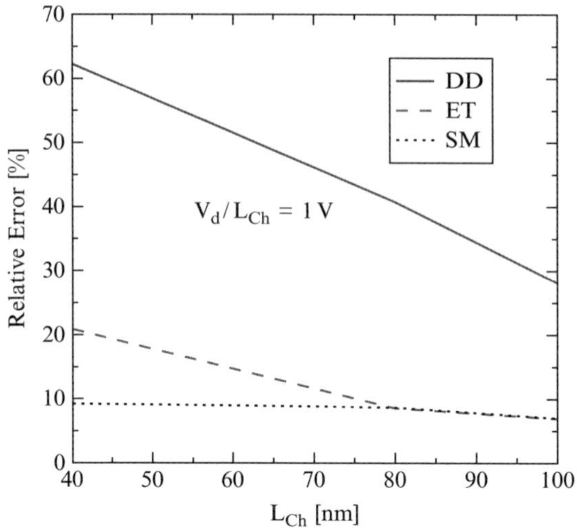

Fig. 12.5 Relative error of the current computed with macroscopic transport models. While the drift-diffusion and the energy-transport models are gradually losing their validity with the channel length reduced, the six-moments model maintains its accuracy down to the channel length of 40 nm

time, and the non-parabolicity factor for non-parabolic bands. The six-moments model requires two additional relaxation times for the second order temperature and the kurtosis flux. Having too many adjustable parameters is a particular inconvenience of the six-moments model. A solution to this problem is based on tabulating some parameters of the model using Monte Carlo simulations. The parameter dependences on temperature, doping, and driving field are determined from the condition that the six-moments transport model reproduces exactly all the six moments obtained from the Monte Carlo simulator under homogeneous conditions. Due to calibration of all these parameters, a generalization of the six-moments transport model to include strain is still pending.

12.3 Mobility in Strained Silicon

The transport models discussed in the previous section contain several parameters which have to be accurately modeled for the purpose of predictive simulations. The drift-diffusion model described by (12.16) and the current relation (12.20) requires the carrier mobility (12.23) as such a parameter. The mobility relates the drift velocity to the electric field and depends on scattering mechanisms encoded into the momentum relaxation time τ as well as on the band structure via the effective mass m. Knowledge of the mobility dependence on temperature, doping, stress, substrate orientation, and driving electric field allows to investigate device performance in different regimes for arbitrary parameters.

12.3.1 Mobility and Piezoresistance

At low strain value the modification of mobility can be obtained by using piezoresistance coefficients. The piezoresistance coefficients π_{ijkl} are defined by

$$\frac{\Delta\rho_{ij}}{\rho} = \pi_{ijkl}\sigma_{kl}, \tag{12.26}$$

where σ_{kl} is the stress tensor defined in Sect. 3.1, and $\Delta\rho_{ij}$ is the stress-induced variation of the resistivity tensor $\rho_{ij} = \rho\delta_{ij}$ which is diagonal in the crystallographic coordinate system in relaxed silicon. Like in the case of the elastic stiffness tensor (3.18), the number of independent entries for to the cubic semiconductors is reduced due to symmetry, and the piezoresistance tensor is characterized by only three independent components conveniently written as π_{11}, π_{12}, and π_{44}. In this case the relation (12.26) between stress and resistivity is written as:

$$\begin{pmatrix} \frac{\Delta\rho_{xx}}{\rho} \\ \frac{\Delta\rho_{yy}}{\rho} \\ \frac{\Delta\rho_{zz}}{\rho} \\ \frac{\Delta\rho_{yx}}{\rho} \\ \frac{\Delta\rho_{xz}}{\rho} \\ \frac{\Delta\rho_{xy}}{\rho} \end{pmatrix} = \begin{pmatrix} \pi_{11} & \pi_{12} & \pi_{12} & 0 & 0 & 0 \\ \pi_{12} & \pi_{11} & \pi_{12} & 0 & 0 & 0 \\ \pi_{12} & \pi_{12} & \pi_{11} & 0 & 0 & 0 \\ 0 & 0 & 0 & \pi_{44} & 0 & 0 \\ 0 & 0 & 0 & 0 & \pi_{44} & 0 \\ 0 & 0 & 0 & 0 & 0 & \pi_{44} \end{pmatrix} \cdot \begin{pmatrix} \sigma_{xx} \\ \sigma_{yy} \\ \sigma_{zz} \\ \sigma_{yz} \\ \sigma_{xz} \\ \sigma_{xy} \end{pmatrix}. \tag{12.27}$$

Due to a simple relation $\Delta\mu_{ij}/\mu = -\Delta\rho_{ij}/\rho$ the stress-induced variation $\Delta\mu_{ij}$ of the mobility tensor $\mu_{ij} = \mu\delta_{ij}$ is expressed through the piezoresistance coefficients as:

$$\begin{pmatrix} \frac{\Delta\mu_{xx}}{\mu} \\ \frac{\Delta\mu_{yy}}{\mu} \\ \frac{\Delta\mu_{zz}}{\mu} \\ \frac{\Delta\mu_{yx}}{\mu} \\ \frac{\Delta\mu_{xz}}{\mu} \\ \frac{\Delta\mu_{xy}}{\mu} \end{pmatrix} = \begin{pmatrix} -\pi_{11} & -\pi_{12} & -\pi_{12} & 0 & 0 & 0 \\ -\pi_{12} & -\pi_{11} & -\pi_{12} & 0 & 0 & 0 \\ -\pi_{12} & -\pi_{12} & -\pi_{11} & 0 & 0 & 0 \\ 0 & 0 & 0 & -\pi_{44} & 0 & 0 \\ 0 & 0 & 0 & 0 & -\pi_{44} & 0 \\ 0 & 0 & 0 & 0 & 0 & -\pi_{44} \end{pmatrix} \cdot \begin{pmatrix} \sigma_{xx} \\ \sigma_{yy} \\ \sigma_{zz} \\ \sigma_{yz} \\ \sigma_{xz} \\ \sigma_{xy} \end{pmatrix}. \tag{12.28}$$

Silicon and germanium were the first materials used as piezoresistors. Smith [161] reported first measurements of large piezoresistive coefficients in bulk silicon. For a lightly doped p-silicon with resistivity of 7.8 Ω-cm the piezoresistance tensor is

$$\pi = \begin{pmatrix} 6.6 & -1.1 & -1.1 & 0 & 0 & 0 \\ -1.1 & 6.6 & -1.1 & 0 & 0 & 0 \\ -1.1 & -1.1 & 6.6 & 0 & 0 & 0 \\ 0 & 0 & 0 & 138.1 & 0 & 0 \\ 0 & 0 & 0 & 0 & 138.1 & 0 \\ 0 & 0 & 0 & 0 & 0 & 138.1 \end{pmatrix} \times 10^{-11} Pa^{-1}, \tag{12.29}$$

while for n-silicon the piezoresistance coefficients are:

$$\pi = \begin{pmatrix} -102.2 & 53.4 & 53.4 & 0 & 0 & 0 \\ 53.4 & -102.2 & 53.4 & 0 & 0 & 0 \\ 53.4 & 53.4 & -102.2 & 0 & 0 & 0 \\ 0 & 0 & 0 & -13.6 & 0 & 0 \\ 0 & 0 & 0 & 0 & -13.6 & 0 \\ 0 & 0 & 0 & 0 & 0 & -13.6 \end{pmatrix} \times 10^{-11} Pa^{-1}, \tag{12.30}$$

We notice that the coefficient π_{44} determines the mobility modification due to shear strain.

Although convenient, piezoresistance coefficients are found to depend on doping and temperature. The piezoresistance coefficients in the surface layers are different from bulk values [19] and depend on such values as an effective field in surface layers. A more refined theory is thus required in order to model mobility.

12.3.2 Compact Mobility Modeling

In some cases, the physically-based consideration helps to generalize the model and to include additional effects. In relaxed silicon, the lowest conduction band consists of six valleys, aligned along the crystallographic axes. In any valley, the mobility is anisotropic due to anisotropy of the inverse effective mass tensor: mobility is higher in the direction corresponding to a smaller effective mass. However, net electron mobility being the average of the components along the three valley orientations is independent of direction, or isotropic. To develop a model for stress dependent mobility, the peculiarity of the band structure modifications has to be taken into account. In silicon the conduction band consists of three pairs of equivalent valleys with their minima located close to the X-points in the Brillouin zone. Uniaxial compression in [001] direction decreases the band energy of the valley parallel to the strain and transfers electrons into these valleys. Since the [001] valleys have high mobility in the (001) plane average mobility in this plane is increased. Alternatively, mobility in the direction of compression is lowered. Elongation in [001] direction has the opposite effect.

Let us consider silicon biaxially stressed in the [001] plane. This type of stress is practically realized in Si grown on [001] SiGe substrate [195]. In this case the degeneracy between the three pairs of valleys is lifted, and the relative shift $\Delta E_C^{(i)}$ between the in-plane and out-of-plane valleys appear. The shift is given as [10]

$$\Delta E_C^{(i)} = \Xi_d(\varepsilon_{xx} + \varepsilon_{yy} + \varepsilon_{zz}) + \Xi_u \varepsilon_{ii}, \quad i = x, y, z, \tag{12.31}$$

where Ξ_d and Ξ_u denote the dilation and shear deformation potentials for the conduction band and the ε_{ii} denote the diagonal components of the strain tensor expressed in the principal coordinate system.

The electron occupation in the valleys depends on their relative energy shifts and must be properly accounted for [30, 116]. As suggested in [116], the electron mobility can be computed by taking the weighted average of the electron mobility tensors $\widehat{\mu}_{\mathrm{n,uns}}^{(i)}$ of the ith pair of valleys in Si with the corresponding electron concentration $n_{\mathrm{str}}^{(i)}$:

$$\widehat{\mu}_{\mathrm{n}}^{\mathrm{tot}} = \sum_{i=1}^{3} p^{(i)} \cdot \widehat{\mu}_{\mathrm{n,uns}}^{(i)}, \quad p^{(i)} = \frac{n_{\mathrm{str}}^{(i)}}{\sum_{i=1}^{3} n_{\mathrm{str}}^{(i)}}, \tag{12.32}$$

$$n_{\mathrm{str}}^{(i)} = N_C^{(i)} \cdot \exp\left[\frac{\Delta E_C^{(i)}}{k_B T}\right]. \tag{12.33}$$

Here $n_{\mathrm{str}}^{(i)}$ is calculated using Boltzmann statistics with $N_C^{(i)}$ as the effective density of states and $\Delta E_C^{(i)}$ as the energy shift, for the ith valley, k_B and T denote the Boltzmann's constant and ambient temperature, respectively. In (12.32) the mobility tensor is a product of a scalar mobility μ_L and the scaled inverse mass tensor:

$$\widehat{\mu}_{\mathrm{n,str}}^{(i)} = \mu_L \cdot \widehat{m}_{(i)}^{-1}, \quad i = x, y, z, \tag{12.34}$$

$$\widehat{m}_x^{-1} = \begin{pmatrix} \frac{m_c}{m_l} & 0 & 0 \\ 0 & \frac{m_c}{m_t} & 0 \\ 0 & 0 & \frac{m_c}{m_t} \end{pmatrix}, \quad \widehat{m}_y^{-1} = \begin{pmatrix} \frac{m_c}{m_t} & 0 & 0 \\ 0 & \frac{m_c}{m_l} & 0 \\ 0 & 0 & \frac{m_c}{m_t} \end{pmatrix}, \quad \widehat{m}_z^{-1} = \begin{pmatrix} \frac{m_c}{m_t} & 0 & 0 \\ 0 & \frac{m_c}{m_t} & 0 \\ 0 & 0 & \frac{m_c}{m_l} \end{pmatrix}. \tag{12.35}$$

The tensors in (12.35) are the inverse effective mass tensors with m_t, m_l denoting the transversal and lateral masses for the ellipsoidal X-valleys in Si. The tensor is scaled to a dimensionless form by the conductivity mass, m_c:

$$m_c = \frac{3}{\frac{2}{m_t} + \frac{1}{m_l}}. \tag{12.36}$$

Electrons in bulk silicon are scattered by phonons and impurities. Due to the later scattering mechanism the electron mobility depends strongly on the doping level. After scattering with phonons an electron may stay within the same valley or scatter into an equivalent valley along the same axis (g-type). The after-scattering state can be also located in a non-equivalent valley (f-type scattering) [76]. By applying the Matthiesson's rule one can model isotropic electron mobility in relaxed Si, including the effects of g-type, f-type phonon scattering, and dependence on the doping concentration N_I via the corresponding relaxation times τ_{equiv}, $\tau_{\text{neq}}(\Delta E_C^{(i)})$, and $\tau_I(N_I)$ as

$$\mu(N_I, \Delta E_C^{(i)}) = \frac{e}{m_c \left(\dfrac{1}{\tau_{\text{equiv}}} + \dfrac{1}{\tau_{\text{neq}}(\Delta E_C^{(i)})} + \dfrac{1}{\tau_I(N_I)} \right)}. \tag{12.37}$$

The electron mobility for the ith valley in strained Si can be written as [25]

$$\widehat{\mu}^{(i)}_{\text{n,str}}(N_I, \Delta E_C^{(i)}) = \frac{\beta \cdot \mu^L}{1 + (\beta - 1) \cdot h^{(i)} + \beta \cdot \left(\dfrac{\mu^L}{\mu^{LI}} - 1 \right)} \cdot \widehat{m}^{-1}_{(i)}. \tag{12.38}$$

where $\widehat{m}^{-1}_{(i)}$ denotes the scaled effective mass tensor for the ith valley in (12.35), and h is a known function of T, $\Delta E_C^{(i)}$ [25]. The parameter $\beta = \dfrac{f \cdot m_t}{m_c}$, where f is the mobility enhancement factor, is defined as the ratio of the saturation value of the transversal mobility in the valley i at high values of strain to the unstrained mobility μ^L. Equation (12.38) is plugged into (12.32) to give the total mobility tensor for electrons in strained Si as a function of doping concentration N_I and strain.

The model was generalized to describe the low-field mobility in silicon under arbitrary stress conditions to include shear strain [26]. For tensile stress along [110] the energy dispersion of the lowest conduction band is influenced as follows (see Sect. 9):

- The valleys located along the [100] and [010] directions move up in energy with respect to the valleys located along the [001] direction, in accordance with (8.2);
- The band minima of the valleys oriented along the [001] direction move down in energy, in accordance with (9.2) and (9.8), and towards the X-point located at $\frac{2\pi}{a_0}[0, 0, \pm 1]$, in accordance with (9.7);
- The shape of the valleys along the [001] direction is warped which results in a modification of the transversal effective masses (9.9)–(9.10) and the longitudinal mass (9.11).

In the case of the stress-dependent masses (9.9)–(9.10) the inverse effective mass tensor m_z^{-1} for the [001] valleys in the crystallographic coordinate system is written as [26]:

$$\widehat{m}_z^{-1} = \begin{pmatrix} \dfrac{m_c}{m_t(\eta)} & \dfrac{m_c}{\Delta m_t(\eta)} & 0 \\ \dfrac{m_c}{\Delta m_t(\eta)} & \dfrac{m_c}{m_t(\eta)} & 0 \\ 0 & 0 & \dfrac{m_c}{m_l(\eta)} \end{pmatrix}, \tag{12.39}$$

where

$$m_t^{-1}(\eta) = \frac{m_{t1}(\eta)^{-1} + m_{t2}(\eta)^{-1}}{2}, \tag{12.40}$$

$$\Delta m_t^{-1}(\eta) = \frac{m_{t2}(\eta)^{-1} - m_{t1}(\eta)^{-1}}{2}, \tag{12.41}$$

and m_c^{-1} for the [001] valleys is defined as

$$m_c^{-1}(\eta) = \frac{m_{t1}(\eta)^{-1} + m_{t2}(\eta)^{-1} + m_l(\eta)^{-1}}{3}.$$

In order to validate compact mobility models and calibrate their parameters, results must be used which are obtained by more precise methods, which, however, require considerably more computational resources. A feasible approach to obtain an accurate solution of the Boltzmann transport equation is based on the Monte Carlo techniques briefly discussed next.

12.3.3 Monte Carlo Methods for Transport Calculations

Methods based on Monte Carlo techniques are well established for studying transport in semiconductors [73]. The motion of charge carriers is simulated in the appropriate phase space formed by position and momentum. In the presence of external fields, the carriers which are considered as point-like objects with well defined momenta and positions move according to Newton's law on classical trajectories. A dispersion relation expressing the carrier energy dependence on the crystal momentum is determined by the semiconductor's energy band structure. The free flight of carriers along the trajectory is interrupted by scattering events which are assumed local in space and instantaneous in time. Scattering is modeled as a random process. The duration of a free flight, the type of scattering mechanism, and the state after scattering are selected randomly from given probability distributions characteristic to the microscopic scattering process. The method of generating sequences of free flights and scattering events appears to be so intuitively transparent, that it is frequently interpreted as a direct emulation of the physical transport process rather than a numerical method. The first Monte Carlo algorithms used in device simulations were originally derived from merely physical considerations, viewing a Monte Carlo simulation as a computer experiment. These algorithms are called Ensemble Monte Carlo (EMC) [68, 120] and One Particle Monte Carlo (OPMC) [76, 138].

Alternatively, one can reformulate the transport equation as an integral equation first and then develop a Monte Carlo algorithm for its solution [75, 128]. When the Boltzmann equation is transformed to an integral equation which is then iteratively substituted into itself [75], the iteration series results in the technique called Monte Carlo Backward. The algorithm is useful, if rare events have to be simulated or the distribution function is needed only in a small phase space domain [94]. If the Boltzmann equation is reformulated in an adjoint integral form [90], a link between a physically-based Monte Carlo method and an iterative procedure of the solution of an integral equation is established.

The work of Kurosawa in 1966 [101] is considered to be the first account of an application of the Monte Carlo method to high-field transport in semiconductors. The following decade has seen considerable improvement of the method and application to a variety of materials [73]. Early papers deal with gallium arsenide [35] and germanium [36]. In the mid 1970s a physical model of silicon has been developed, capable of explaining major macroscopic transport characteristics [17, 74]. The used energy band structure models were represented by simple analytic expressions accounting for non-parabolicity and anisotropy. With the increase of the carriers' energy the need for accurate, numerical energy band structure models arose [42, 100, 152, 197]. For electrons in silicon, the most thoroughly investigated case, it is believed that a satisfactory understanding of the basic scattering mechanisms gives rise to a new "standard model" [38]. With the introduction of strain to enhance the performance of MOSFETs, however, the need for accurate full-band transport analysis has regained considerable interest [41, 43, 79].

We demonstrate the importance of full-band consideration by using an analysis of the electron low-field mobility in strained Si as an example. We apply the simulator VMC [187], which offers simulation algorithms for both bulk semiconductors and devices with models based either on analytical band structure or on the full-band structure analyses. VMC includes a comprehensive set of scattering models with phonons, ionized impurities, alloy scattering, as well as impact ionization both for electrons and holes.

Figure 12.6 demonstrates mobility enhancement of holes due to uniaxial stress applied along [111] direction to lightly doped p-silicon. Band structure was obtained with the empirical pseudopotential method described in Chaps. 5 and 8. Compressive stress favors significant mobility enhancement while tensile stress reduces hole mobility. The model based on the piezoresistance coefficients is also shown by the straight line. The piezoresistance coefficient value $\pi_{<111>} = 71.8 \times 10^{-11}$ Pa^{-1} along the [111] stress direction [19] calculated from the values from [161] is in good agreement with the results of the full-band calculations for stress absolute value below 0.1 GPa. For larger stress values the mobility enhancement factor deviates significantly from that predicted by the piezoresistance mode, especially for tensile stress above 0.5 GPa where the mobility modification saturates.

For electrons in n-silicon, a phonon scattering model based on Jacoboni and Reggiani [76] is used. The model takes into account long-wavelength acoustic phonons causing intravalley transitions and three f- and g-type phonons for intervalley transitions. Intravalley scattering from acoustic phonons is treated as elastic

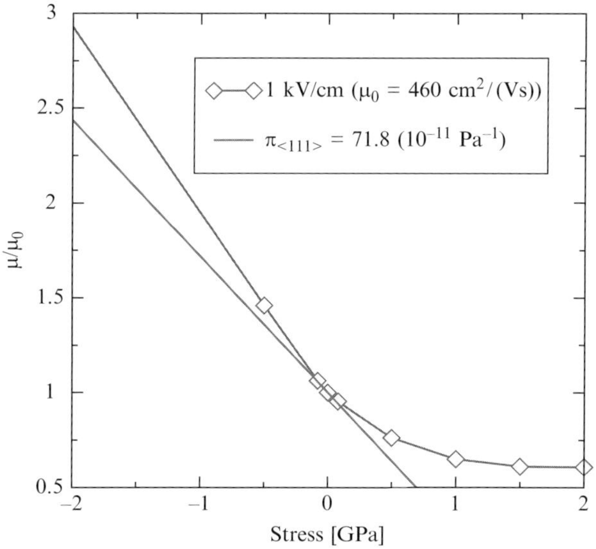

Fig. 12.6 Hole mobility enhancement factor along the [111] stress direction obtained with the full-band Monte Carlo simulations (*symbols*). Prediction based on the piezoresistance model with parameters for lightly doped p-type silicon [19, 161] is shown by a straight line

Table 12.1 Modes, coupling constants, phonon energies, and selection rule of inelastic phonon scattering

Mode	Δ [MeV/cm]	$\hbar\omega$ [meV]	Selection rule
Transversal Acoustic	47.2	12.1	f
Longitudinal Acoustic	75.5	18.5	f
Longitudinal Optical	1042.0	62.0	f
Transversal Acoustic	34.8	19.0	g
Longitudinal Acoustic	232.0	47.4	g
Transversal Optical	232.0	58.6	g

process. Following [25] we slightly adjust the original values for the coupling constants for intervalley phonon scattering [76] to achieve a bulk mobility enhancement factor of 70% in biaxially strained Si layers. The coupling constants for acoustic and optical intervalley phonons, as well as the phonon energies are listed in Table 12.1. In full-band simulations the scattering rates are proportional to the density of states calculated from the band structure [79].

Figure 12.7 demonstrate the orientation-dependent electron mobility enhancement factor for tensile stress applied in [111] direction obtained from the full-band Monte Carlo simulations. The current direction is defined by the unit vector **n**

$$\mathbf{n} = [\cos(\Theta + \Phi_0)/\sqrt{2}, \cos(\Theta + \Phi_0)/\sqrt{2}, \sin(\Theta + \Phi_0)],$$

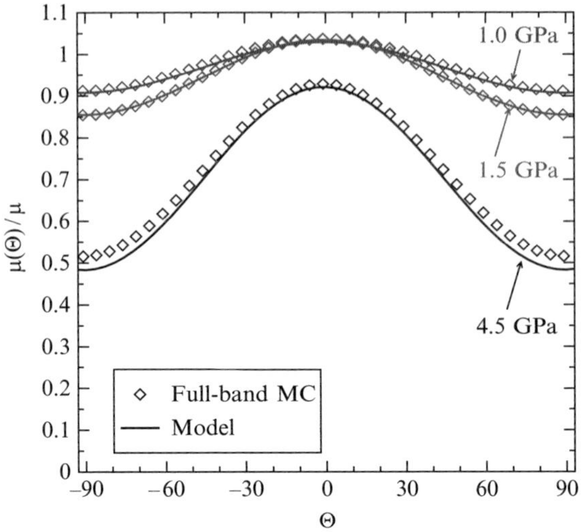

Fig. 12.7 Orientation dependent electron mobility variation in silicon under tensile stress along [111] direction, for several stress values. Results of full-band Monte Carlo are well explained by dependence of the effective masses on shear strain. Note a mobility degradation at high stress values caused by an increase of scattering due to stress-dependent increase of the density-of-states mass

where $\Phi_0 \approx 36°$ is the angle between [111] and [110] directions. The angle Θ determines the current direction within the plane defined by the [111] and [1$\bar{1}$0] is counted from the [111] direction.

Results of the full-band simulations coincide well with the analytical model presented in the previous subsection, even for high stress value. Stress along [111] direction affects all six valleys in a similar way and thus does not produce relative energy shift between the valleys. [111] stress produces shear strain component, which in this case affects all six valleys in a similar way. In this case mobility of an ellipsoidal valley in an arbitrary orientation **n** with respect to the crystallographic coordinate system in a certain direction **n** is determined by the (inverse) conductivity mass $1/m_{\mathbf{n}}$ in this direction

$$m_{\mathbf{n}}^{-1} = \mathbf{n}^T \hat{m}^{-1}(\eta)\mathbf{n}$$

where $\mathbf{m}^{-1}(\eta)$ is the inverse effective mass tensor for a valley in the crystallographic system. Taking into account the dependence of the density-of-state effective mass $m_{dos}(\eta) = (m_l(\eta)m_{t1}(\eta)m_{t2}(\eta))^{1/3}$ on dimensionless strain η, the mobility dependence on [111] strain in a direction **n** is obtained as

$$\frac{\mu(\eta, \mathbf{n})}{\mu_0} = \frac{m_c}{m_{\mathbf{n}}}\left[\frac{m_{dos}}{m_{dos}(\eta)}\right]^{3/2}, \tag{12.42}$$

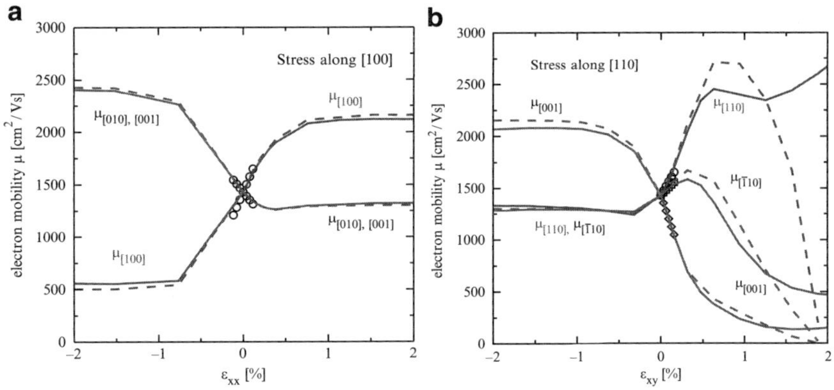

Fig. 12.8 Simulated bulk mobility of intrinsic Si as a function of strain for stress direction [100] (**a**) and [110] (**b**). Mobility is plotted along the stress direction and along two orthogonal directions from full-band Monte Carlo (*solid lines*) and analytical band Monte Carlo (*dotted lines*). Symbols indicate the change of mobility calculated using the piezoresistance coefficients [161]

where μ_0 is the mobility of relaxed silicon, $m_c = 3(1/m_l + 2/m_t)^{-1}$. The strain dependence of the inverse effective mass tensor $\mathbf{m}^{-1}(\eta)$ on shear strain is determined by (9.9, 9.10, 9.11). For the [001] valleys $\mathbf{m}^{-1}(\eta)$ is determined by (12.39, 12.40, 12.41).

Comparing simulation results from analytical band structure Monte Carlo with full-band Monte Carlo allows determining the limits of validity of the analytical band model. In Fig. 12.8 simulation results for the electron mobility of strained Si obtained with full-band and analytical band structures for the stress directions [100] and [110] as well as predictions from a model based on the linear piezoresistance coefficients [161] are shown. Mobility is plotted in three directions, one parallel and two perpendicular to the stress direction. Good agreement is found for both stress directions at small stress ($<$100 MPa).

In Fig. 12.8a the simulation results from analytical and full-band Monte Carlo for stress along [100] are compared and good agreement is obtained. The resulting mobility is anisotropic in the (001) plane ($\mu_{[100]} \neq \mu_{[010]}$), and can be explained by strain induced X-valley shifts. Mobility saturates at approximately 1% strain, regardless of the sign of strain. The saturated mobility values are larger for compressive strain, since in this case four X-valleys with unfavorable conductivity masses are depopulated.

In Fig. 12.8b simulation results are shown for stress along [110]. At compressive stress, four valleys move down in energy, yielding a decreased mobility in the (001) plane and mobility enhancement along [001]. For tensile stress along [110], the mobility behavior is remarkably different from the previous case. First, mobilities along the directions [110], [$\bar{1}$10], and [001] are different from each other, with the largest mobility enhancement observed in [110] direction. Furthermore, no clear in-plane mobility saturation is observed as the value of stress increases. The mobility enhancement in tensile stress direction is determined by the effective mass change induced by the shear strain component in the primarily populated valleys along [001] [168, 183]. Thus, analytical models solely based on strain induced intervalley electron transfer [67] would fail to predict correctly the mobility enhancement along [110] and to explain the origin of the non-vanishing shear piezoresistance coefficient $\pi_{44} = -13.6 \cdot 10^{-11}\mathrm{Pa}^{-1}$, hence these models are not capable of reproducing the anisotropy of electron mobility in uniaxially stressed channels with [110] channel direction.

It can be seen that the results from analytical band structure Monte Carlo agree well with those from full-band results up to 0.5% shear strain. At larger strain levels the band deformation is so pronounced that the energy band description in terms of an effective mass is no longer accurate, and full-band Monte Carlo simulations must be used to calculate the low-field mobility.

12.4 Mixed Quantum-Semi-Classical Description and Quantum Corrections in Current Transport Models

The characteristics of modern semiconductor devices are strongly influenced by quantum-mechanical effects. Due to this fact, purely classical device simulation may not suffice to accurately reproduce transport. Size quantization of carrier motion in the confining potential of an inversion layer is the most investigated and well understood quantum-mechanical effect in modern MOSFETs. Because of the size quantization, the energy spectrum becomes discrete in the confinement direction, while it is still continuous in the transport direction. Thus the three-dimensional energy band structure is partitioned into a set of two-dimensional quantum subbands. In order to find the subband energy levels and the corresponding wave function, the Schrödinger equation has to be solved. Within each subband transport can be described by a set of Boltzmann equations for the subband distribution functions. Unlike the bulk case, the Boltzmann equations for different subbands are coupled because of intersubband scattering. This complicates the computation of the mobility. Even more, in inversion layers and especially in FETs with ultra-thin body (UTB) the subband occupation may be large at high effective fields, and degeneracy effects due to Pauli exclusion principle start playing an important role.

Indeed, the Pauli exclusion principle forbids the double occupancy of the same quantum state. Therefore, scattering to an already occupied state is prohibited. In order to describe transport correctly, degeneracy effects must be carefully taken into account. Below an efficient Monte Carlo algorithm based on the linearized Boltzmann equation including usually ignored effects of the fermion statistics and degeneracy is presented. Degeneracy effects lead to enhanced intersubband scattering in (001) double-gate (DG) UTB FETs which results in a carrier mobility qualitatively different from the case when degeneracy is ignored.

The potential entering into the Schrödinger equation depends on the charge carrier concentration in the inversion layer. By knowing the wave functions and occupations of the subbands, the charge carrier concentration in the inversion layer can be obtained. Therefore, the potential has to be found self-consistently by solving the Schrödinger and the Poisson equations simultaneously. This procedure is time consuming and should be avoided whenever timely results must be obtained. One option is to exploit the well established semi-classical transport models, while correcting them in such a way that they mimic the behavior of the complete quantum-mechanical system. Various quantum correction models are available [29,65,80,130,131]. Some of these corrections are based on empirical fits with numerous parameters [80, 130]. In other models the convergence [29] is strongly affected by the dependence on the electrical field. Some models [65] have to be re-calibrated for each particular device. A comprehensive comparison of these models was recently performed in [189].

The quantum correction which mimics the local density of states close to the interface can also be understood as an additional quantum potential [37] which has to be added to the classical self-consistent potential in order to describe the decrease of carrier concentration at the interface correctly. Because this additional quantum potential enters into the current relations, it opens an opportunity to introduce quantum-mechanical effects into the drift-diffusion and higher moments transport models. In fact, the appearance of the quantum potential can be easily illustrated, if one just substitutes the wave function represented as the product of its real amplitude and the exponent containing the phase factor

$$\psi(\mathbf{r},t) = A(\mathbf{r},t)\exp(i\phi(\mathbf{r},t)). \tag{12.43}$$

into the Schrödinger equation. The density $n(\mathbf{r},t) = A^2(\mathbf{r},t)$, the velocity $\mathbf{v} = \hbar\nabla\phi(\mathbf{r},t)/m$, and the current $\mathbf{j} = N\mathbf{v}$ are introduced and one obtains the following system of equations [185]:

$$\frac{\partial n(\mathbf{r},t)}{\partial t} + \nabla\left[n(\mathbf{r},t)\mathbf{v}\right] = 0, \tag{12.44}$$

$$\frac{\partial \mathbf{v}}{\partial t} = \frac{v^2}{2m} + V(\mathbf{r},t) + V_{QC}(\mathbf{r},t). \tag{12.45}$$

Here, the additional quantum correction potential $V_{QC}(\mathbf{r},t)$ is given by the following expression:

$$V_{QC}(\mathbf{r},t) = -\frac{\hbar^2}{2m}\frac{\Delta\sqrt{n}}{\sqrt{n}}. \tag{12.46}$$

A similar expression for the effective quantum correction potential can be obtained from the quantum-mechanical theory based on the evolution of the moments for the Wigner function [69]:

$$V_{QC}(\mathbf{r},t) = -\frac{\hbar^2}{3r}\Delta\ln n, \tag{12.47}$$

where r is the parameter, taking values between $r = 1$ in the extreme case with only a single subband filled and $r = 3$ in case of many subbands filled.

The concept of quantum correction potentials opens the way for using the particle-based Monte Carlo algorithm to investigate transport in systems with strong size quantization. Therefore, the full power of three-dimensional full-band Monte Carlo methods which include accurate band-structure and scattering processes to obtain transport characteristics in inversion layers and SOI structures can be applied [83, 134]. Recently, mobility of stressed Si/SiGe inversion layers was investigated [34, 99, 184]. Interestingly, due to a concentration decrease at the interface, the surface roughness scattering is underestimated, which requires special corrections [15]. Due to statistical noise in the second derivative of the carrier concentration, smooth approximations for the quantum correction potential are used in practice. The popular form of the effective potential introduced by Ferry [37] relates it to the self-consistent potential via an integral smoothing relation with a Gaussian kernel. Using this quantum correction potential, transport characteristics in ultra-thin body structures with short channels were recently investigated and compared against analytical predictions [133].

The form of the quantum potential (12.47) is commonly referred to as the density gradient correction and is extensively used in quantum hydrodynamic calculations [3–6]. The expression for the current density including the quantum correction reads [8]:

$$\mathbf{J} = en\mu\mathbf{E} + eD\nabla n + en\mu\frac{\hbar^2}{2r}\nabla\frac{\Delta\sqrt{n}}{\sqrt{n}}. \tag{12.48}$$

Substitution of the current relation (12.48) into the continuity equation (12.2) results in the differential equation for the particle concentration n of the fourth order. Such an equation needs two boundary conditions. If one considers the interface between the semiconductor and the dielectric, the first boundary condition is the standard one put on the normal derivative of the concentration requiring the normal current component to be zero at the interface. The second boundary condition allows to set the carrier concentration to zero at the interface. Thus, the quantum drift-diffusion theory based on (12.48) supplemented with the corresponding boundary conditions automatically reproduces the concentration decrease at the interface, mimicking the quantum-mechanical behavior.

For numerical transport calculations it is convenient to avoid the discretization of the fourth order equation and to include the quantum potential correction term into a generalized electro-chemical potential [8]. After a careful calibration of the resulting density gradient model the transport calculations including the source-drain tunneling in ultra-scaled MOSFETs becomes possible [8]. Quantum hydrodynamic models are reviewed in [78].

We are not aware of any generalization of the quantum correction method to include strain. The possible reason is that, although the calibration of quantum correction to include strain seems possible, it is not straightforward to convert the modification of the band structure in stressed silicon to a particular form of the quantum correction potential. Indeed, according to (12.43), the wave function as a solution of the Schrödinger equation is required to guess the form of the quantum correction potential. Since the accurate evaluation of the quantum-mechanical carrier density is based on the calculation of the eigenstates of a Hamiltonian, it is beneficial to solve the Schrödinger and Poisson equations self-consistently on slices in the quantization direction perpendicular to the transport direction. In this case carriers belong to quantum subbands. The motion along the channel in this quantum subbands is characterized by the subband dispersion relations $E_n(\mathbf{k})$. This dispersion relation can be straightforwardly integrated into a Monte Carlo algorithm to solve a transport equation within the subband. Thus, in the MOSFET devices with a channel length larger than the mean free path, a mixed quantum-mechanical description in the quantization direction and a classical description along the transport direction can be used. Strain-induced band structure modifications and scattering rates can be easily introduced in the method allowing an accurate analyses of transport properties in stressed MOSFETs.

12.4.1 Subband Monte Carlo and Degeneracy Effects

Monte Carlo is a well-established numerical method to solve the Boltzmann transport equation. Within this approach, particles are moving on classical trajectories determined by Newton's law. The motion along the trajectory is interrupted by scattering processes with phonons and impurities. Scattering is modeled as a random process. The duration of a free flight, the scattering mechanism and the state after scattering are selected randomly from a given probability distribution which is determined by the scattering process.

The Pauli exclusion principle prohibits scattering into an occupied state. Therefore, scattering rates depend on the occupation probability of the after-scattering state, given by the distribution function, which is the solution of the Boltzmann equation. Dependence of scattering rates on the solution makes the Boltzmann transport equation nonlinear. In many cases the occupation numbers are small and can be safely neglected in transport simulations. However, in the silicon-on-insulator based MOSFETs utilizing thin silicon films the occupation probabilities become high and can reach values close to unity at high carrier concentrations. In this case, a self-consistent solution of the Boltzmann transport equation with respect to the occupation probabilities given by the subband distribution functions is needed.

Different approaches are known to include degeneracy effects into Monte Carlo algorithms. When the distribution function is close to the equilibrium solution, the blocking factor can be approximated with the Fermi-Dirac distribution function [42]. Another method method is to compute the occupation numbers self-consistently [12, 111]. This approach is applicable not only to low-field mobility simulations but also for higher driving fields [110]. Recently, the method was generalized to include the subband structure obtained from a self-consistent solution of a $\mathbf{k}\cdot\mathbf{p}$-based Schrödinger and Poisson equations. The method describes hole transport in a silicon device subject to arbitrary stress conditions. Because the subband distribution functions have to be obtained self-consistently, the method is computationally very demanding.

In practice, engineers are interested in obtaining the device current-voltage characteristics timely, therefore, a simplified approach based on a semi-classical transport model is frequently applied. In this case, one needs to model the transport model parameters including their dependence on stress. Due to the presence of an interface, the surface layer mobility is affected by extra scattering mechanisms. Surface roughness scattering, phonon scattering, and Coulomb scattering at the interfacial states are the most important scattering mechanisms limiting surface mobility at high, intermediate, and low carrier concentration, respectively. Each scattering mechanism depends strongly on the value of the effective field confining the carriers at the interface. The effective field depends strongly on the applied gate voltage. Stress alters the relative occupation of the valleys, subbands, and their dispersion at the gate voltage fixed, however, the degree to which these changes appear depend on the value of the effective field. Therefore, stress-induced mobility modification is a function of the gate voltage. It then follows that even at low stress values, when the mobility modification can be characterized by piezoresistance coefficients, it is not very convenient to use these coefficients for a particular device because of their dependence not only on doping level but on the effective field as well. Thus, a reliable method to accurately model mobility in stressed surface layers and SOI structures is required.

A Monte Carlo algorithm for arbitrary stress conditions which includes the degeneracy effects for mobility calculations is developed [160]. The method can be generalized to include subband structure [167] and magnetic field [81]. The method is based on the linearized version of the Boltzmann transport equation and thus incorporates degeneracy effects *exactly* in the limit of vanishing driving fields and is valid for arbitrary scattering mechanisms and for general band structure.

In order to obtain the low-field mobility, one compute the response to a small electric field $\mathbf{E}(t)$. Generated by this field a small perturbation $\delta f_n(\mathbf{k},t)$ of the equilibrium Fermi-Dirac distribution $f_0(E_n(\mathbf{k}))$ in each quantum subband satisfies the system of coupled linearized subband equations:

$$\frac{\partial \delta f_n(\mathbf{k},t)}{\partial t} = -e\mathbf{E}(t)\nabla_{\mathbf{k}} f_0(E_n(\mathbf{k})) + Q_n[\delta f], \tag{12.49}$$

where Q_n is the scattering operator of the linearized Boltzmann equation

$$Q_n[\delta f] = \sum_m \int \frac{d\mathbf{k}'}{(2\pi)^2}(\Lambda_{nm}(\mathbf{k},\mathbf{k}')\delta f_m(\mathbf{k}',t) - \Lambda_{mn}(\mathbf{k}',\mathbf{k})\delta f_n(\mathbf{k},t)). \tag{12.50}$$

The scattering rates $\Lambda_{mn}(\mathbf{k},\mathbf{k}')$ in (12.50) are related to the rates $S_{mn}(\mathbf{k},\mathbf{k}')$ of the original Boltzmann equation via

$$\Lambda_{mn}(\mathbf{k}',\mathbf{k}) = (1 - f_0(E_m(\mathbf{k}')))S_{mn}(\mathbf{k}',\mathbf{k}) + f_0(E_m(\mathbf{k}'))S_{mn}(\mathbf{k},\mathbf{k}'), \tag{12.51}$$

where $E_n(\mathbf{k})$ is the total energy in the n-th subband. The equation for the perturbation has a form similar to the Boltzmann equation, with two important differences:

(i) the presence of a source term which depends on the small electric field and is proportional to the derivative of the equilibrium function, and (ii) renormalized scattering rates which enforce the equilibrium solution of the homogeneous equation (12.49) to be $f_0(E_n(\mathbf{k}))(1 - f_0(E_n(\mathbf{k})))$, and not $f_0(E_n(\mathbf{k}))$.

In order to calculate the mobility, a subband Monte Carlo method is used to solve the system (12.49). Following the procedure outlined in [160], we assume the time dependence of the driving field to be a set of instantaneous delta-like pulses:

$$\mathbf{E}(t) = \mathbf{E}_0 \tau \sum_i \delta(t - t_i). \tag{12.52}$$

In (12.52) τ is the average period between the delta-pulses and $\mathbf{E}_0$ is the value of the field averaged over a long simulation time T:

$$\mathbf{E}_0 = \frac{1}{T} \int_0^T dt \mathbf{E}(t).$$

We compute the current response $\mathbf{I}_i(t)$ produced by an electric field pulse at the moment t_i as

$$\mathbf{I}_i = eH(t - t_i) \sum_n \int \frac{d\mathbf{k}}{(2\pi)^2} \mathbf{v}_n \delta f_n(t - t_i), \tag{12.53}$$

where $\mathbf{v}_n$ is the velocity in the n-th subband, and $H(t)$ is the Heaviside function. The instantaneous current density $\mathbf{J}(t) = \sum_i \mathbf{I}_i$ is calculated as the sum over current densities $\mathbf{I}_i$ produced by all pulses i. The current density value averaged over some long time T is then expressed as

$$\mathbf{J} = \frac{1}{T} \sum_i \int_0^T dt \mathbf{I}_i(t).$$

The low field mobility is defined as $\mu_{\alpha\beta} = J_\alpha / (enE_\beta)$, where the direction of the β-axis coincides with the direction of $\mathbf{E}_0$, and n is the carrier concentration. Now the mobility can be easily computed using a single-particle Monte Carlo technique.

The method can be illustrated as follows. The diffusion tensor $D_{\alpha\beta}$ is calculated as an integral of the velocity auto-correlation function [145]

$$D_{\alpha\beta} = \int_0^\infty d\tau \left\langle v_\alpha(t) v_\beta(t + \tau) \right\rangle, \tag{12.54}$$

where angular brackets denote the time averaging over the stochastic dynamics determined by the rates $\Lambda_{mn}(\mathbf{k}, \mathbf{k}')$ of the *linearized* multi-subband Boltzmann scattering integral in case of degenerate statistics.

The mobility tensor $\tilde{\mu}_{\alpha\beta}$ is related to the diffusion tensor via the Einstein relation for degenerate statistics

$$\tilde{\mu}_{\alpha\beta} = eD_{\alpha\beta}\frac{1}{n}\frac{\mathrm{d}n}{\mathrm{d}E_F}, \tag{12.55}$$

where E_F is the Fermi level.

In order to compute the mobility, we accumulate three temporary estimators t, w_β, and $\nu_{\alpha\beta}$ during the Monte Carlo simulations:

1. Initialize $t = 0$, $w_\beta = 0$, $\nu_{\alpha\beta} = 0$, and start the particle trajectory with the stochastic dynamics determined by the scattering rates $\Lambda_{mn}(\mathbf{k}, \mathbf{k}')$ from (12.51) of the *linearized* multi-subband Boltzmann equations;
2. Before each scattering event update $\nu_{\alpha\beta}$, w_β, and t:

$$t = t + \frac{\tau(j)}{1 - f(E(j))},$$

$$w_\beta = w_\beta + v_\beta(j)\tau(j),$$

$$\nu_{\alpha\beta} = \nu_{\alpha\beta} + \tau(j)v_\alpha(j)w_\beta(j);$$

3. When t is sufficiently large, compute the mobility tensor as

$$\tilde{\mu}_{\alpha\beta} = \frac{e}{k_B T}\frac{\nu_{\alpha\beta}}{t},$$

where $v_\alpha(j)$ denotes the α-component of the velocity, $E(j)$ is the particle energy, $f(E)$ is the Fermi-Dirac function, and $\tau(j)$ is the time of j-th free flight. The convergence of the method is improved by resetting $w_\beta = 0$ each time a velocity randomizing scattering event occurs.

The multi-subband generalization of the method requires the subband energies and wave functions to be pre-calculated self-consistently, for each value of the effective field. The transport calculations account for electron-phonon interaction [76, 139] and surface roughness scattering, which are the dominant mechanisms determining the mobility in the region of high effective fields. This captures the main difference of the mobility behavior in Double-Gate (DG) and Single-Gate (SG) structures at high concentrations for different orientations, as we demonstrate below.

In order to validate the computation procedure, we compare the results of mobility simulations with data available in the literature [31], see Fig. 12.9. Our simulations are in good agreement with previous calculations for a FET with 20 nm and 2.5 nm thick (100) thick silicon body, assuming similar surface roughness scattering parameters.

With downscaling of semiconductor devices continuing, the introduction of double-gate silicon-on-insulator field-effect transistors with ultra-thin silicon body seems increasingly likely [106]. The second gate allows to achieve excellent electrostatic channel control [106] which makes these devices perfect candidates for the far-end of ITRS scaling [1].

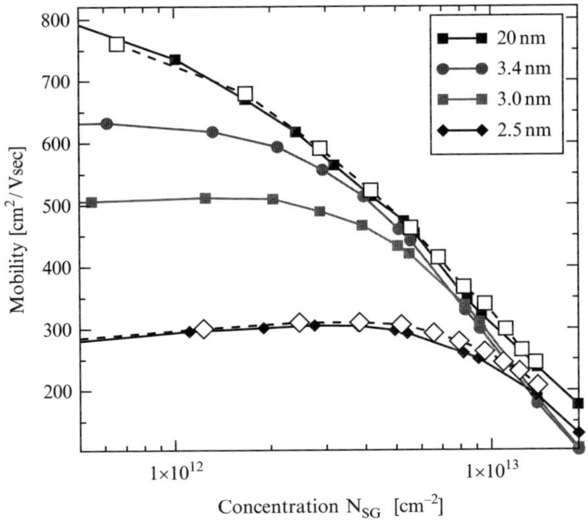

Fig. 12.9 Mobility simulations for SG (100) SOI FETs (*filled symbols*). Results are in good agreement with those reported previously [31] (*open symbols*)

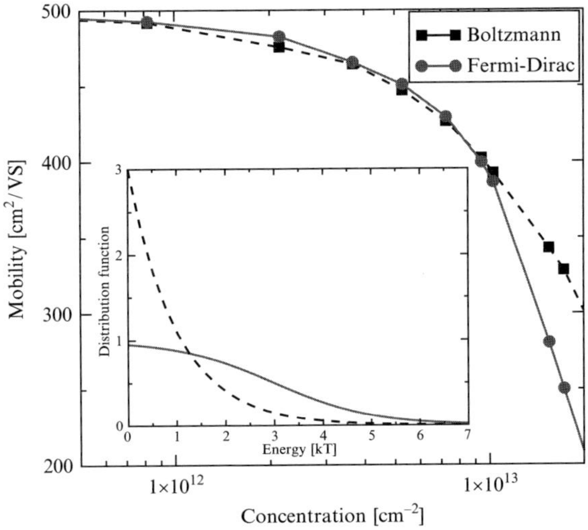

Fig. 12.10 Mobility simulations for a 3 nm thick SG structure, using Boltzmann and Fermi-Dirac statistics. Mobility calculated without Pauli exclusion principle taken into account is overestimated at high carrier concentrations. It is due to different subband population for the same N_s, with and without degeneracy effects included. Inset: Boltzmann and Fermi-Dirac distributions for the same temperature and N_s

Mobility in a single-gate (100) FET is compared for degenerate and non-degenerate statistics of carriers. Figure 12.10 shows that neglecting the Fermi blocking factor leads to a significant overestimation of mobility, especially at higher carrier concentrations. The difference between the mobility values for degenerate and non-degenerate statistics looks surprising. Indeed, at high carrier concentrations the principal scattering mechanism limiting the low-field mobility is elastic surface roughness scattering. For elastic scattering the forward and inverse scattering rates are equal, and the Pauli blocking factor cancels out from the equations for the elastic scattering rates. However, the Pauli blocking factor is also present in the inelastic electron-phonon part of the total scattering integral and ensures the equilibrium solution to be the Fermi-Dirac distribution function. In the degenerate case the higher value of the Fermi level may open additional channels for inter-subband scattering, which reduces the mobility.

12.4.2 *Simulation Results for Mobilities in Single- and Double-Gate FETs*

After the surface roughness parameters were calibrated to reproduce the bulk (100) surface mobility, the same model parameters are used to simulate mobilities of single- and double-gate (100) structures. Results are compared in Fig. 12.11 for several Si body thicknesses. For a 20 nm thick Si body the two inversion layers formed at the opposite interfaces of the DG structure are well separated and independent. The double-gate FET mobility plotted as a function of the concentration per single channel is in good agreement with the results for a SG structure. Both curves are in reasonable agreement with experimental data for a 32 nm thick Si body [180]

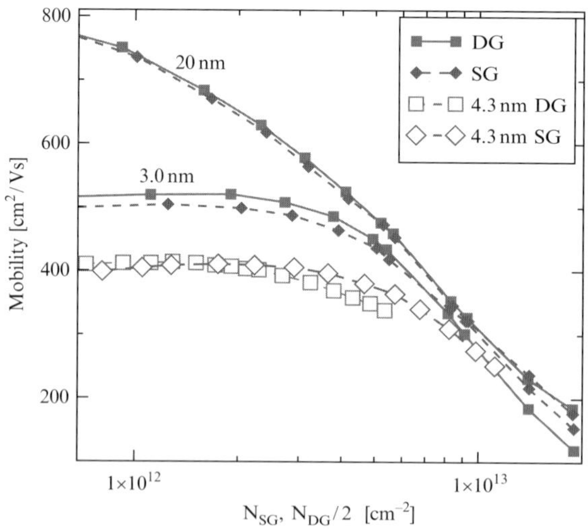

Fig. 12.11 Mobility for (100) UTB FETs. Mobility is plotted as a function of concentration for the SG structure and as a function of half of the concentration for the DG structure. Mobility in the DG structure is smaller than that in the SG structure at high concentration, which is in qualitative agreement with the experimental data [181] and in contrast to the bulk inversion concept [9]

and reproduce the universal mobility curve [175]. For 3 nm body thickness the DG mobility, plotted as a function of $N_s/2$, tends to be slightly higher at small effective fields. However, for high N_s the DG mobility for (100) orientation becomes lower than the SG mobility. This is in qualitative agreement with experimental data, also shown in Fig. 12.11. This behavior contradicts the volume inversion concept [9]. According to this concept in volume inversion carriers are located in the center of the Si body and are further away from the interfaces. Due to reduced surface roughness scattering the mobility in a DG structure is expected to be higher than the mobility in a SG structure.

It should be noted that in order to obtain a better quantitative agreement with the experimental mobility, additional mechanisms such as surface optical phonons interaction, or scattering on fluctuations of δT_{SOI} [32] could be included. Introduction of acoustic phonon confinement relevant to FET structures with thin silicon bodies also improves agreement with the experimental data [27]. However, our model captures the main difference in the mobility behavior in DG and SG structures at high concentrations, namely that the DG mobility is lower than the SG mobility. It should also be noted that in DG structures the mobility plotted as a function of the total concentration is always higher than in SG structures, as was pointed out in [33].

Using a thick body DG FET, we demonstrate the importance of inter-subband scattering for mobility calculations. For a thick DG structure the wave functions $\psi_i(z)$ corresponding to two independent inversion layers $i = 1, 2$ are located at the opposite interfaces. Due to the large separation between the interfaces, the value of the wave function and its derivatives is zero at the interface opposite to its location. By making symmetric $\psi_+(z)$ and asymmetric $\psi_-(z)$ combinations of $\psi_1(z)$ and $\psi_2(z)$, one obtains the wave functions corresponding to the first and the second subband in the Si body:

$$\psi_{\pm}(z) = \frac{1}{\sqrt{(2)}}(\psi_1(z) \pm \psi_2(z)). \tag{12.56}$$

For the sake of simplicity let us consider surface roughness scattering only at the left interface, $z = 0$. The surface roughness scattering rate $\Gamma_{ij}(q)$ between the two subbands k and j is proportional to [7, 137]

$$\Gamma_{kj} = \left(\frac{\hbar^2}{2m_z}\right)^2 \left(\frac{d\psi_k(0)}{dz}\right)^2 \left(\frac{d\psi_j(0)}{dz}\right)^2, \tag{12.57}$$

where the derivative $d\psi_k(0)/dz$ of the k-th wave function is computed at the interface $z = 0$, and m_z is the quantization mass. For a thick body FET it holds $d\psi_2(0)/dz = 0$, and the total scattering rate within the first subband $\psi_+(z)$ is:

$$\Gamma_+ = \Gamma_{++} + \Gamma_{+-} = \frac{1}{2}\left(\frac{\hbar^2}{2m_z}\right)^2 \left(\frac{d\psi_1(0)}{dz}\right)^4, \tag{12.58}$$

where both Γ_{++} and Γ_{-+} are:

$$\Gamma_{++} = \Gamma_{-+} = \frac{1}{4}\left(\frac{\hbar^2}{2m_z}\right)^2 \left(\frac{d\psi_1(0)}{dz}\right)^4. \qquad (12.59)$$

Therefore, a half of the total scattering rate would be missed if the intersubband scattering were neglected. A similar conclusion also holds for electron-phonon scattering [157].

Now we turn our attention to (110) UTB FETs. Due to the surface orientation, the effective equation for the envelope functions in quantization direction contains also the first order derivative of the wave functions. The derivative can be removed from the equation by a wave function transformation [163], which results in a standard Schrödinger-like equation for the envelope function in the quantization direction, allowing the determination of quantization mass, conductivity mass and density-of-state effective mass. For (110) orientation the subband structure consists of a fourfold and a twofold degenerate subband ladder. Subbands in both ladders are anisotropic. Simulation results for 20 nm thick SG and DG structures are shown in Fig. 12.12. The results for SG and DG structures, which are plotted as a function of the concentration per inversion layer, are in good agreement with each other and with experimental data [180]. Due to the high anisotropy of the subbands of each ladder the surface mobility is anisotropic. Diagonal values of the mobility tensor along $\langle 001\rangle$ and $\langle 1\bar{1}0\rangle$ directions are in agreement with those observed experimentally [180].

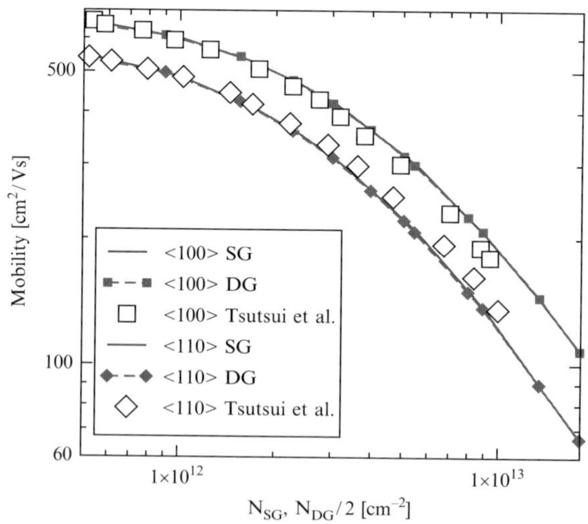

Fig. 12.12 Low field mobility in different directions calculated for 20 nm (110) SG and DG structures (*filled symbols*) closely follows experimental data [180] (*open symbols*). Mobility tensor components per inversion layer in thick DG and SG structures are the same

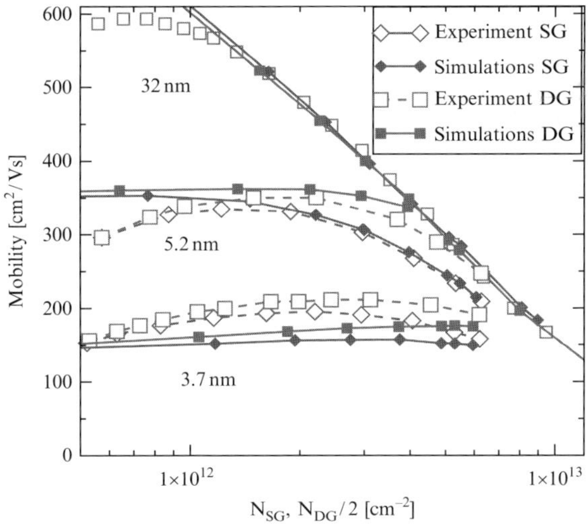

Fig. 12.13 Mobility in $\langle 1\bar{1}0\rangle$ direction for SG and DG structures with (110) thin body orientation. Mobility is plotted as a function of concentration for SG structures and as function of half the concentration for DG structures. DG (110) UTB mobility is higher than the SG mobility in whole range of concentrations, in good agreement with the experimental data [180], supporting the bulk inversion concept [9]

Figure 12.13 shows results of mobility simulations in $\langle 1\bar{1}0\rangle$ direction performed for several Si body thicknesses using the same scattering parameters as for thick Si films. The mobility of (110) UTB FETs coincides reasonably well with the experimental data [180] shown in Fig. 12.13. The mobility for (110) DG structures is higher than the mobility of (110) SG structures in the whole concentration range. It is consistent with recent experimental studies [156, 180] and confirms the concept of the mobility enhancement in DG UTB FETs due to volume inversion [9].

Mobility in a (100) UTB double-gate FET plotted as function of carrier concentration per inversion layer becomes lower than the mobility in a corresponding SG FET. This is in sharp contrast with results for (110) FETs mobilities and is in obvious contradiction with the concept of the mobility enhancement due to volume inversion. In order to explain the apparent contradiction, high occupation of primed subbands in (100) DG FETs with a significantly lower mobility caused by heavier conductivity mass was assumed in [180]. The higher occupation of primed subbands in DG FETs was justified by smaller energy splitting between the lowest primed and unprimed subbands in a (100) DG FET induced by the interaction of the inversion layers [180]. However, in UTB FETs the subband structure is mainly determined by strong geometrical confinement. Therefore, the subband structure and the occupation of subbands should be quite similar in UTB DG and SG FETs provided the total carrier concentration is the same in both structures. At the same time the SG and DG mobilities should be compared for the same gate voltages. In case of an intrinsic body this corresponds to the mobilities compared as functions of the concentration per single inversion layer. Since in a DG FET there are two inversion layers created at the front and the back interface of the Si body, the total concentration in a DG structure is twice as high as in a SG structure for the same gate voltage. Also due to the double carrier concentration the occupation of the primed subbands can be higher in DG structures.

In order to compare relative contributions of two different mechanisms to increase of the primed subband occupation in a DG FET with a SG FET, in Fig. 12.14 we plot the occupation of the primed subband ladder as a function of the total concentration in SG and DG structures. The contribution Δ_2 computed at the same total concentration is due to the difference in energy splitting between primed and unprimed ladders and corresponds to the mechanism proposed in [180]. This contribution plays a minor role compared to the increase Δ_1 of the primed subband occupation due to the doubled carrier concentration in DG structures for similar gate voltages.

If we now apply similar argumentation to (110) UTB FETs, we find that the occupation of primed subbands increases rapidly in DG (110) UTB FETs at high effective fields, which should also lead to a more rapid decrease of DG mobility in (110) FETs due to the lower mobility in primed subbands. However, this contradicts the experimental data and the results of simulations. Higher occupancy of primed subbands in DG structures is present in both (100) and (110) UTB FETs and cannot be the only reason for the mobility lowering observed in DG (100) structures.

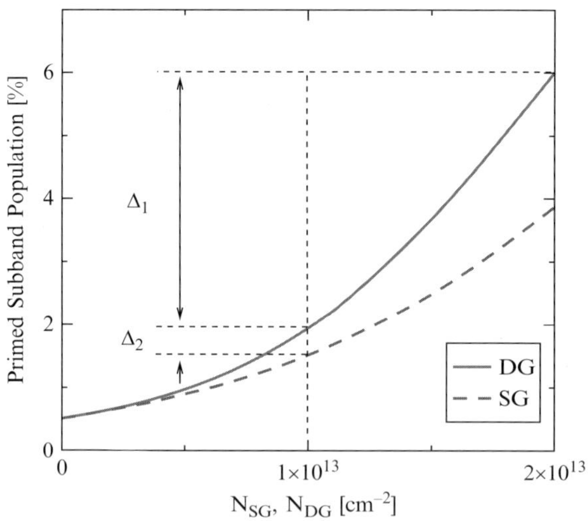

Fig. 12.14 Occupation of primed ladder in a 3 nm thick (100) UTB FET as a function of total concentration for either SG or DG operation mode. Higher primed ladder occupation in DG FET is mainly due to the Δ_2 contribution resulting from twice-as-high carrier concentration in DG as compared to SG FETs, for similar gate voltage. Contribution Δ_2 due to subband structure difference [180] is small

Due to different quantization masses, the subband structures in (100) and (110) UTB FETs differ significantly. In (100) UTB FETs the subband energy of the second unprimed subband is nearly equal to the energy of the first primed subband, while in (110) FETs the second unprimed fourfold degenerate subband lies significantly higher than the lowest primed subband. Even though the occupation of primed subbands increases for both (100) and (110) orientations with increasing effective field, higher unprimed subbands remain virtually de-populated in (110) UTB FETs. Contrary, the occupation of higher unprimed subbands increases rapidly in DG (100) FETs for high effective fields. Figure 12.15 illustrates the higher unprimed subband occupation in (100) compared to (110) DG FETs.

Figure 12.16 shows a strong correlation between the rapid mobility degradation and the occupation increase of the higher unprimed subbands in a (100) DG structure. However, estimations demonstrate that the fraction of carriers occupying the higher subbands is not sufficient to cause such a significant drop in the DG (100) FET mobility. A more substantial correction comes from the mobility lowering in the ground subband itself. Since the higher unprimed subband is occupied in (100) DG structures, an extra channel of scattering between the ground and the second unprimed subbands opens. This mechanism plays a significant role in the (100) DG mobility lowering: If we switch off the inter-subband scattering in the simulations, thus forbidding the scattering between the lowest and higher unprimed subbands, the DG mobility exceeds the corresponding (100) SG mobility, even at high carrier concentrations, in compliance with the volume inversion concept.

Degeneracy effects therefore play a significant role in compensating the volume inversion induced mobility enhancement in (100) DG structures. They lead to a significant occupation of higher subbands in the unprimed ladder, which results in increased intersubband scattering and mobility lowering reported in [156].

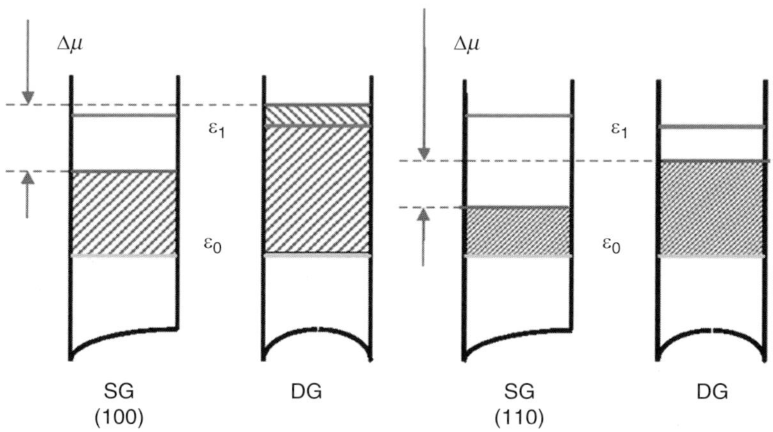

Fig. 12.15 Illustration of higher unprimed subband occupation in (100) and (110) UTB FETs. Contrary to (110) FETs, the higher unprimed subbands are occupied significantly in DG (100) FETs, for the same concentrations, due to a heavier quantization mass and a lighter in-plane density-of-states mass

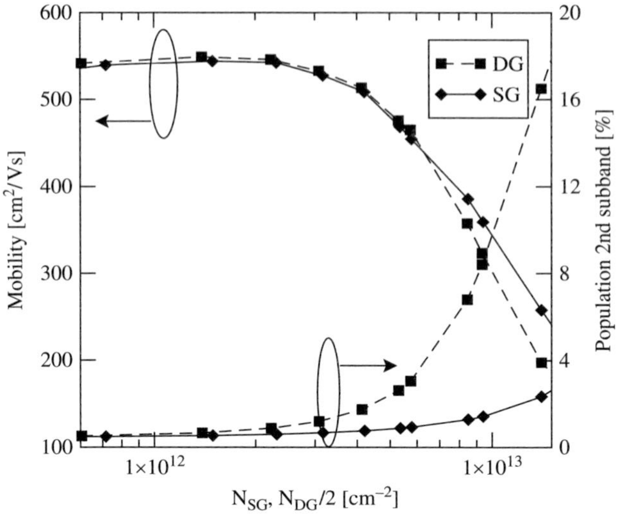

Fig. 12.16 Correlation between mobility degradation and occupation of higher unprimed subbands in a 3 nm thick DG (100) SOI FET

12.4.3 Electron Mobility Enhancement in FETs with Ultra-Thin Silicon Body

The (001) substrate orientation intrinsically yields the largest effective mobility for electrons [71, 180]. Let us now apply uniaxial stress to a (001) oriented substrate. Two cases are considered: stress and channel both parallel to the [110] direction or to the [100] direction. In addition, the impact of the silicon body thickness on the mobility enhancement is investigated.

In Fig. 12.17a the mobility parallel and perpendicular to the [110] channel direction [110] under 1 GPa tensile stress along [110] is compared to the unstrained mobility for two Si body thicknesses. At relatively large body thicknesses ($t > 20$ nm) tensile stress along [110] direction has two beneficial effects on the [110] mobility. First, the splitting between the unprimed and primed subband ladders is increased, leading to higher occupation of the unprimed ladder with lower conductivity effective mass. Second, the transport mass in stress direction is reduced with respect to the relaxed case, according to (9.10). These two effects determine the mobility enhancement at all inversion layer densities.

Contrary, in the direction $[\bar{1}10]$ perpendicular to stress, the effective mass is increased (9.9). This resulting to a smaller mobility enhancement in this direction at small inversion layer concentration, which further diminishes at larger inversion layer concentrations.

In ultra-thin body FETs with the thickness ($t < 5$ nm), the energy separation between the subband ladders is already high due to strong quantum confinement. Thus, the additional valley splitting due to stress have a negligible effect on the

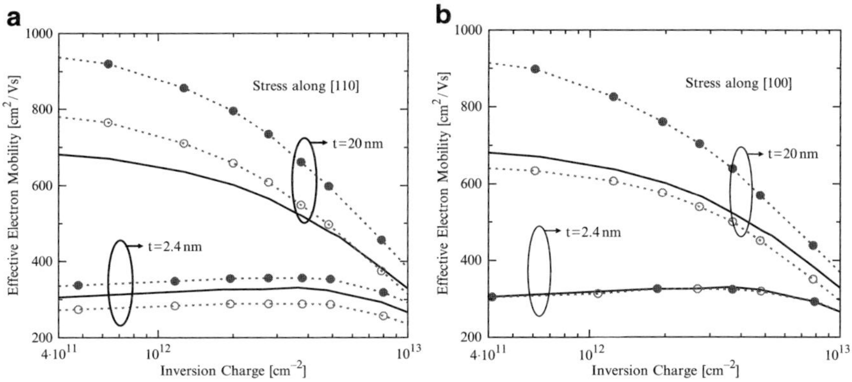

Fig. 12.17 Simulated electron mobility for substrate orientation (001) of unstressed (*solid lines*) and 1 GPa stressed (*dashed lines*) silicon for two body thicknesses. The mobility components are plotted with filled symbols in the direction parallel to stress and with open symbols in the direction perpendicular to stress along (**a**) [110] and (**b**)[100] direction

occupation of unprimed subbands and mobility. Therefore, larger (smaller) mobility parallel (perpendicular) to the [110] stress direction is because of the effective mass change (9.9,9.10) alone, in good agreement with experimental data [71, 182].

In Fig. 12.17b the effect of [110] uniaxial stress on [100] channel mobility is shown. Stress along [100] lifts the degeneracy of the fourfold (primed) ladder by pushing the subbands resulting from the [100] oriented valleys up in energy. Thus, the occupation of the primed subbands from the [100] ladder is reduced. Since these subbands have a larger transport mass in [100] direction, by depopulating them we enhance total mobility in [100] direction in thick films. Alternatively, mobility perpendicular to stress is reduced due to an increased occupation of [010] primed subbands with unfavorable mobility along [010] direction.

In ultra-thin films, however, the occupation of primed subbands is negligible due to high energy separation between primed and unprimed subbands. Therefore, stress-induced shift between the [100] and [010] primed subbands does not result in subband repopulation. Since no effective mass change occurs the mobility enhancement vanishes at $t = 2.4$ nm.

12.4.4 Stress-Induced Mobility and Drive Current Enhancement

We now summarize the latest results on stress-induced transport enhancement obtained by different groups. It is predicted [165] that the highest enhancement factor for surface mobility of holes is achieved, when uniaxial compressive stress is applied along the [110] channel direction to the hole surface channel at the (001) silicon interface. Physics of mobility enhancement is determined by the strain-induced valence band warping [191] shown in Fig. 8.4 which results in the smaller conductivity mass and enhanced current in the transport direction. The enhancement factor rises almost linearly with stress up to 3 GPa, where it reaches the value close to four. At higher stresses a saturation in the hole mobility enhancement is reported [165]. For the (110) substrate orientation the enhancement factor is lower, however, at high stress the hole mobility value is found to be the same as for the case of (001) orientation for [110] stress/channel direction, independent on whether compressive stress/channel is in $\langle 110 \rangle$ or $\langle 111 \rangle$ direction [165]. These observations are confirmed in [97], where it is demonstrated that for the same [110] direction of the channel the response at high stresses is determined by the stress and transport orientation and not by the wafer orientation. By applying deterministic Boltzmann/multisubband $\mathbf{k}\cdot\mathbf{p}$-based device solver [135] it is shown that values for the drive current enhancement achieved are comparable to that of the mobility enhancement, in the broad range of drive voltages.

In case of electrons, the highest mobility enhancement factor due to tensile uniaxial stress is predicted for (110) substrate orientation, when stress is applied along the $\langle 110 \rangle$ channel direction [97, 98]. This behavior is consistent with that expected from the subband structure modification discussed in Sect. 11.8. It is also supported by recent mobility simulations in FinFETs where the highest mobility enhancement was reported for a longitudinally elongated $\langle 110 \rangle$ oriented FinFET [151], due to a substantial contribution in the total mobility of [110] oriented interfaces.

It is shown [97] that, even at high level of stress close to 3 GPa mobility on (110) substrate remains inferior to [110]/(001) channel mobility of uniaxially elongated along the [110] direction. Therefore, the standard [110]/(001) channel orientation allows achieving superior electron mobility under practically relevant values of uniaxial stress [97, 98, 178, 183]. The reason for the highest mobility in this case is the depopulation of primed subband ladder with unfavorable conductivity mass augmented by the favorable strain-reduced transport mass reduction along the stress direction in the unprimed subband ladder.

Recently, alternative materials with higher mobility like germanium [18, 88] and III-V semiconductors [142, 143] were considered as possible substitutions to silicon to boost the channel transport properties. It is interesting to note that the transport properties of these new channel materials can be significantly enhanced by stress [88, 121, 143] extending applicability of currently developed stress technologies to several technology generations ahead.

12.5 Quantum Transport Models

As device sizes become comparable to the electron wave length, quantum effects start playing a significant role. At the same time scattering with phonons, impurities, and surface roughness may still be strong and cannot be neglected. At this point a full quantum transport formalism which consistently describes quantum coherent propagation and dissipative scattering processes, must be employed.

Established techniques used to address dissipative quantum transport can be classified according to the functions they are based upon: the nonequilibrium Green's functions, the density matrix, and the Wigner function. All three approaches are based on fundamental equations of motion and are equivalent at the most general level of formal description of a dissipative quantum system. The resulting system of integral-differential equations for the Green's function $G(\mathbf{r}_1, \mathbf{r}_2, t_1, t_2)$, or the density matrix $\rho(\mathbf{r}_1, \mathbf{r}_2, t_1, t_2)$, or the Wigner function $f(\mathbf{r}, \mathbf{p}, \omega, t)$ is in many cases too complex to allow for a direct numerical solution. Each function depends on two vector and two scalar arguments. For a numerical solution, each argument of the function must be discretized. In the case of a three-dimensional system ($d = 3$) the total number of unknowns is $N_{\text{tot}} = (N_x \cdot N_y \cdot N_z \cdot N_t)^2$. Assuming 100 grid points for each argument this results in the astronomical number $N_{\text{tot}} = 10^{16}$, which results in a prohibitively large memory requirements. Approximations and simplifications must be incorporated in order to make the problem numerically tractable. It is mainly these simplifying assumptions that make the differences between the different approaches to address quantum transport.

Although there are evidences that transport in MOSFETs with 10 nm channels may still be governed by scattering, it is instructive to begin with a purely coherent carrier propagation in order to highlight the importance of quantum-mechanical effects in transport direction in short channel devices.

12.5.1 Ballistic Transport and Tunneling

A MOSFET utilizing ballistic transport was first suggested and studied by Natori [123]. It consists of a thin Si film connected to two reservoirs. One or two gate electrodes are adjusted close to the side film interfaces. The gate electrodes are electrically separated from the Si film by a dielectric material. In order to reduce scattering in the channel the Si film is not doped.

The occupation of left- or right-propagating transversal modes is determined by the chemical potential of the left and the right reservoir, respectively. In equilibrium, the chemical potentials of both reservoirs are the same and the currents of the left- and right-propagating modes are equal in value and opposite in sign and thus compensate each other. The total current in equilibrium is zero. The source-drain voltage shifts the chemical potentials and creates a unbalance in occupation between the left- and right-propagating modes. The number of propagating modes is controlled by the potential barrier in the silicon film, which is determined by the gate voltage. This picture is valid as long as there is no (back)scattering of the propagating modes in the silicon channel. We note that, except for the filling of the propagating modes in accordance with the reservoirs' Fermi-Dirac distribution, we did not need quantum mechanics to describe the ballistic transport [124].

For ballistic transport the channel must be shorter than the mean free path. In Si at room temperature the mean free path is a few nanometers. With the silicon channel scaled down, the potential barrier profile $V(x)$ along the channel becomes steeper and close to its maximum at $x = x_0$ it can be approximated by a parabolic dispersion $V(x) = V_m - m_l\omega(x - x_0)^2/2$. The characteristic curvature of the potential ω increases with the channel length L_c decreased approximately as [106]

$$\omega \propto \frac{V_m}{m_l L_c^2}. \tag{12.60}$$

For large curvatures, quantum-mechanical tunneling of electrons under the barrier becomes possible. The total current through the channel is determined by the sum of currents from all propagating modes and of the contribution of the modes tunneling under the barrier.

In order to estimate the curvature ω beyond which transport is becoming determined by quantum-mechanical tunneling, we use the Kemble formula for the transmission coefficient close to the maximum of the potential (12.60), following [106]:

$$T(E) = \left[1 + \exp\left(2\pi \frac{V_m - E}{\hbar\omega}\right)\right]^{-1}. \tag{12.61}$$

Electrons are injected according to the Fermi-Dirac distribution

$$f_0(E) = \left[1 + \exp\left(\frac{V_m - E}{k_B T}\right)\right]^{-1}. \tag{12.62}$$

The current is determined by carriers with $E \geq V_m$ flying above the potential maximum, if $\hbar\omega/2\pi \leq T$. In this case tunneling can be neglected and transport is computed semi-classically. At the opposite limit $\hbar\omega/2\pi \geq k_B T$, the current must be computed quantum-mechanically, because the main contribution is due to carriers with $E \leq V_m$ tunneling under the barrier. For typical parameters of Si and $V_m = 50$ meV the channel length beyond which tunneling under the barrier becomes important is around 10 nm.

Due to the absence of scattering in the channel and excellent channel control, the double-gate ballistic MOSFET can be considered as an ultimate MOSFET. In order to describe transport accurately, we assume (100) orientation of the Si channel at the Si/SiO_2 interface. As already discussed, a self-consistent solution of the two- or three-dimensional Schrödinger equation together with the Poisson equation represents a significant computational challenge, cf. [104]. Because of the strong lateral confinement in z direction due to the channel thickness $t = 2$ nm, only the first subband with the heavy mass m_h is populated with carriers which participate in transport for moderately high drain-source voltages. The wave function can then be written in the following approximate form:

$$\Psi(x, y, z) = \psi(x)e^{iky}\sqrt{\frac{2}{t}}\cos(\frac{\pi z}{t}). \tag{12.63}$$

This approximation simplifies the calculations and reduces the computational effort significantly [20, 66, 166, 192]. The effective Schrödinger equation for the function $\psi(x)$ along the channel has the usual form $H\psi = E\psi$, with the Hamiltonian H

$$H = -\frac{\hbar^2 d^2}{dx^2}\psi(x) + \frac{2}{t}\int_0^t dz \cos\left(\frac{\pi z}{t}\right) eV(x,z), \tag{12.64}$$

where the potential $V(x,z)$ is found from the Poisson equation. The Schrödinger equation must be supplemented with the boundary conditions. To describe an open system, one uses the boundary conditions corresponding to plane waves propagating in electrodes far from the channel. The Schrödinger equation can be solved numerically [190] or using the WKB semi-classical approximation [170]. The solution of the Schrödinger equation with open boundary conditions can also be achieved by means of the quantum transmitting boundary method [49, 105]. Simulators accounting for a full two-dimensional solution of the open-boundary Schrödinger equation have been reported and applied to 10 nm double-gate MOSFETs [104, 147].

A Schrödinger-Poisson solver [188] with open boundary conditions can be used to find the wave functions self-consistently. As a result, the transmission function through the channel for each propagating mode is found. The total current is obtained by summing up the contributions from all modes propagating in forward direction from source to drain and in backward direction from drain to source. In an UTB FET when only a single subband is occupied, the current as function of a source-drain voltage V is [123]

$$J(V) = e\frac{\sqrt{2mT}}{\pi^{3/2}\hbar^2}\int dET(E)\left(F_{-1/2}\left(\frac{E_F - E}{k_B T}\right) - F_{-1/2}\left(\frac{E_F - E - eV}{k_B T}\right)\right), \tag{12.65}$$

where $F_{1/2}(x)$ is the Fermi-Dirac integral of the order 1/2. The structure of (12.65) is quite transparent. The two terms in the brackets correspond to supply functions from the source and from the drain electrode. Their difference, multiplied by the transmission function $T(E)$, gives the current through the device.

Figure 12.18 shows output characteristics of a transistor with $t_{ox} = 1.5$ nm, $t = 2$ nm, for several values of gate voltage [170]. Two sets of IVs are presented corresponding to the gate lengths of 5 nm and 10 nm. For 10 nm channel length IV characteristics are very similar to the characteristics of the ideal transistor, see Fig. 12.18a, with a perfect saturation, suggesting an almost vanishing Drain-Induced Barrier Lowering (DIBL) effect. The 10 nm transistor is therefore very close to the ideal double-gate ballistic transistor, described in [123]. With the decreased gate length the current saturation becomes less pronounced. The reason for this degradation is twofold: tunneling under the barrier formed by the potential in the channel and drain-induced barrier lowering. The characteristics are acceptable for $L = 5$ nm, see Fig. 12.18b, while for $L = 2.5$ nm the saturation practically vanishes. The output characteristics simulated for an ultra-thin body double-gate

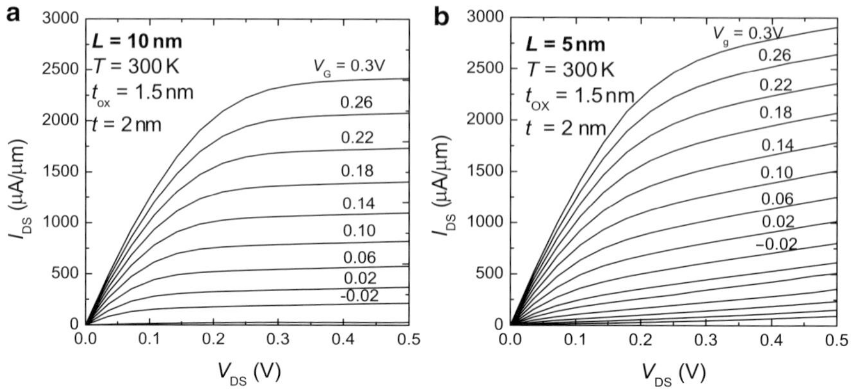

Fig. 12.18 Output characteristics of a DG FET with 2 nm Si body and 1.5 nm silicon dioxide thickness. (**a**) Gate length 10 nm and (**b**) gate length 5 nm. With the gate length decreased, the current saturation is less pronounced

MOSFET with a gate length L as short as 5 nm shows that even such a small device possesses an I_{on}/I_{off} ratio sufficient for logic applications and displays a reasonable short-channel effect and acceptable DIBL, a conclusion recently reached from more detailed atomistic calculations [146]. It should be noted that the sensitivity to small MOSFET dimension variations, the control of doping as well as the whole manufacturing process represent significant challenges for multi-gate MOSFETs with a gate length below 10 nm.

Using the ballistic transport model and neglecting tunneling under the barrier (length $L > 10$ nm), one can evaluate influence of strain-induced electron subband structure modification discussed in Chap. 11 on transport in ultra-thin body [110]/(001) MOSFETs. We are using stress-dependence of the subband energies and the effective masses evaluated numerically in Sect. 11.4.3.

For the thick film ($t = 10.9$ nm) the effective masses closely follow the analytical expressions (9.9, 9.10) [169], obtained by neglecting the valley coupling. The valley coupling caused by strain and reduced thickness introduces a substantial difference in the effective masses of the two unprimed subbands with the same quantum number, which are usually treated as completely equivalent. Even more, for $t = 2.7$ nm a clear difference in masses is already observed in relaxed films visible in Fig. 11.16, which becomes high in an ultra-thin film with the thickness $t = 1.4$ nm (Fig. 11.17).

Examples of the IVs for $t = 10.9$ nm and $t = 1.4$ nm at the gate voltage $V_G = 0.5$ V are shown in Figs. 12.19 and 12.20, correspondingly. The enhancement of the drive current for all V_{DS} with shear strain is observed. In order to characterize the enhancement quantitatively, we have evaluated the relative current increase as $(I(\varepsilon_{xy}, V) - I_0(V))/I_0(V)$ for each source-drain voltage. Results for different film thicknesses are shown in Figs. 12.21–12.24.

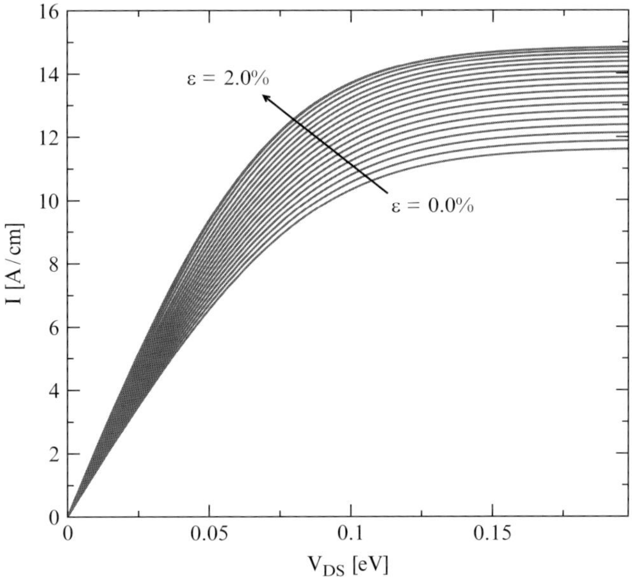

Fig. 12.19 The *IV* characteristics of a double-gate ballistic MOSFET with a silicon body of 10.9 nm thickness as function of shear strain at $V_G = 0.5$ eV. Strain is increased from 0 to 2% in steps of 0.1%

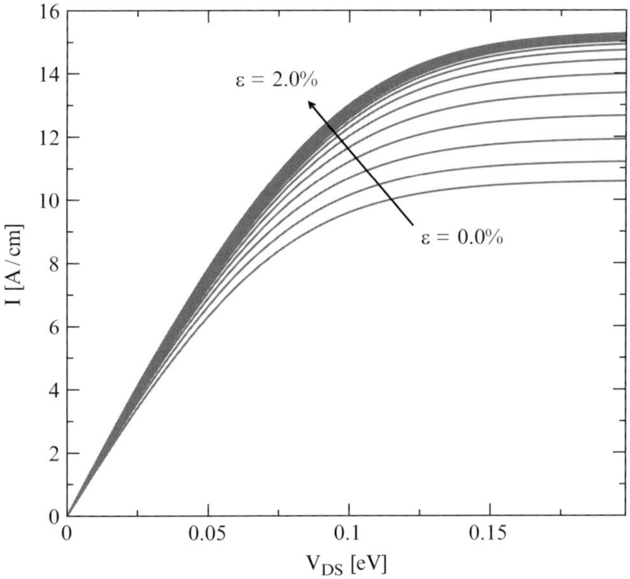

Fig. 12.20 The *IV* characteristics of a double-gate ballistic MOSFET with a silicon body of 1.4 nm thickness as function of shear strain at $V_G = 0.5$ eV. Strain is increased from 0 to 2% in steps of 0.1%

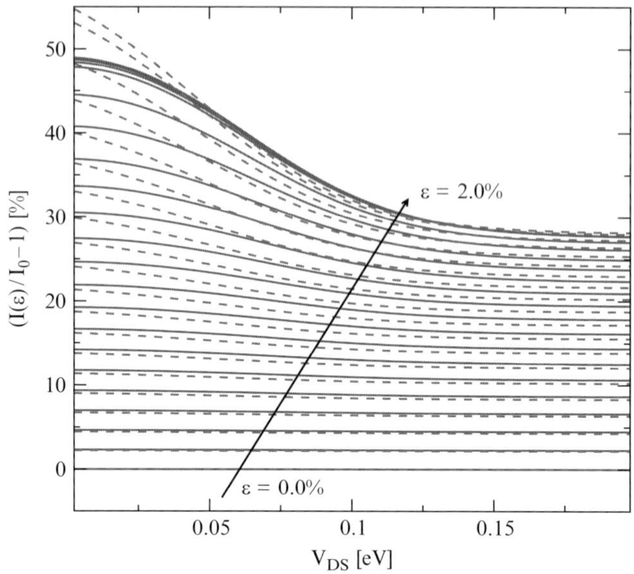

Fig. 12.21 Relative current enhancement $(I(\varepsilon_{xy}, V) - I_0(V))/I_0(V)$ as function of shear strain for a double-gate MOSFET with the body thickness $t = 10.9\,\text{nm}$. Dotted lines are obtained using the analytical model (9.9), (9.10) for the effective masses

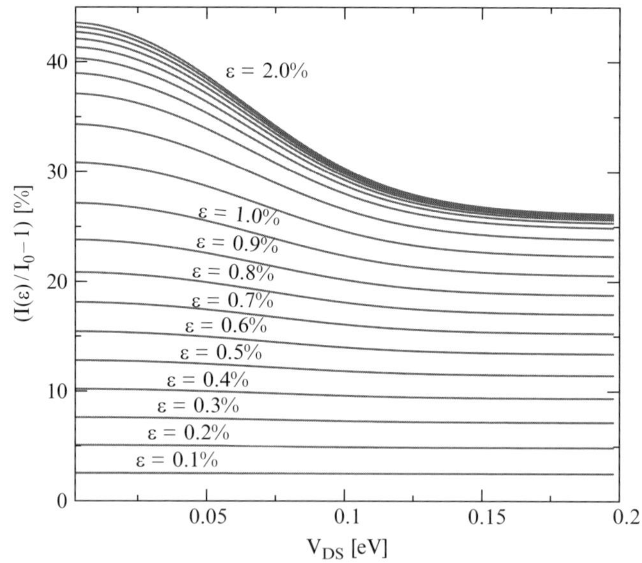

Fig. 12.22 Relative current enhancement $(I(\varepsilon_{xy}, V) - I_0(V))/I_0(V)$ as function of shear strain for a double-gate MOSFET with body thickness $t = 5.4\,\text{nm}$. As for $t = 10.9\,\text{nm}$ body thickness, the enhancement is larger in the linear regime

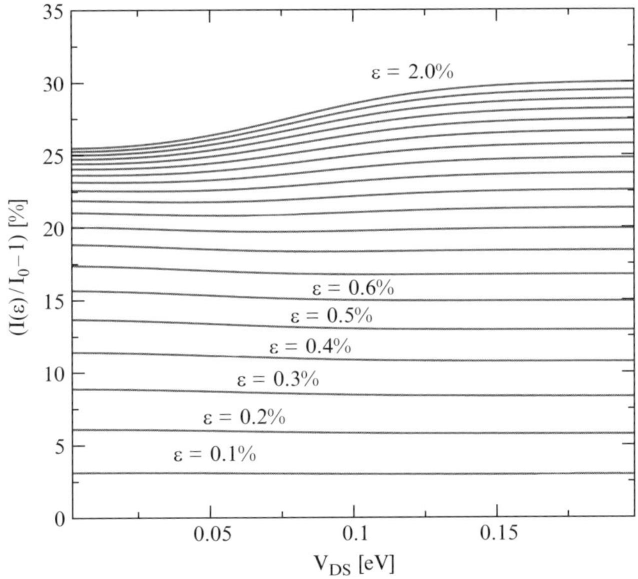

Fig. 12.23 Relative current enhancement $(I(\varepsilon_{xy}, V) - I_0(V))/I_0(V)$ as function of shear strain for a double-gate MOSFET with body thickness $t = 2.7\,\text{nm}$. Contrary to the MOSFETs with thicker bodies,the enhancement becomes larger at saturation

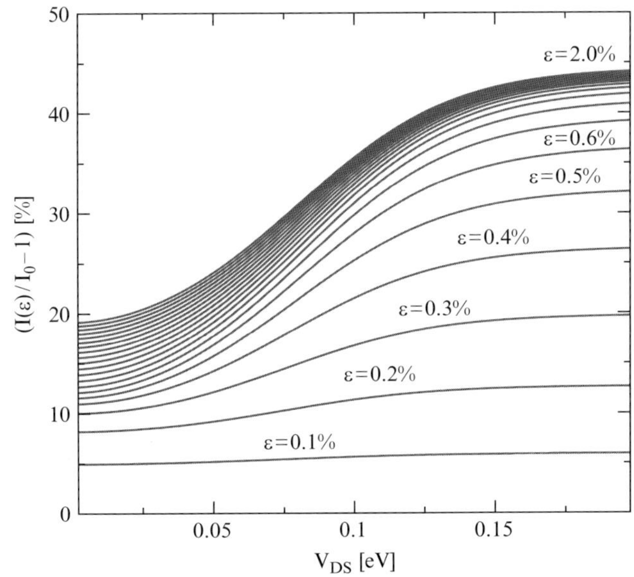

Fig. 12.24 Relative current enhancement $(I(\varepsilon_{xy}, V) - I_0(V))/I_0(V)$ as function of shear strain for a double-gate MOSFET with body thickness $t = 1.4\,\text{nm}$. Due to strain-induced subband splitting shear strain is extremely efficient for the enhancement of the on-current in MOSFETs with ultra-thin bodies

For small shear strain values the current enhancement is bias independent for all studied film thicknesses. For extremely high strain the saturation of the current enhancement is observed. For intermediate strain values the enhancement behaves differently in thick and thin silicon films at high and low voltages.

In thicker films the enhancement in the linear regime is larger than in saturation (Figs. 12.21 and 12.22). Figure 12.21 shows the current increase with the values from Fig. 11.14 for the effective masses and with the masses (9.9) and (9.10) in bulk silicon, which demonstrates that the enhancement in a $t = 10.9$ nm film is due to the effective mass modifications with strain. For small strain values the current enhancement is due to the m_y increase. However, for larger strain the density-of-states effective mass $\sqrt{m_x m_y}$ starts increasing. For fixed gate voltage it lowers the chemical potential ϕ, thus reducing the current.

At saturation only half of the states flowing from source to drain is filled. For a fixed gate voltage it results in a higher chemical potential ϕ compared to the linear current regime, where the difference in states filling flowing from source to drain and in opposite direction is small. The decrease of ϕ due to the strain-induced increase in the density-of-states is thus more pronounced in saturation leading to a substantially smaller current enhancement.

In thin silicon films the behavior of the enhancement is reversed (Figs. 12.23 and 12.24). This cannot be explained by the change in the subband effective masses alone. The main reason is the strain-induced energy splitting between the unprimed subbands with the same quantum number [169]. The splitting increases with strain and is particularly large in thin films. Because of this splitting the density of states decreases with strain prompting an increase in the chemical potential and current. In saturation the increase of ϕ is larger than in the linear regime guaranteeing the large drive current enhancement.

The methods described so far are either based on the assumption of semi-classical or pure quantum-mechanical ballistic transport. The former modeling approach has proven to be adequate to describe transport in previous generations of microelectronic devices. The latter one may be used for transport description, when the carrier coherence length is larger than the device size. Recent studies show that even for devices with a channel length as short as 15 nm scattering may still play a significant role [134]. An adequate transport model for ultra-scaled MOSFETs must therefore account for quantum-mechanical and dissipative effects simultaneously.

12.5.2 Quantum Transport Models with Scattering

Alternative approaches which can handle both quantum-mechanical and dissipative scattering effects are based on the reduced density matrix [39, 40] and on the Wigner function [48, 89, 196]. The evolution for the reduced electron density matrix is described by a quantum Liouville equation. It is obtained from the equation of motion for the total density matrix by tracing out the degrees of freedom of the environment. To simplify the equation, the limit of weak interaction between the device and the environment is applied. Memory effects can be neglected, if one is interested in system evolution for times larger than the reservoir correlation time [109]. This coarse-graining in time results in a simplified, Markovian description, when the future dynamics of the system is determined by its current state and is independent from the past. The resulting quantum Liouville equation for the reduced density matrix is of Lindblad form and describes the quantum evolution of a system with loss [107]. The Lindblad form guarantees positive definiteness, normalization, and hermiticity of the reduced density matrix.

Coupling to contacts introduces carrier exchange between device and reservoirs. The difference between the electrochemical potentials of the reservoirs causes current flow through the device. In the approach developed by Fischetti, based on the Pauli master equation, this coupling is introduced in a phenomenological manner [39, 40]. Assuming a flat potential profile within each reservoir, electron wave functions are plane waves with well defined momentum. All electron states may be separated in two distinct groups, representing electrons traveling towards and away from the device, respectively. The role of each reservoir is to supply the incident wave functions with occupation determined by the equilibrium distribution within the reservoir. Therefore, the basis of choice to analyze the dynamics of the reduced density matrix is formed by the states incoming from each reservoir. These scattering states are obtained by a numerical solution of the Schrödinger equation subject to open boundary conditions. Without dissipation in the system, i.e., when the evolution is governed by the system Hamiltonian, the density matrix is diagonal in the basis of scattering states. Occupation of each scattering state is determined by the equilibrium distribution within the reservoirs. Once the occupations of scattering states are known, the quantum ballistic current is readily determined. In the presence of weak dissipation, when the device length is shorter than the phase coherence length, the state occupations are found from a solution of the Pauli master equation, which contains only diagonal elements of the density matrix [39, 40]. Carrier density and current density are calculated from the occupations of the scattering states. The Pauli master equation and the Poisson equation are solved repeatedly in a self-consistent iteration loop. In this way subband quantization and rapid potential variations in the transport direction are taken into account self-consistently. The eigenfunctions are used to compute transition rates among the corresponding states. Application of the Pauli master equation is restricted to stationary systems, since in the non-stationary case current continuity would be violated [48].

An interesting solution free from the above mentioned shortcoming of phenomenological coupling of the device to the reservoirs was recently suggested by Gebauer and Car [50, 51]. They suggest to impose periodic boundary conditions upon the non-perturbed system. A constant electric field is introduced into the system via the vector-potential term, which linearly depends on time. The master equation is solved in two steps. First, the Hamiltonian dynamics of the density matrix is evaluated for a certain time-step t. During this time evolution the vector potential grows linearly with time. It can be scaled out from the Hamiltonian by means of a guage transformation [72] at the expense of an additional phase factor in the wave function. The time-step at which the guage transformation is performed is chosen such that the phase added to the wave function satisfies the periodic boundary condition. In the second step the change of the density matrix due to the Lindblad scattering operator is evaluated. This procedure is repeated until a steady state is achieved. Since the eigenfunctions of the periodic system do not carry any current, the current in the Gebauer and Car formulation is due to off-diagonal elements. This approach can also be used to describe transients. The only limitation is that the time scale is much larger than the reservoir coherence time, a condition arising from coarse graining. It has been shown that in addition to the usual Hamiltonian current component a dissipative current component due to interaction with the environment appears. The total current is then conserved also for transients.

Another approach capable of handling both quantum coherent propagation and dissipative scattering effects is based on the Wigner function. The Wigner function is defined as the density matrix in a mixed coordinate/momentum representation [89, 196]. A practically used approximation to incorporate realistic scattering processes into the Wigner equation is to utilize a properly adapted Boltzmann scattering operator [92]. In this way well established scattering models already calibrated within semi-classical transport approaches can be employed in quantum transport calculations. The inclusion of dissipation through the Boltzmann scattering operator, although intuitively appealing, raises some concerns about the validity of such a procedure. The Boltzmann scattering operator is semi-classical by its nature, and represents a good approximation for sufficiently smooth device potentials.

To account for scattering more rigorously, spectral information has to be included into the Wigner function, resulting in an energy-dependence in addition to the momentum dependence [114]. An alternative approach to construct quantum-mechanical extensions of the semi-classical electron-phonon scattering operator has been reported in [13, 23]. These quantum collision operators satisfy a quantum H-theorem and relax systems towards quantum equilibria.

The kinetic equation for the Wigner function is similar to the semi-classical Boltzmann equation, except for a non-local potential term. In the case of a slowly varying potential this non-local term reduces to the local classical force term, and the semi-classical description given by the Boltzmann equation is obtained from the Wigner equation. This semi-classical limit of the Wigner transport equation allows to link seamlessly a semi-classical description of the extended contact regions with the quantum description of the active region of a device using the same formalism [92].

Reports on finite difference solutions of the one-particle Wigner equation for device applications were given by Ravaioli [144], Kluksdahl [87], and coworkers, and date back to the mid 1980s. Frensley [44–46] was the first who introduced boundary conditions on the Wigner function to model open quantum systems. Later, self-consistency was added to the Wigner equation solvers [47, 86]. Main and Haddad included a reduced Boltzmann scattering operator in transient Wigner function-based simulations [115]. Research on finite difference methods for the Wigner equation culminated in 1990 when review articles of Frensley [48] and Buot and Jensen [16] appeared. In 2002, implementations of Monte Carlo methods for solving the Wigner device equation have been reported [125, 153]. With the finite difference method scattering was restricted to the relaxation time approximation and the momentum space to one dimension.

The Monte Carlo method allows scattering processes to be included on a more detailed level, e.g., through the Boltzmann scattering operator, assuming a three-dimensional momentum-space [93,154]. Realistic scattering processes can be easily embedded into the Wigner equation via Boltzmann-like scattering integrals, which turns out to be a good approximation. The Wigner function approach reduces to a semi-classical transport description in contacts providing an important advantage of a seamless treatment between classical and quantum-mechanical regions in device simulations [92].

The Wigner function is given by the density matrix in mixed representation [89, 196] defined by the Wigner-Weyl transform

$$f_{\mathrm{w}}(\mathbf{r},\mathbf{k},t) = \int \rho\left(\mathbf{r}+\frac{\mathbf{s}}{2},\mathbf{r}-\frac{\mathbf{s}}{2},t\right)\exp(-i\mathbf{k}\cdot\mathbf{s})\,\mathrm{d}\mathbf{s}.$$

The kinetic equation for the Wigner function is similar to the Boltzmann equation:

$$\left(\frac{\partial}{\partial t}+\mathbf{v}\cdot\nabla_r\right) f_{\mathrm{w}} = \int V_{\mathrm{w}}(\mathbf{r},\mathbf{k}'-\mathbf{k}) f_{\mathrm{w}}(\mathbf{k}',\mathbf{r},t)\mathrm{d}\mathbf{k}' + \left(\frac{\partial f_{\mathrm{w}}}{\partial t}\right)_{\mathrm{coll}}. \tag{12.66}$$

The Wigner potential entering into the non-local operator in the right-hand side is defined as

$$V_{\mathrm{w}}(\mathbf{r},\mathbf{k}) = \frac{1}{\imath\hbar\,(2\pi)^3}\int \left(V\left(\mathbf{r}-\frac{\mathbf{s}}{2}\right)-V\left(\mathbf{r}+\frac{\mathbf{s}}{2}\right)\right)\exp\left(-i\mathbf{k}\cdot\mathbf{s}\right)\,\mathrm{d}\mathbf{s}. \tag{12.67}$$

In case of slowly varying potentials the difference term in the right-hand side of (12.67) can be developed into the series. Keeping the first non-vanishing terms, one rewrites (12.67) as

$$V_{\mathrm{w}}(\mathbf{r},\mathbf{k}'-\mathbf{k}) = -\frac{1}{\hbar\,(2\pi)^3}\nabla_{\mathbf{r}}V\left(\mathbf{r}\right)\frac{\partial}{\partial\mathbf{k}}\int\exp\left(-i\left(\mathbf{k}'-\mathbf{k}\right)\cdot\mathbf{s}\right)\,\mathrm{d}\mathbf{s}. \tag{12.68}$$

After substituting (12.68) into (12.66) the term with the non-local potential reduces to the classical force term. Following [53], one can introduce a spectral decomposition of the potential profile $V(x)$ into a slowly varying, classical component and a rapidly changing component treated quantum-mechanically.

$$V(\mathbf{r}) = V_{\mathrm{cl}}(\mathbf{r}) + V_{\mathrm{qm}}(\mathbf{r}). \tag{12.69}$$

This separation of the total potential into a smooth classical and a small quantum-mechanical contribution can improve the stability of a numerical solution method. The quantum-mechanical contribution may be moved into the right-hand side of the transport equation. Considered as a perturbation, the quantum-mechanical term can be interpreted as a quantum scattering integral. It allows to treat quantum effects on equal footing with classical scattering:

$$\left(\frac{\partial}{\partial t}+\mathbf{v}\cdot\nabla_r-\frac{q\nabla_r V_{cl}(\mathbf{r})}{\hbar}\cdot\nabla_k\right)f_w=\int V_{qm}(\mathbf{r},\mathbf{k}'-\mathbf{k})f_w(\mathbf{k}',\mathbf{r},t)\mathrm{d}\mathbf{k}'+\left(\frac{\partial f_w}{\partial t}\right)_{coll}. \tag{12.70}$$

The Wigner function formalism treats scattering and quantum-mechanical effects on equal footing through the corresponding scattering integrals [126]. By analogy to the Monte Carlo methods used for the Boltzmann transport equation, it is tempting to solve the quantum Wigner transport equation (12.70) by means of the Monte Carlo technique. Such a program was recently realized in [91, 96, 140, 141, 155, 166]. However, since the kernel of the quantum scattering operator is not positively defined, the numerical weight of the particle trajectory increases rapidly, and the numerical stability of the trajectory-based Monte Carlo algorithm becomes a critical issue. A multiple trajectories method was suggested [92] to overcome this difficulty. In the algorithm developed the problem of a growing statistical weight of a single trajectory is addressed by creating an increasing number of trajectories with constant weights, which may assume positive and negative values. Being formally equivalent to the former method, the algorithm allows the annihilation of particles with similar statistical properties, introducing a possibility to control the number of trajectories.

This method was recently applied to double-gate MOSFETs [166] and quantum wires [127]. In the coherent mode, where scattering is turned off, a comparison to conventional Schrödinger solvers can be performed. In order to estimate the tunneling component of the current the Wigner Monte Carlo simulations were carried out for a MOSFET with a gate length of 10 nm, and a good agreement between the two approaches was found.

The carrier concentration must be used to update the potential in the device by solving the Poisson equation. A superimposed iteration loop makes the Wigner-Poisson solver self-consistent. An example of self-consistent potentials for n-i-n Si structures with an intrinsic region of length W ranging from 20 nm to 2.5 nm, as calculated with the Wigner Monte Carlo method and the classical Monte Carlo method is shown in Fig. 12.25. The doping profile is assumed to increase gradually from the intrinsic channel to the highly doped contacts over the same distance W. Phonon and Coulomb scattering were included. As expected, for long W the classical and quantum calculations yield similar results for the self-consistent potential. For $W = 2.5$ nm an extra space charge due to electrons tunneling under the barrier becomes important, which results in a potential barrier increase. Despite of the potential barrier increase, the current in self-consistent Wigner simulations is approximately 20% higher compared to its classical value found by a self-consistent solution of the Boltzmann and the Poisson equations.

Relative differences between I_{WIG} and the current I_{BALL} computed for a "ballistic" device with scattering inside the intrinsic and transition regions turned off is shown in Fig. 12.26. For $W = 2.5$ nm the relative differences in current due to quantum effects and scattering in the barrier are still of the order of 25% and cannot be neglected.

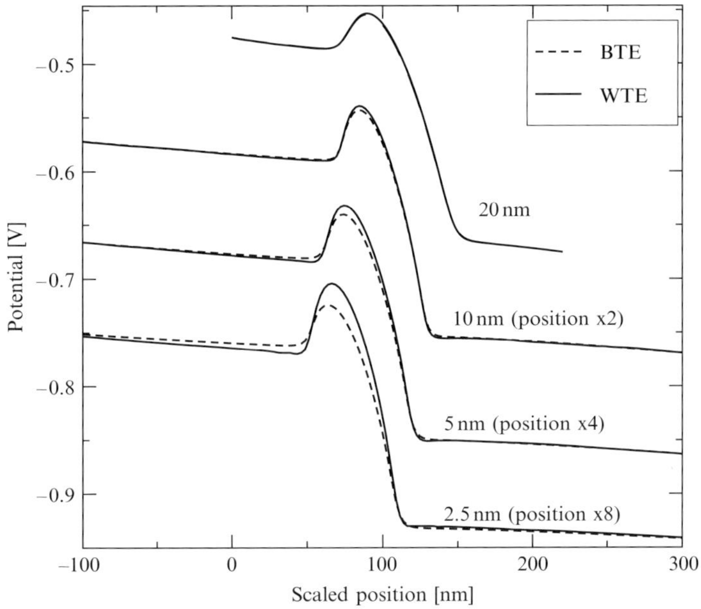

Fig. 12.25 Self-consistent potential profiles calculated for *n-i-n* structures with Wigner (*solid lines*) and Boltzmann (*dashed lines*) transport equations. For long *n-i-n* structures the results are similar. For short *n-i-n* structures the additional charge due to tunneling electrons results in a higher potential barrier

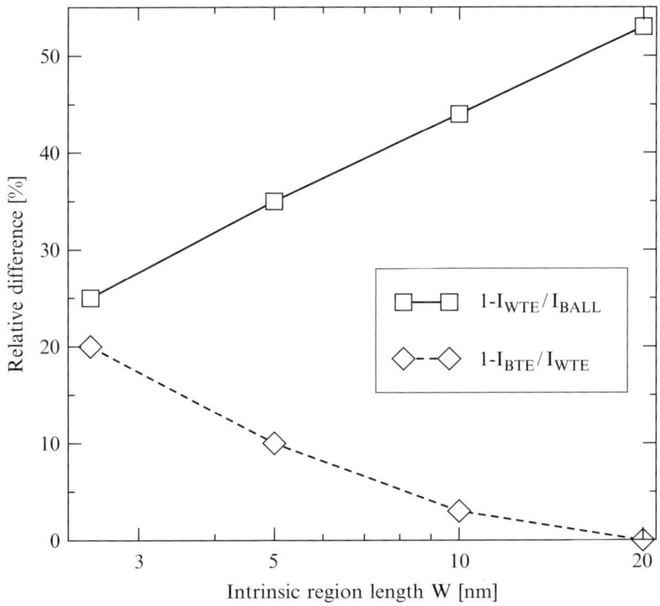

Fig. 12.26 Relative difference between currents calculated with the Wigner and Boltzmann Monte Carlo methods (*diamonds*) and calculated with the Wigner Monte Carlo method for *n-i-n* structures, with and without scattering in the intrinsic region

The Wigner Monte Carlo method gives accurate results not only for single-barrier devices, but it can also be applied to purely quantum-mechanical systems such as resonant tunneling diodes [92]. A typical output characteristic of a GaAs resonant tunneling diode is shown in Fig. 12.27. Scattering with polar optical phonons as well as Coulomb scattering in the contacts is considered. A region of negative differential resistance common to transport via a resonant level is clearly visible after the resonance peak at 250 mV applied voltage. A self-consistent solution of the Wigner transport and Poisson equation is mandatory for the correct determination of the resonance position due to charge accumulation at the cathode side of the resonant tunneling diode. Before the barrier, an accumulation layer forms, depending on the applied voltage, as seen in Fig. 12.28. This results in a voltage shift of the resonance peak of the *IV* characteristics shown in Fig. 12.27. A typical distribution of the concentration in resonance condition and off-resonance is presented in Fig. 12.28. The amount of charge localized in the potential well is much higher at resonance. This leads to a potential barrier increase and also contributes to the shift of the resonance peak. This example demonstrates the importance of quantum-mechanical effects for the simulation of properties of ultra-scaled devices. It also shows that space charge effects are of crucial importance for the accurate prediction of output characteristics of single- and double-barrier devices.

Generalization of the Wigner equation-based transport models to include strain is pending because of additional complications due to inclusion of the band structure in the formalism. The Green's function method briefly discussed below allows to investigate stress dependent transport properties provided that the parameters of the tight-binding model which describes the band structure are properly calibrated to include strain.

12.5.3 Non-Equilibrium Green's Function Method

The Green's function method provides a powerful technique to evaluate the properties of a many-body system both in thermodynamic equilibrium and non-equilibrium situations. The single-particle Green's function of a system allows evaluation of carrier density and current. The many-particle information about interactions is cast into self-energies. The perturbative solution of the equations for the Green's functions and self-energies is the key technique to evaluate the properties of an interacting system in external fields.

Four types of Green's functions can be defined as the non-equilibrium statistical ensemble averages of the single particle correlation operator [114]:

$$\begin{aligned}
G^{>}(\mathbf{r},t;\mathbf{r}',t') &= -i\hbar^{-1}\langle\hat{\psi}(\mathbf{r},t)\hat{\psi}^{\dagger}(\mathbf{r}',t')\rangle,\\
G^{<}(\mathbf{r},t;\mathbf{r}',t') &= +i\hbar^{-1}\langle\hat{\psi}^{\dagger}(\mathbf{r}',t')\hat{\psi}(\mathbf{r},t)\rangle,\\
G^{\mathrm{R}}(\mathbf{r},t;\mathbf{r}',t') &= \theta(t-t')[G^{>}(\mathbf{r},t;\mathbf{r}',t')-G^{<}(\mathbf{r},t;\mathbf{r}',t')],\\
G^{\mathrm{A}}(\mathbf{r},t;\mathbf{r}',t') &= \theta(t'-t)[G^{<}(\mathbf{r},t;\mathbf{r}',t')-G^{>}(\mathbf{r},t;\mathbf{r}',t')],
\end{aligned} \tag{12.71}$$

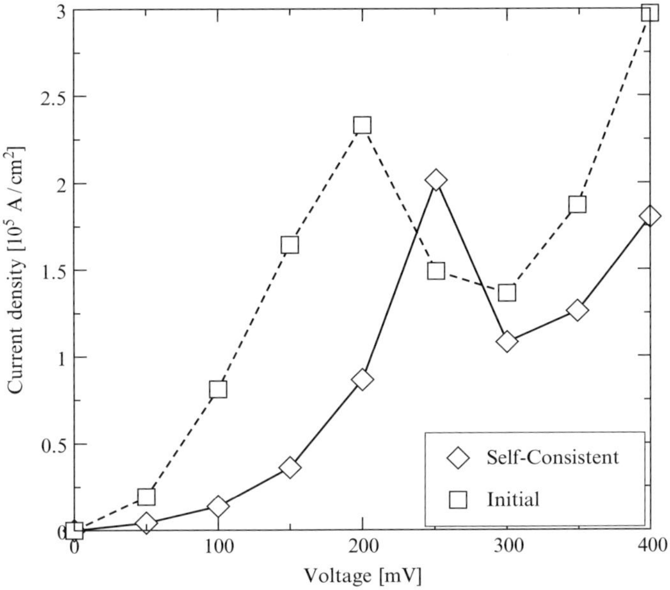

Fig. 12.27 Typical *IV* curve of a resonant tunneling diode, calculated self-consistently (*solid line*), contrasted against a non-self-consistent characteristics. Charge accumulation before the potential barrier as well as a higher charge density inside the RTD between the potential barriers are responsible for the significant shift of the *IV* resonance peak, thus demonstrating the importance of space - charge effects

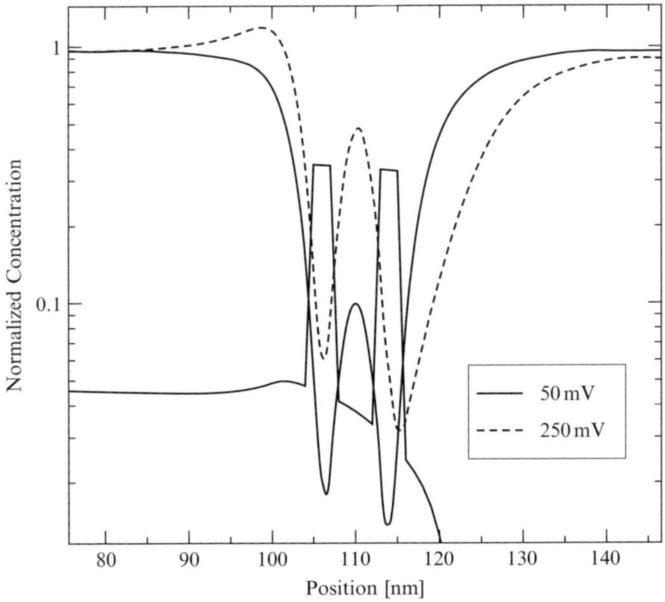

Fig. 12.28 Normalized electron concentration off-resonance (*dashed line*) and at resonance (*solid line*) in RTD

where $\theta(t)$ is the unit step function, and $\hat{\psi}^\dagger(\mathbf{r},t)$ and $\hat{\psi}(\mathbf{r},t)$ are the field operators creating or destroying a particle at point $(\mathbf{r},t)$ in space-time, respectively. Only three of these Green's functions are independent.

Under steady-state conditions the equation of motion for the Green's functions can be written as [21]:

$$[E - H(\mathbf{r})]G^{\mathrm{R}}(\mathbf{r},\mathbf{r}';E) - \int d\mathbf{r}_1 \Sigma^{\mathrm{R}}(\mathbf{r},\mathbf{r}_1;E)G^{\mathrm{R}}(\mathbf{r}_1,\mathbf{r}';E) = \delta(\mathbf{r}-\mathbf{r}'), \quad (12.72)$$

$$G^{\lessgtr}(\mathbf{r},\mathbf{r}';E) = \int d\mathbf{r}_1 \int d\mathbf{r}_2 G^{\mathrm{R}}(\mathbf{r},\mathbf{r}_1;E)\Sigma^{\lessgtr}(\mathbf{r}_1,\mathbf{r}_2;E)[G^{\mathrm{R}}(\mathbf{r}_2,\mathbf{r}';E)]^\dagger, \quad (12.73)$$

where H is the one-particle Hamiltonian, and Σ^{R}, $\Sigma^{<}$, and $\Sigma^{>}$ are the retarded, lesser, and greater self-energies, respectively. The total self energy contains contributions due to the coupling of the device to the contacts and other interactions, such as electron-phonon interaction, $\Sigma = \Sigma_c + \Sigma_{\mathrm{el-ph}}$. The self-energies due to the coupling of the device to the contacts are only non-zero at the boundaries [21]

$$\Sigma_c^{\mathrm{R}}(E) = \tau g_c \tau^\dagger, \quad (12.74)$$
$$\Sigma_c^{<}(E) = -2i\Im m[\Sigma_c^{\mathrm{R}}]f_c(E), \quad (12.75)$$
$$\Sigma_c^{>}(E) = +2i\Im m[\Sigma_c^{\mathrm{R}}](1 - f_c(E)), \quad (12.76)$$

where τ is the coupling matrix between the device and the contact, g_c is the surface Green's function [21], and $f_c(E)$ is the Fermi-Dirac distribution function at the contact c. Within the self-consistent Born approximation, the self-energies for the electron-phonon interaction are

$$\Sigma_{\mathrm{el-ph}}^{<}(\mathbf{r},\mathbf{r}';E) = \sum_j \int \frac{d\mathbf{q}}{(2\pi)^3} e^{i\mathbf{q}.(\mathbf{r}-\mathbf{r}')} D_{\mathbf{q},j}\left(n_{\mathbf{q},j} + \frac{1}{2} \pm \frac{1}{2}\right) G^{<}(\mathbf{r},\mathbf{r}';E \pm \hbar\omega_{\mathbf{q},j}), \quad (12.77)$$

$$\Sigma_{\mathrm{el-ph}}^{>}(\mathbf{r},\mathbf{r}';E) = \sum_j \int \frac{d\mathbf{q}}{(2\pi)^3} e^{i\mathbf{q}.(\mathbf{r}-\mathbf{r}')} D_{\mathbf{q},j}\left(n_{\mathbf{q},j} + \frac{1}{2} \pm \frac{1}{2}\right) G^{>}(\mathbf{r},\mathbf{r}';E \mp \hbar\omega_{\mathbf{q},j}), \quad (12.78)$$

$$\Sigma_{\mathrm{el-ph}}^{\mathrm{R}}(\mathbf{r},\mathbf{r}';E) = -\frac{i}{2}\Gamma_{\mathrm{el-ph}}(\mathbf{r},\mathbf{r}';E) + \mathrm{P}\int \frac{dE'}{2\pi} \frac{\Gamma_{\mathrm{el-ph}}(\mathbf{r},\mathbf{r}';E')}{E-E'}. \quad (12.79)$$

$\hbar\omega_{\mathbf{q},j}$ denotes the phonon energy of branch j at the wave-vector $\mathbf{q}$, $n_{\mathbf{q},j}$ is the average phonon occupation number, $D_{\mathbf{q},j}$ is the electron-phonon interaction strength, $\Gamma_{\mathrm{el-ph}} \equiv i(\Sigma_{\mathrm{el-ph}}^{>} - \Sigma_{\mathrm{el-ph}}^{<})$ defines the broadening, and $\mathrm{P}\int$ represents the principal part of the integration. The imaginary part of the retarded self-energy broadens the density of states, whereas the real part shifts it. The plus and minus signs in (12.77) and (12.78) denote the phonon emission and absorption processes, respectively. Assuming that the environment stays in thermal equilibrium, $n_{\mathbf{q},j}$ is given by

$$n_{\mathbf{q},j} = \frac{1}{\exp(\hbar\omega_{\mathbf{q},j}/k_B T) - 1}. \tag{12.80}$$

The self-energy due to electron-phonon interaction comprises the contributions of elastic and inelastic scattering mechanisms, $\Sigma^{\nu}_{\text{e-ph}} = \Sigma^{\nu}_{\text{el}} + \Sigma^{\nu}_{\text{inel}}$. By definition the particle energy is conserved for elastic scattering. For inelastic scattering, the after-scattering energy is different from its initial value. The energy difference is carried away by a scattering mediator, e.g., by a phonon. For optical phonons the frequency dependence $\omega_{\mathbf{q}}$ on the wave vector q is weak and it has a finite value ω_0 as $\mathbf{q} \to 0$, therefore carrier scattering with optical phonons is inelastic. Inelastic interaction of carriers with optical phonons is usually considered local [102]. In this case the self-energies can be written as

$$\Sigma^{<,\nu}_{\text{inel}}(E) = \sum_j D_{\text{inel},j}\left(n_B(\hbar\omega_j) + \frac{1}{2} \pm \frac{1}{2}\right) G^{<,\nu}(E \pm \hbar\omega_j), \tag{12.81}$$

$$\Sigma^{>,\nu}_{\text{inel}}(E) = \sum_j D_{\text{inel},j}\left(n_B(\hbar\omega_j) + \frac{1}{2} \pm \frac{1}{2}\right) G^{>,\nu}(E \mp \hbar\omega_j)\,. \tag{12.82}$$

The electron-phonon interaction strength is given by

$$D_{\text{inel},j} = \frac{\hbar |M_j^{\text{OP}}|^2}{2nm_c\omega_j}, \tag{12.83}$$

where m_c is the mass of a carbon atom.

For acoustic phonons $\omega_{\mathbf{q},j} = v_j q$ is proportional to q. Therefore, the acoustic phonon energy at small q is negligible compared to the carrier energy, and carrier scattering with acoustic phonons is frequently considered elastic. The corresponding self-energies for acoustic phonon interaction are

$$\Sigma^{\lessgtr,\nu}_{\text{el}}(E) = D^{\nu}_{\text{el}} G^{\lessgtr,\nu}(E), \tag{12.84}$$

$$D_{\text{el},j} = \frac{k_B T |M_j^{\text{AP}}|^2}{2nm_c v_j}. \tag{12.85}$$

The non-equilibrium Green's function method addresses the quantum transport problem in the most consistent and complete way, however, the method is computationally complex and, with a few exceptions [117, 118, 173], usually applied to one-dimensional problems [103] and for a restricted set of scattering mechanisms [172] only. Scattering requires the knowledge of the corresponding self-energies and thus complicates computations significantly [172]. The self-consistent

Born approximation for the self-energy is an extremely time consuming but necessary step, because it guarantees the current continuity. The convergence of the self-consistent iteration is a critical issue where fine resonances at some energies have to be resolved accurately [77, 136]. For that purpose an adaptive method for selecting the energy grid is essential [136].

The one-dimensional equation of motion for the Green's functions, (12.72) and (12.73), are solved for $G^{R,\nu}$ and $G^{\lessgtr,\nu}$ using a recursive Green's function algorithm [173]. Coupling to reservoirs [21, 171] is usually described by the contact self-energies (12.74). For numerical solutions it is convenient to transform the equations for the Green's function in the eigen-mode representation [186]. Alternatively, a discretized version on a mesh in real space can be used [22]. This results in a tight-binding-like Hamiltonian defined on the mesh nodes. The carrier concentration at a node l of the spatial grid and the current density at the edge between the nodes l and $l+1$ are given by

$$n_l = -2i \sum_{\nu} \int \frac{dE}{2\pi} G_{l,l}^{<,\nu}(E), \tag{12.86}$$

$$j_{l,l+1} = \frac{2q}{\hbar} \sum_{\nu} \int \frac{dE}{2\pi} 2\Re e\{G_{l,l+1}^{<,\nu}(E) t_{l+1,l}^{\nu}\}, \tag{12.87}$$

where the factor 2 is due to the spin degeneracy. The equations for the Green's functions must be solved self-consistently with the Poisson equation [198].

Recently an atomistic and full-band quantum transport simulator designed for CMOS devices based on ultra-thin body silicon films and nanowires is being developed [112]. The simulator is able to resolve two- and three-dimensional Schrödinger-Poisson equations based on the $sp^3d^5s^*$ semi-empirical tight-binding method. Carrier and current densities are obtained by injecting electrons and holes at different energies into the device and by solving the resulting system of equations in the non-equilibrium Green's function formalism. The $sp^3d^5s^*$ model was only recently calibrated [14] to include shear strain and results of current calculations in uniaxially stressed nanowires are expected to be available soon.

12.5.4 Conclusion and Trends

Transport modeling for TCAD applications has grown into a mature field of research, software development, and applications. Models of different complexity, precision and accuracy are offered and implemented in various commercial and academic TCAD tools. Starting with the drift-diffusion model and higher moments models, a complete hierarchy of transport models for semiconductor device simulation has been gradually constructed. Depending on parameter values and device scales, either semi-classical or quantum-mechanical transport description can be adopted.

Monte Carlo techniques are the basis for ultimate tools to obtain the solution of the Boltzmann transport equation with arbitrary scattering mechanisms. These methods require significant CPU resources and are relatively rarely used for industrial TCAD applications, when timely but perhaps less accurate results are of primary importance. Monte Carlo methods can easily be extended and generalized to incorporate strain-induced modifications in band structure and scattering rates. With quantum corrections carefully added, full-band Monte Carlo methods can provide accurate results in strained ultra-scaled devices with strong size quantizations. Monte Carlo solvers are indispensable for calibration of the parameters of higher order moment models to include strain.

With the channel size of MOSFETs in the decananometer region, the development of conceptually new future devices and architectures is becoming increasingly important. New nanoelectronic structures, utilizing carbon nanotubes, silicon nanowires, and molecules are considered to be prominent candidates for the post-CMOS era. Nanoelectronic devices are expected to complement and substitute some of the current CMOS functions. New physical principles for carrier propagations must be included in transport models for nanodevices. Indeed, at this small device size the geometrical spread of the carrier wave packet can no longer be ignored. Thus, the complete information about carrier dynamics inside the device including the phase of the wave function is needed. Although many particles with different energies and phases are injected into a device, the phases of an individual carrier may not average out completely, if the potential in transport direction is changing rapidly. The full quantum-mechanical description is needed, when the curvature of the potential close to its maximum is so big that tunneling under the potential barrier contributes significantly to the current.

When the device channel length is smaller than the scattering length, current transport inside the device becomes ballistic. The scattering length is only several nanometers in modern MOSFETs operating at room temperature, therefore, even in ultra-scaled modern devices transport is not completely ballistic. If the potential changes rapidly enough on the scale of the carrier wave packet, a quantum-mechanical description of electron motion between two scattering events must be adopted. In the case when both the scattering length and the spread of the wave packet are comparable to the channel length, the interplay between coherent propagation and scattering determines transport. Methods for dissipative quantum transport are based on the non-equilibrium Green's function formalism, the Liouville/von-Neumann equation for the density matrix, and the kinetic equation for the Wigner function. A proper generalisation of the methods to incorporate stress is still pending. All the quantum methods for transport are time consuming, and highly accurate simulations of emerging nanoscale devices represent an outstanding modeling and computational challenge. Development of new efficient numerical algorithms as well as a comprehensive comparison between different quantum-mechanical models describing dissipative transport in open systems is mandatory and is currently on the research agenda.

In modern microelectronic devices quantum effects are usually dominant in a small active region connected to relatively large, heavily doped contact areas, where the carrier dynamics is essentially classical. Modern TCAD simulators must be able to incorporate both semi-classical and (dissipative) quantum-mechanical modeling approaches within the same simulation core. It is, therefore, necessary to have an accurate multiscale simulation technique, which is able to bridge the gap between semi-classical and quantum-mechanical simulations.

With shrinking device dimensions, the demand for full three-dimensional accurate solvers for the coupled transport/Poisson equations [117, 179, 193] and atomistic based simulations [85, 112, 194] has grown significantly. With the advances in computer architectures, computational power and memory capabilities increasing, state-of-the art software, development of fast numerical algorithms and conceptually new generic simulation platforms a fundamental breakthrough in speed, reliability, and accuracy of multiscale three-dimensional TCAD simulation tools is anticipated. This will facilitate an accurate design of the macroscopic transport properties of emerging nanoelectronic and molecular devices based on their microscopic atomistic electronic characteristics.

References

1. International Technology Roadmap for Semiconductors: 2005 Edition (2005). http://www.itrs.net/Links/2009ITRS/Home2009.htm
2. Online Simulations and More (2010). http://www.nanohub.org
3. Ancona, M.G.: Macroscopic description of quantum-mechanical tunneling. Phys. Rev. B **42**(2), 1222–1233 (1990)
4. Ancona, M.G., Tiersten, H.F.: Quantum correction to the equation of state of an electron gas in a semiconductor. Phys. Rev. B **39**(13), 9536–9540 (1989)
5. Ancona, M.G., Yu, Z., Dutton, R.W., Voorde, P.J.V., Cao, M., Vook, D.: Density-gradient analysis of tunneling in MOS structures with ultra-thin oxides. In: Proc. Intl. Conf. Simulation of Semiconductor Processes and Devices, pp. 235–238 (1999)
6. Ancona, M.G., Yu, Z., Dutton, R.W., Voorde, P.J.V., Cao, M., Vook, D.: Density-gradient analysis of MOS tunneling. IEEE Trans. Electron Devices **47**(12), 2310–2319 (2000)
7. Ando, T., Fowler, A.B., Stern, F.: Electronic properties of two-dimensional systems. Rev. Mod. Phys. **54**(2), 437–672 (1982)
8. Asenov, A., Brown, A.R., Watling, J.R.: Quantum corrections in the simulation of decanano MOSFETs. Solid State Electron. **47**(7), 1141–1145 (2003)
9. Balestra, F., Cristoloveanu, S., Benachir, M., Brini, J., Elewa, T.: Double-gate silicon-on-insulator transistor with volume inversion: a new device with greatly enhanced performance. IEEE Electron Device Lett. **8**(9), 410–412 (1987)
10. Balslev, I.: Influence of uniaxial stress on the indirect absorption edge in silicon and germanium. Phys. Rev. **143**, 636–647 (1966)
11. Blotekjaer, K.: Transport equations for electrons in two-valley semiconductors. IEEE Trans. Electron Devices **17**(1), 38–47 (1970)
12. Bosi, S., Jacoboni, C.: Monte Carlo high field transport in degenerate GaAs. J. Phys. C: Solid State Phys. **9**, 315–319 (1976)
13. Bourgade, J.P., Degond, P., Mehats, F., Ringhofer, C.: On quantum extensions to classical spherical harmonics expansion/Fokker-Planck models. J. Math. Phys. **47**(4), 043302 (2006)
14. Boykin, T.B., Luisier, M., Salmani-Jelodar, M., Klimeck, G.: Strain-induced, off-diagonal, same-atom parameters in empirical tight-binding theory suitable for [110] uniaxial strain applied to a silicon parametrization. Phys. Rev. B **81**(12), 125,202 (2010)

15. Bufler, F.M., Hudé, R., Erlebach, A.: On a simple and accurate quantum correction for Monte Carlo simulations. In: Intl. Workshop Comput. Electroncis, pp. 101–102. Wien (2006)
16. Buot, F., Jensen, K.: Lattice Weyl-Wigner formulation of exact many-body quantum-transport theory and applications to novel solid-state quantum-based devices. Phys. Rev. B **42**(15), 9429–9457 (1990)
17. Canali, C., Jacoboni, C., Nava, F., Ottaviani, G., Quaranta, A.: Electron drift velocity in silicon. Phys. Rev. B **12**(4), 2265–2284 (1975)
18. Caymax, M., Eneman, G., Bellenger, F., Merckling, C., Delabie, A., Wang, G., Loo, R., Simoen, E., Mitard, J., De Jaeger, B., Hellings, G., De Meyer, K., Meuris, M., Heyns, M.: Germanium for advanced CMOS anno 2009: A SWOT analysis. In: Intl. Electron Devices Meeting, pp. 1–4 (2009)
19. Colman, D., Bate, R.T., Mize, J.P.: Mobility anisotropy and piezoresistance in silicon p-type inversion layers. J. Appl. Phys. **39**(4), 1923–1931 (1968)
20. Curatola, G., Fiori, G., Iannaccone, G.: Modeling and simulation challenges for nanoscale mosfets in the ballistic limit. Solid State Electron. **48**(4), 581–587 (2004)
21. Datta, S.: Electronic Transport In Mesoscopic Systems. Cambridge University Press, Cambridge (1995)
22. Datta, S.: Quantum Transport: Atom To Transistor. Cambridge University Press, Cambridge (2005)
23. Degond, P., Ringhofer, C.: Quantum moment hydrodynamics and entropy principle. J. Stat. Phys. **112**(3), 587–628 (2003)
24. DeMari, A.: An accurate numerical steady-state one-dimensional solution of the p-n junction. Solid State Electron. **11**, 33–58 (1968)
25. Dhar, S., Kosina, H., Palankovski, V., Ungersboeck, E., Selberherr, S.: Electron mobility model for strained-Si devices. IEEE Trans. Electron Devices **52**(4), 527–533 (2005)
26. Dhar, S., Ungersboeck, E., Kosina, H., Grasser, T., Selberherr, S.: Electron mobility model for $\langle 110 \rangle$ stressed silicon including strain-dependent masses. IEEE Trans. Nanotechnol. **6**(1), 97–100 (2007)
27. Donetti, L., Gámiz, F., Rodriguez, N., Jamenez, F., Sampedro, C.: Influence of acoustic phonon confinement on electron mobility in ultrathin silicon on insulator layers. Appl. Phys. Lett. **88**(1), 122108(1–3) (2006)
28. Doris, B., Ieong, M., Kanarsky, T., Zhang, Y., Roy, R.A., Documaci, O., Ren, Z., Jamin, F.F., Shi, L., Natzle, W., Huang, H.J., Mezzapelle, J., Mocuta, A., Womack, S., Gribelyuk, M., Jones, E.C., Miller, R.J., Wong, H.S.P., Haensch, W.: Extreme scaling with ultra-thin si channel MOSFETs. In: Intl. Electron Devices Meeting, pp. 267–270 (2002)
29. van Dort, M.J., Woerlee, P.H., Walker, A.J.: A simple model for quantization effects in heavily-doped silicon MOSFETs at inversion conditions. Solid State Electron. **37**(3), 411–414 (1994)
30. Egley, J., Chidambarao, D.: Strain effects on devide characteristics: Implementation in drift-difusion simulators. Solid State Electron. **36**(12), 1653–1664 (1993)
31. Esseni, D.: On the modeling of surface roughness limited mobility in SOI MOSFETs and its correlation to the transistor effective field. IEEE Trans. Electron Devices **51**(3), 394–401 (2004)
32. Esseni, D., Abramo, A.: Mobility modelling of SOI MOSFETs. Semicond. Sci. Technol. **19**, S67–S70 (2004)
33. Esseni, D., Mastrapasqua, M., Celler, G., Fiegna, C., Selmi, L., Sangiorgi, E.: An experimental study of mobility enhancement in ultrathin SOI transistors operated in double-gate mode. IEEE Trans. Electron Devices **50**(3), 802–808 (2003)
34. Fan, X.F., Register, L.F., Winstead, B., Foisy, M.C., Chen, W.Q., Zheng, X., Ghosh, B., Banerjee, S.K.: Hole mobility and thermal velocity enhancement for uniaxial stress in Si up to 4 GPa. IEEE Trans. Electron Devices **54**(2), 291–296 (2007)
35. Fawcett, W., Boardman, A., Swain, S.: Monte Carlo determination of electron transport properties in gallium arsenide. J. Phys. Chem. Solids **31**, 1963–1990 (1970)

36. Fawcett, W., Paige, E.: Negative differential mobility of electrons in germanium: A Monte Carlo calculation of the distribution function, drift velocity and carrier population in the <111> and <100> minima. J. Phys. C: Solid State Phys. **4**, 1801–1821 (1971)
37. Ferry, D., Akis, R., Vasileska, D.: Quantum effects in MOSFETs: Use of an effective potential in 3D Monte Carlo simulations in ultra-short channel devices. In: Intl. Electron Devices Meeting, pp. 287–290 (2000)
38. Fischetti, M., Laux, S.: Monte Carlo simulation of electron transport in Si: The first 20 years. In: Baccarani, G., Rudan, M. (eds.) 26th European Solid State Device Research Conference, pp. 813–820. Editions Frontiers, Bologna, Italy (1996)
39. Fischetti, M.V.: Theory of electron transport in small semiconductor devices using the Pauli master equation. J. Appl. Phys. **83**(1), 270–291 (1998)
40. Fischetti, M.V.: Master-equation approach to the study of electronic transport in small semiconductor devices. Phys. Rev. B **59**(7), 4901–4917 (1999)
41. Fischetti, M.V., Gámiz, F., Hänsch, W.: On the enhanced electron mobility in strained-silicon inversion layers. J. Appl. Phys. **92**(12), 7320–7324 (2002)
42. Fischetti, M.V., Laux, S.E.: Monte Carlo analysis of electron transport in small semiconductor devices including band-structure and space-charge effects. Phys. Rev. B **38**(14), 9721–9745 (1988)
43. Fischetti, M.V., Ren, Z., Solomon, P.M., Yang, M., Rim, K.: Six-band **k·p** calculation of the hole mobility in silicon inversion layers: Dependence on surface orientation, strain, and silicon thickness. J. Appl. Phys. **94**(2), 1079–1095 (2003)
44. Frensley, W.: Quantum transport simulation of the resonant tunneling diode. In: Intl. Electron Devices Meeting, Los Angeles, pp. 571–574 (1986)
45. Frensley, W.: Transient response of a tunneling device obtained from the Wigner function. Phys. Rev. Lett. **57**(22), 2853–2856 (1986)
46. Frensley, W.: Wigner-function model of a resonant-tunneling semiconductor device. Phys. Rev. B **36**(3), 1570–1580 (1987)
47. Frensley, W.: Effect of inelastic processes on the self-consistent potential in the resonant-tunneling diode. Solid State Electron. **32**(12), 1235–1239 (1989)
48. Frensley, W.: Boundary conditions for open quantum systems driven far from equilibrium. Rev. Mod. Phys. **62**(3), 745–791 (1990)
49. Frensley, W.: Numerical evaluation of resonant states. Superlattices Microstructures **11**(3), 347–350 (1992)
50. Gebauer, R., Car, R.: Current in open quantum qystems. Phys. Rev. Lett. **93**(16), 160,404 (2004)
51. Gebauer, R., Car, R.: Kinetic theory of quantum transport at the nanoscale. Phys. Rev. B **70**(12), 125,324 (2004)
52. Gehring, A., Grasser, T., Kosina, H., Selberherr, S.: Simulation of hot-electron oxide tunneling current based on a non-Maxwellian electron energy distribution function. J. Appl. Phys. **92**(10), 6019–6027 (2002)
53. Gehring, A., Kosina, H.: Wigner-function based simulation of quantum transport in scaled DG-MOSFETs using the Monte Carlo method. J. Comput. Electron. **4**(1–2), 67–70 (2005)
54. Gilbert, M., Akis, R., Ferry, D.: Phonon-assisted ballistic to diffusive crossover in silicon nanowire transistors. J. Appl. Phys. **98**(9), 094,303–1–8 (2005)
55. Grasser, T., Jungemann, C., Kosina, H., Meinerzhagen, B., Selberherr, S.: Advanced transport models for sub-micrometer devices. In: Proc. Intl. Conf. Simulation of Semiconductor Processes and Devices, pp. 1–8 (2004)
56. Grasser, T., Kosik, R., Jungemann, C., Kosina, H., Selberherr, S.: Nonparabolic macroscopic transport models for device simulation based on bulk Monte Carlo data. J. Appl. Phys. **97**(9), 0937,101–09371,012 (2005)
57. Grasser, T., Kosina, H., Gritsch, M., Selberherr, S.: Using six moments of Boltzmann's transport equation for device simulation. J. Appl. Phys. **90**(5), 2389–2396 (2001)
58. Grasser, T., Kosina, H., Heitzinger, C., Selberherr, S.: Characterization of the hot electron distribution function using six moments. J. Appl. Phys. **91**(6), 3869–3879 (2002)

59. Grasser, T., Kosina, H., Selberherr, S.: An impact ionization model including non-maxwellian and non-parabolicity effects. In: Proc. Intl. Conf. Simulation of Semiconductor Processes and Devices, pp. 46–49 (2001)
60. Grasser, T., Kosina, H., Selberherr, S.: Hot carrier effects within macroscopic transport models. Intl. J. High Speed Electron. **13**(3), 873–901 (2003)
61. Gritsch, M.: Numerical modeling of SOI MOSFETs. Dissertation, Technische Universität Wien (2002). http://www.iue.tuwien.ac.at/phd/gritsch
62. Gritsch, M., Kosina, H., Grasser, T., Selberherr, S.: Influence of generation/recombination effects in simulations of partially depleted SOI MOSFETs. Solid State Electron. **45**(4), 621–627 (2001)
63. Gritsch, M., Kosina, H., Grasser, T., Selberherr, S.: Revision of the standard hydrodynamic transport model for SOI simulation. IEEE Trans. Electron Devices **49**(10), 1814–1820 (2002)
64. Gummel, H.: A self-consistent iterative scheme for one-dimensional steady state transistor calculations. IEEE Trans. Electron Devices **11**, 455–465 (1964)
65. Hänsch, W., Vogelsang, T., Kircher, R., Orlowski, M.: Carrier transport near the Si/SiO_2 interface of a MOSFET. Solid State Electron. **32**(10), 839–849 (1989)
66. Heinz, F., Schenk, A., Scholze, A., Fichtner, W.: Full quantum simulation of silicon-on-insulator single-electron devices. J. Comput. Electron. **1**(1), 161–164 (2002)
67. Herring, C., Vogt, E.: Transport and deformation-potential theory for many-valley semiconductors with anisotropic scattering. Phys. Rev. **101**(3), 944–961 (1956)
68. Hockney, R., Eastwood, J.W.: Computer Simulation Using Particles. Adam Hilger, Bristol and Philadelphia (1988)
69. Iafrate, G.J., Grubin, H.L., Ferry, D.K.: Utilization of quantum-distribution function for ultra-submicron device transport. J. Phys. **42**, 307–312 (1981)
70. Institut für Mikroelektronik: MINIMOS-NT 2.1 User's Guide. Technische Universität Wien, Austria (2010)
71. Irie, H., Kita, K., Kyuno, K., Toriumi, A.: In-plane mobility anisotropy and universality under uni-axial strains in nand p-MOS inversion layers on (100), [110], and (111) Si. In: Intl. Electron Devices Meeting, pp. 225–228 (2004)
72. Jackson, J.: Classical Electrodynamics, Third Edition. Academic Press, New York (1998)
73. Jacoboni, C.: A new approach to Monte Carlo simulation. In: Intl. Electron Devices Meeting, pp. 469–472. IEEE Electron Devices Society, Washington, D.C. (1989)·
74. Jacoboni, C., Minder, R., Majni, G.: Effects of band non-parabolicity on electron drift velocity in silicon above room temperature. J. Phys. Chem. Solids **36**, 1129–1133 (1975)
75. Jacoboni, C., Poli, P., Rota, L.: A new Monte Carlo technique for the solution of the Boltzmann transport equation. Solid State Electron. **31**(3/4), 523–526 (1988)
76. Jacoboni, C., Reggiani, L.: The Monte Carlo method for the solution of charge transport in semiconductors with applications to covalent materials. Rev. Mod. Phys. **55**(3), 645–705 (1983)
77. John, D.L., Castro, L.C., Pereira, P.J.S., Pulfrey, D.L.: A Schrödinger-Poisson solver for modeling carbon nanotube FETs. In: Proc. of Nanotech 2004 (2004)
78. Jungel, A.: Quasi-hydrodynamic semiconductor equations, In: Progress in Nonlinear Differential Equations and Their Applications, vol. 41. A Birkhauser book, Switzerland (2001)
79. Jungemann, C., Meinerzhagen, B.: Hierarchical Device Simulation. The Monte Carlo Perspective. Springer, New york (2003)
80. Jungemann, C., Nguyen, C.D., Neinhüs, B., Decker, S., Meinerzhagen, B.: Improved modified local density approximation for modeling of size quantization in nMOSFETs. In: Proc. Intl. Conf. Modeling and Simulation of Microsystems, pp. 458–461 (2001)
81. Jungemann, C., Pham, A.T., Meinerzhagen, B.: A linear response Monte Carlo algorithm for inversion layers and magnetotransport. In: Proc. Intl. Workshop Comput. Electronics, pp. 13–14 (May, 2006)
82. Kadanoff, L.P., Baym, G.: Quantum Statistical Mechanics. Benjamin, New York (1962)
83. Kathawala, G., Winstead, B., Ravaioli, U.: Monte Carlo simulations of double-gate MOSFETs. IEEE Trans. Electron Devices **50**(12), 2467–2473 (2003)

84. Kennedy, D.: On the ambipolar diffusion of impurities into silicon. Proc. IEEE **54**(6), 1202–1203 (1969)
85. Klimeck, G., Luisier, M.: From nemo1d and nemo3d to omen: Moving towards atomistic 3-d quantum transport in nano-scale semiconductors. In: Intl. Electron Devices Meeting, pp. 1–4 (2008)
86. Kluksdahl, N., Kriman, A., Ferry, D., Ringhofer, C.: Self-consistent study of the resonant-tunneling diode. Phys. Rev. B **39**(11), 7720–7735 (1989)
87. Kluksdahl, N., Pötz, W., Ravaioli, U., Ferry, D.: Wigner function study of a double quantum barrier resonant tunneling diode. Superlattices Microstructures **3**(1), 41–45 (1987)
88. Kobayashi, M., Irisawa, T., Magyari-Kope, B., Saraswat, K., Wong, H.S., Nishi, Y.: Uniaxial stress engineering for high-performance Ge NMOSFETs. IEEE Trans. Electron Devices **57**(5), 1037 –1046 (2010)
89. Kosina, H., Nedjalkov, M.: Handbook Of Theoretical And Computational Nanotechnology, vol. 10, chap. Wigner function based device modeling, pp. 731–763. American Scientific, Los Angeles (2006)
90. Kosina, H., Nedjalkov, M., Selberherr, S.: Theory of the Monte Carlo method for semiconductor device simulation. IEEE Trans. Electron Devices **47**(10), 1899–1908 (2000)
91. Kosina, H., Nedjalkov, M., Selberherr, S.: A Monte Carlo method seamlessly linking quantum and classical transport calculations. J. Comput. Electron. **2**(2–4), 147–151 (2002)
92. Kosina, H., Nedjalkov, M., Selberherr, S.: A Monte Carlo method seamlessly linking classical and quantum transport calculations. J. Comp. Electron. **2**(2–4), 147–151 (2003)
93. Kosina, H., Nedjalkov, M., Selberherr, S.: Quantum Monte Carlo Simulation Of A Resonant Tunneling Diode Including Phonon Scattering. In: Laudon, M., Romanowicz, B. (eds.) Nanotech, Computational Publications, San Francisco, pp. 190–193 (2003)
94. Kosina, H., Nedjalkov, M., Selberherr, S.: A stable backward Monte Carlo method for the solution of the Boltzmann equation. In: Lecture Notes in Computer Science 2907: Large-Scale Scientific Computing, Springer, Berlin, pp. 170–177 (2003)
95. Kosina, H., Selberherr, S.: Device simulation demands of upcoming microelectronics devices. Intl. J. High Speed Electron. **16**(1), 115–136 (2006)
96. Kosina, H., Sverdlov, V., Grasser, T.: Wigner Monte Carlo simulation: Particle annihilation and device applications. In: Proc. Intl. Conf. on Simulation of Semiconductor Processes and Devices, pp. 357–360 (2006)
97. Kotlyar, R., Giles, M., Cea, S., Linton, T., Shifren, L., Weber, C., Stettler, M.: Modeling the effects of applied stress and wafer orientation in silicon devices: From long channel mobility physics to short channel performance. J. Comput. Electron. **8**(2), 110–123 (2009)
98. Kotlyar, R., Weber, C., Shifren, L., Cea, S., Giles, M., , Stettler, M.: Effect of band warping and wafer orientation on NMOS mobility under arbitrary applied stress. J. Comput. Electron. **7**(3), 95–98 (2007)
99. Krishnamohan, T., Jungemann, C., Kim, D., Ungersboeck, E., Selberherr, S., Wong, P., Nishi, Y., Saraswat, K.: Theoretical investigation of performance in uniaxially- and biaxially-strained Si, SiGe and Ge double-gate p-MOSFETs. In: Intl. Electron Devices Meeting, pp. 937–940 (2006)
100. Kunikiyo, T., Takenaka, M., Kamakura, Y., Yamaji, M., Mizuno, H., Morifuji, M., Taniguchi, K., Hamaguchi, C.: A Monte Carlo simulation of anisotropic electron transport in silicon including full band structure and anisotropic impact-ionization model. J. Appl. Phys. **75**(1), 297–312 (1994)
101. Kurosawa, T.: Monte Carlo calculation of hot electron problems. In: Proc. Intl. Conf. on Physics of Semiconductors, pp. 424–426 (1966)
102. Lake, R., Datta, S.: Nonequilibrium Green's-function method applied to double-barrier resonant-tunneling diodes. Phys. Rev. B **45**(12), 6670–6685 (1992)
103. Lake, R., Klimeck, G., Bowen, R.C., Jovanovic, D.: Single and multiband modeling of quantum electron transport through layered semiconductor devices. J. Appl. Phys. **81**(12), 7845–7869 (1997)
104. Laux, S., Kumar, A., Fischetti, M.: Ballistic FET modeling using QDAME: Quantum device analysis by modal evaluation. IEEE Trans. Nanotechnol. **1**(4), 255–259 (2002)

105. Lent, C., Kirkner, D.: The quantum transmitting boundary method. J. Appl. Phys. **67**(10), 6353–6359 (1990)
106. Likharev, K.K.: Sub-20-nm electron devices. In: Morkoc, H. (ed.) Advanced Semiconductor and Organic Nano-Techniques, Academic Press, New York, pp. 239–302 (2003)
107. Lindblad, G.: On the generators of quantum dynamical semigroups. Comm. Math. Phys. **48**, 119–130 (1976)
108. Loeb, H., Andrew, R., Love, W.: Application of 2-dimensional solutions of the Shockley-Poisson equation to inversion-layer M.O.S.T. devices. Electron. Lett. **4**, 352–354 (1968)
109. Louisell, W.H.: Quantum Statistical Properties Of Radiation. Willey, New York (1973)
110. Lucci, L., Palestri, P., D.Esseni, Selmi, L.: Multi-subband Monte-Carlo modeling of nano-MOSFETs with strong vertical quantization and electron gas degeneration. In: Intl. Electron Devices Meeting, pp. 531–534 (2005)
111. Lugli, P., Ferry, D.K.: Degeneracy in the ensemble Monte Carlo method for high field transport in semiconductors. IEEE Trans. Electron Devices **32**(11), 2431–2437 (1985)
112. Luisier, M., Schenk, A., Fichtner, W., Klimeck, G.: Atomistic simulations of nanowires in the $sp^3d^5s^*$ tight-binding formalism: From boundary conditions to strain calculations. Phys. Rev. B **74**, 205323(1–12) (2006)
113. Lundstrom, M.: Fundamentals Of Carrier Transport. Cambridge University Press, Cambridge (2000)
114. Mahan, G.: Many-Particle Physics. Premium Press, New York (1990)
115. Mains, R.K., Haddad, G.I.: Time-dependent modeling of resonant-tunneling diodes from direct solution of the Schrödinger equation. J. Appl. Phys. **64**(7), 3564–3569 (1988)
116. Manku, T., Nathan, A.: Electron drift mobility model for devices based on unstrained and coherently strained $Si_{1-x}Ge_x$ grown on $< 001 >$ silicon substrate. IEEE Trans. Electron Devices **39**(9), 2082–2089 (1992)
117. Martinez, A., Barker, J.R., Anantram, M.P., Svizhenko, A., Asenov, A.: Developing a full 3D NEGf simulator with random dopant and interface roughness. In: Intl. Workshop Comput. Electroncis, Wien, pp. 275–276 (2006)
118. Martinez, A., Svizhenko, A., Anantram, M.P., Barker, J.R., Brown, A.R., Asenov, A.: A study of the effect of the interface roughness on a DG-MOSFET using a full 2D NEGF technique. In: Intl. Electron Devices Meeting, pp. 627–630 (2005)
119. Mistry, K., Allen, C., Auth, C., Beattie, B., Bergstrom, D., Bost, M., Brazier, M., Buehler, M., Cappellani, A., Chau, R., Choi, C.H., Ding, G., Fischer, K., Ghani, T., Grover, R., Han, W., Hanken, D., Hattendorf, M., He, J., Hicks, J., Huessner, R., Ingerly, D., Jain, P., James, R., Jong, L., Joshi, S., Kenyon, C., Kuhn, K., Lee, K., Liu, H., Maiz, J., Mclntyre, B., Moon, P., Neirynck, J., Pae, S., Parker, C., Parsons, D., Prasad, C., Pipes, L., Prince, M., Ranade, P., Reynolds, T., Sandford, J., Shifren, L., Sebastian, J., Seiple, J., Simon, D., Sivakumar, S., Smith, P., Thomas, C., Troeger, T., Vandervoorn, P., Williams, S., Zawadzki, K.: A 45 nm logic technology with high-k+metal gate transistors, strained silicon, 9 Cu interconnect layers, 193 nm dry patterning, and 100% Pb-free packaging. In: Intl. Electron Devices Meeting, pp. 247–250 (2007)
120. Moglestue, C.: Monte Carlo particle modelling of small semiconductor devices. Comput. Methods Appl. Mech. Eng. **30**, 173–208 (1982)
121. Nainani, A., Raghunathan, S., Witte, D., Kobayashi, M., Irisawa, T., Krishnamohan, T., Saraswat, K., Bennett, B., Ancona, M., Boos, J.: Engineering of strained III-V heterostructures for high hole mobility. In: Intl. Electron Devices Meeting, pp. 1 –4 (2009)
122. Natarajan, S., Armstrong, K., Bost, M., Brain, R., Brazier, M., Chang, C.H., Chikarmane, V., Childs, M., Deshpande, H., Dev, K., Ding, G., Ghani, T., Golonzka, O., Han, W., He, J., Heussner, R., James, R., Jin, I., Kenyon, C., Klopcic, S., Lee, S.H., Liu, M., Lodha, S., McFadden, B., Murthy, A., Neiberg, L., Neirynck, J., Packan, P., Pae, S., Parker, C., Pelto, C., Pipes, L., Sebastian, J., Seiple, J., Sell, B., Sivakumar, S., Song, B., Tone, K., Troeger, T., Weber, C., Yang, M., Yeoh, A., Zhang, K.: A 32 nm logic technology featuring 2nd-generation high-k + metal-gate transistors, enhanced channel strain and $0.171 \mu m^2$ SRAM cell size in a 291Mb array. In: Intl. Electron Devices Meeting, pp. 941–943 (2008)

123. Natori, K.: Ballistic metal-oxide-semiconductor field-effect transistor. J. Appl. Phys. **78**(8), 4879–4890 (1994)
124. Naveh, Y., Likharev, K.K.: Modeling of 10 nm-scale ballistic MOSFETs. IEEE Electron Device Lett. **21**(5), 242–244 (2000)
125. Nedjalkov, M., Kosik, R., Kosina, H., Selberherr, S.: Wigner transport through tunneling structures - scattering interpretation of the potential operator. In: Simulation of Semiconductor Processes and Devices, Publication Office Business Center for Academic Societies Japan, Kobe, Japan, pp. 187–190 (2002)
126. Nedjalkov, M., Kosina, H., Selberherr, S., Ringhofer, C., Ferry, D.K.: Unified particle approach to wigner-boltzmann transport in small semiconductor devices. Phys. Rev. B **70**(11), 115,319 (2004). DOI 10.1103/PhysRevB.70.115319
127. Nedjalkov, M., Vasileska, D., Ferry, D.K., Jacoboni, C., Ringhofer, C., Dimov, I., Palankovski, V.: Wigner transport models of the electron-phonon kinetics in quantum wires. Phys. Rev. B **74**(3), 035,311 (2006). DOI 10.1103/PhysRevB.74.035311
128. Nedjalkov, M., Vitanov, P.: Iteration approach for solving the Boltzmann equation with the Monte Carlo method. Solid State Electron. **32**(10), 893–896 (1989)
129. Nguyen, B.Y., Mazure, C., Delprat, D., Aulnette, C., Daval, N., Andrieu, F., Faynot, O.: Overview of FDSOI technology from substrate to device. In: Semiconductor Device Research Symposium, 2009. ISDRS '09. Intl., pp. 1 –2 (2009)
130. Nguyen, C.D., Jungemann, C., Meinerzhagen, B.: Modeling of size quantization in strained Si-nMOSFETs with the improved modified local density approximation. In: Proc. Nanotech 2005 Vol. 3, pp. 33–36 (2005)
131. Paasch, G., Übensee, H.: Carrier density near the semiconductor-insulator interface - local density approximation for non-isotropic effective mass. Phys. Stat. Sol. (b) **118**(1), 255–266 (1983)
132. Packan, P., Akbar, S., Armstrong, M., Bergstrom, D., Brazier, M., Deshpande, H., Dev, K., Ding, G., Ghani, T., Golonzka, O., Han, W., He, J., Heussner, R., James, R., Jopling, J., Kenyon, C., Lee, S.H., Liu, M., Lodha, S., Mattis, B., Murthy, A., Neiberg, L., Neirynck, J., Pae, S., Parker, C., Pipes, L., Sebastian, J., Seiple, J., Sell, B., Sharma, A., Sivakumar, S., Song, B., St. Amour, A., Tone, K., Troeger, T., Weber, C., Zhang, K., Luo, Y., Natarajan, S.: High performance 32 nm logic technology featuring 2nd generation high-k + metal gate transistors. IEDM Proc. pp. 1–4 (2009)
133. Palestri, P., Eminente, S., Esseni, D., Fiegna, C., Sangiorgi, E., Selmi, L.: An improved semi-classical Monte-Carlo approach for nano-scale MOSFET simulation. Solid State Electron. **49**, 727–732 (2005)
134. Palestri, P., Esseni, D., Eminente, S., Fiegna, C., Sangiorgi, E., Selmi, L.: Understanding quasi-ballistic transport in nano-MOSFETs: Part I - scattering in the channel, and in the drain. IEEE Trans. Electron Devices **52**(12), 2727–2735 (2005)
135. Pham, A., Jungemann, C., Meinerzhagen, B.: Deterministic multisubband device simulations for strained double gate PMOSFETs including magnetotransport. In: Intl. Electron Devices Meeting, pp. 895–898 (2008)
136. Pourfath, M., Kosina, H.: Fast convergent Schrödinger-Poisson solver for the static and dynamic analysis of carbon nanotube field effect transistors. Lecture Notes in Computer Science **3743**, 578–585, (2006)
137. Prange, R.E., Nee, T.W.: Quantum spectroscopy of the low-field oscillations in the surface impedance. Phys. Rev. **168**(3), 779–786 (1968)
138. Price, P.J.: Monte Carlo calculation of electron transport in solids. Semiconductors Semimetals **14**, 249–308 (1979)
139. Price, P.J.: Resonant tunneling via an accumulation layer. Ann. Phys. **133**, 217 (1981)
140. Querlioz, D., Dollfus, P.: The Wigner Monte Carlo Method For Nanoelectronic Devices - A Particle Description Of Quantum Transport And Decoherence. Wiley, New York (2010)
141. Querlioz, D., Saint-Martin, J., Do, V.N.: A study of quantum transport in end-of-Roadmap DG-MOSFETs using a fully self-consistent Wigner Monte Carlo approach. IEEE Trans. Nanotechnol. **5**(6), 737–744 (2006)

142. Radosavljevic, M., Ashley, T., Andreev, A., Coomber, S., Dewey, G., Emeny, M., Fearn, M., Hayes, D., Hilton, K., Hudait, M., Jefferies, R., Martin, T., Pillarisetty, R., Rachmady, W., Rakshit, T., Smith, S., Uren, M., Wallis, D., Wilding, P., Chau, R.: High-performance 40nm gate length insb p-channel compressively strained quantum well field effect transistors for low-power ($v_{CC} = 0.5$V) logic applications. In: Intl. Electron Devices Meeting, pp. 1–4 (2008)
143. Radosavljevic, M., Chu-Kung, B., Corcoran, S., Dewey, G., Hudait, M., Fastenau, J., Kavalieros, J., Liu, W., Lubyshev, D., Metz, M., Millard, K., Mukherjee, N., Rachmady, W., Shah, U., Chau, R.: Advanced high-k gate dielectric for high-performance short-channel in0.7ga0.3as quantum well field effect transistors on silicon substrate for low power logic applications. In: Intl. Electron Devices Meeting, pp. 1–4 (2009)
144. Ravaioli, U., Osman, M., Pötz, W., Kluksdahl, N., Ferry, D.: Investigation of ballistic transport through resonant-tunneling quantum wells using Wigner function approach. Physica B **134**, 36–40 (1985)
145. Reggiani, L., Lugli, P., Gantsevich, S., Gurevich, V., Katilius, R.: Diffusion and fluctuations in a nonequilibrium electron gas with electron-electron collisions. Phys. Rev. B **40**(18), 12,209–12,214 (1989). DOI 10.1103/PhysRevB.40.12209
146. Risch, L.: Pushing CMOS beyond the roadmap. In: Proc. European Solid-State Device Research Conf., pp. 63–68 (2005)
147. Sabathil, M., Hackenbuchner, S., Majewski, J.A., Zandler, G., Vogl, P.: Towards fully quantum mechanical 3D device simulations. J. Comput. Electron. **1**, 81–85 (2002)
148. Scharfetter, D., Gummel, H.: Large-signal analysis of a silicon read diode oscillator. IEEE Trans. Electron Devices **16**(1), 64–77 (1969)
149. Schroeder, J., Muller, R.: IGFET analysis through numerical solution of Poisson's equation. IEEE Trans. Electron Devices **15**(12), 954–961 (1968)
150. Selberherr, S.: Analysis and Simulation of Semiconductor Devices. Springer, Heidelberg (1984)
151. Serra, N., Esseni, D.: Mobility enhancement in strained n-FinFETs: Basic insight and stress engineering. IEEE Trans. Electron Devices **57**(2), 482 –490 (2010)
152. Shichijo, H., Hess, K.: Band-structure-dependent transport and impact ionization in GaAs. Phys. Rev. B **23**(8), 4197–4207 (1981)
153. Shifren, L., Ferry, D.K.: A Wigner function based ensemble Monte Carlo approach for accurate incorporation of quantum effects in device simulation. J. Comput. Electron. **1**, 55–58 (2002)
154. Shifren, L., Ringhofer, C., Ferry, D.: Inclusion of nonlocal scattering in quantum transport. Phys. Lett. A **306**, 332–336 (2003)
155. Shifren, L., Ringhofer, C., Ferry, D.: A Wigner function-based quantum ensemble Monte Carlo study of a resonant tunneling diode. IEEE Trans. Electron Devices **50**(3), 769–773 (2003)
156. Shimizu, K., Saraya, T., Hiramoto, T.: Suppression of electron mobility degradation in (100)-oriented double-gate ultrathin body nMOSFETs. IEEE Electron Device Lett. **31**(4), 284–286 (2010)
157. Shoji, M., Horiguchi, S.: Electronic structure and phonon-limited electron mobility of double-gate silicon-on-insulator si inversion layers. J. Appl. Phys. **85**(5), 2722–2731 (1999)
158. Silvaco, Santa Clara, CA: ATLAS user's manual (2010)
159. Slotboom, J.: Iterative scheme for 1- and 2-dimensional d.c.-transistor simulation. Electron. Lett. **5**, 677–678 (1969)
160. Smirnov, S., Kosina, H., Nedjalkov, M., Selberherr, S.: Monte Carlo method for modeling of small signal response including the Pauli exclusion principle. J. Appl. Phys. **94**(9), 5791–5799 (2003)
161. Smith, C.S.: Piezoresistance effect in germanium and silicon. Phys. Rev. **94**(1), 42–49 (1954)
162. Sonoda, K.I., Yamaji, M., Taniguchi, K., Hamaguchi, C., Dunham, S.T.: Moment expansion approach to calculate impact ionization rate in submicron silicon devices. J. Appl. Phys. **80**(9), 5444–5448 (1996)

163. Stern, F., Howard, W.E.: Properties of semiconductor surface inversion layers in the electric quantum limit. Phys. Rev. **163**(3), 816–835 (1967)
164. Stratton, R.: Diffusion of hot and cold electrons in semiconductor barriers. Phys. Rev. **126**(6), 2002–2014 (1962)
165. Sun, G., Sun, Y., Nishida, T., Thompson, S.E.: Hole mobility in silicon inversion layers: Stress and surface orientation. J. Appl. Phys. **102**(8), 084501 (2007)
166. Sverdlov, V., Gehring, A., Kosina, H., Selberherr, S.: Quantum transport in ultra-scaled double-gate MOSFETs: A Wigner function-based Monte Carlo approach. Solid State Electron. **49**(9), 1510–1515 (2005)
167. Sverdlov, V., Ungersboeck, E., Kosina, H., Selberherr, S.: Volume inversion mobility in SOI MOSFETs for different thin body orientations. Solid State Electron. **51**, 299–305 (2007)
168. Sverdlov, V., Ungersboeck, E., Kosina, H., Selberherr, S.: Influence of uniaxial [110] stress on silicon band structure and electron low-field mobility in ultra-thin body SOI FETs. In: Proc. EUROSOI 2007, pp. 39–40 (January, 2007)
169. Sverdlov, V.A., Selberherr, S.: Electron subband structure and controlled valley splitting in silicon thin-body SOI FETs: Two-band **k·p** theory and beyond. Solid State Electron. **52**(12), 1861–1866 (2008)
170. Sverdlov, V.A., Walls, T.J., Likharev, K.K.: Nanoscale silicon MOSFETs: A theoretical study. IEEE Trans. Electron Devices **50**(9), 1926–1933 (2003)
171. Svizhenko, A., Anantram, M.: Effect of Scattering and Contacts on Current and Electrostatics in Carbon Nanotubes. Phys. Rev. B **72**, 085,430–085,440 (2005)
172. Svizhenko, A., Anantram, M.P.: Role of scattering in nanotransistors. IEEE Trans. Electron Devices **50**, 1459–1466 (2003)
173. Svizhenko, A., Anantram, M.P., Govindan, T.R., Biegel, B., Venugopal, R.: Two-dimensional quantum mechanical modeling of nanotransistors. J. Appl. Phys. **91**, 2343–2354 (2002)
174. Synopsys, Mountain View, CA: Sentaurus device user's manual (2010)
175. Takagi, S.I., Toriumi, A., Iwase, M., Tango, H.: On the universality of inversion layer mobility in Si MOSFETs: Part I - effects of substrate impurity concentration. IEEE Trans. Electron Devices **41**(12), 2357–2362 (1994)
176. Thompson, S.E., Armstrong, M., Auth, C., Alavi, M., Buehler, M., Chau, R., Cea, S., Ghani, T., Glass, G., Hoffmann, T., Jan, C.T., Kenyon, C., , Klaus, J., Kuhn, K., Ma, Z., McIntyre, B., Mistry, K., Murthy, A., Obradovic, B., Nagisetty, R., Nguyen, P., Sivakumar, S., Shaheed, R., Shifren, L., Tufts, B., Tyagi, S., Bohr, M., El-Mansy, Y.: A 90-nm logic nanotechnology featuring strained-silicon. IEEE Trans. Electron Devices **51**(11), 1790–1797 (2004)
177. Thompson, S.E., Armstrong, M., Auth, C., Cea, S., Chau, R., Glass, G., Hoffmann, T., Klaus, J., Ma, Z., McIntyre, B., Murthy, A., Obradovic, B., Shifren, L., Sivakumar, S., Tyagi, S., Ghani, T., Mistry, K., Bohr, M., El-Mansy, Y.: A logic nanotechnology featuring strained-silicon. IEEE Electron Device Lett. **25**(4), 191–193 (2004)
178. Thompson, S.E., Suthram, S., Sun, Y., Sun, G., Pathasarathy, S., Chu, M., Nishida, T.: Future of strained Si/semiconductors in nanoscale MOSFETs. In: Intl. Electron Devices Meeting, pp. 681–684 (2006)
179. Trellakis, A., Zibold, T., Andalauer, T., Smith, S.B.A.K., Morschal, R., Vogl, P.: The 3D nanometer device project nextnano^3: Concepts, methods, results. In: Intl. Workshop Comput. Electroncis, Wien, pp. 173–174 (2006)
180. Tsutsui, G., Saitoh, M., Saraya, T., Nagumo, T., Hiramoto, T.: Mobility enhancement due to volume inversion in (110)-oriented ultra-thin body double-gate nMOSFETs with body thickness less than 5nm. In: Intl. Electron Devices Meeting, pp. 747–750 (2005)
181. Uchida, K., Koga, J., Takagi, S.: Experimental study on carrier transport mechanisms in double- and single-gate ultrathin-body MOSFETs - Coulomb scattering, volume inversion, and δt_{SOI}-induced scattering. In: Intl. Electron Devices Meeting, pp. 805–808 (2003)
182. Uchida, K., Krishnamohan, T., Saraswat, K.C., Nishi, Y.: Physical mechanisms of electron mobility enhancement in uniaxial stressed MOSFETs and impact of uniaxial stress engineering in ballistic regime. In: Intl. Electron Devices Meeting, pp. 129–132 (2005)

183. Ungersboeck, E., Dhar, S., Karlowatz, G., Sverdlov, V., Kosina, H., Selberherr, S.: The effect of general strain on band structure and electron mobility of silicon. IEEE Trans. Electron Devices **54**(9), 2183–2190 (2007)
184. Aubry-Fortuna, V., Dollfus,P., Galdin-Retailleau, S.: Electron effective mobility in strained-Si/$Si_{1-x}Ge_x$ MOS devices using Monte Carlo simulation. Solid State Electron. **49**(8), 1320–1329 (2005)
185. Vasileska, D., Ferry, D., Goodnick, S.: Handbook Of Theoretical And Computational Nanotechnology, vol. 10, chap. Computational Nanoelectronics, American Scientific, Los Angeles, pp. 1–135 (2006)
186. Venugopal, R., Ren, Z., Datta, S., Lundstrom, M.S., Jovanovic, D.: Simulation of quantum transport in nanoscale transistors: Real versus mode-space approach. J. Appl. Phys. **92**(7), 3730–3739 (2002)
187. VMC2.0: Vienna Monte Carlo 2.0 user's guide. Institut für Mikroelektronik, http://www.iue.tuwien.ac.at/software, Technische Universität Wien, Austria (2006)
188. VSP1.0: Vienna Schrödinger-Poisson solver 1.0 user's guide. Institut für Mikroelektronik, http://www.iue.tuwien.ac.at/software, Technische Universität Wien, Austria (2007)
189. Wagner, M., Karner, M., Grasser, T.: Quantum correction model for modern semiconductor devices. In: Proc. of the XIII Intl. Workshop Semiconductor Devices, pp. 458–459 (2005)
190. Walls, T.J., Sverdlov, V.A., Likharev, K.K.: Nanoscale SOI MOSFETs: A comparison of two options. Solid State Electron. **48**, 857–865 (2004)
191. Wang, E., Matagne, P., Shifren, L., Obradovic, B., Kotlyar, R., Cea, S., Stettler, M., Giles, M.D.: Physics of hole transport in strained silicon MOSFET inversion layers. IEEE Trans. Electron Devices **53**(8), 1840–1851 (2006)
192. Wang, J., Polizzi, E., Ghosh, A., Datta, S., Lundstrom, M.: Theoretical investigation of surface roughness scattering in silicon nanowire transistor. J. Appl. Phys. **87**, 0431,011–0431,013 (2005)
193. Wang, J., Polizzi, E., Lundstrom, M.: A three-dimensional quantum simulation of silicon nanowire transistors with the effective-mass approximation. J. Appl. Phys. **96**(4), 2192–2203 (2004)
194. Wang, J., Rahman, A., Ghosh, A., Klimeck, G., Lundstrom, M.: On the validity of the parabolic effective-mass approximation for the $i - v$ calculation of silicon nanowire transistors. IEEE Trans. Electron Devices **52**(7), 1589–1595 (2005)
195. Welser, J., Hoyt, J., Gibbons, J.: NMOS and PMOS transistors fabricated in strained silicon/relaxed silicon-germanium structures. In: Intl. Electron Devices Meeting, pp. 1000–1002 (1992)
196. Wigner, E.: On the Quantum Correction for Thermodynamic Equilibrium. Phys. Rev. **40**, 749–759 (1932)
197. Yoder, P., Higman, J., Bude, J., Hess, K.: Monte Carlo simulation of hot electron transport in Si using a unified pseudopotential description of the crystal. Semicond. Sci. Technol. **7**(3B), 357–359 (1992)
198. Zahid, F., Ghosh, A., Paulsson, M., Polizzi, E., Datta, S.: Charging-induced asymmetry in molecular conductors. Phys. Rev. B **70**, 245,317 (2004)

Author Index

Aamari, K., 15
Abbadie, A., 9
Absil, P.P., 10, 11
Agnello, P., 6
Ahmed, S., 8
Ajmera, A., 8
Akamatsu, H., 15
Akbar, S., 6, 170
Akis, R., 171, 193, 194
Alavi, M., 170
Allen, C., 6, 11, 12, 170
Allen, S., 6
Alshareef, H., 10, 11
An, J., 14
Anantram, M.P., 109, 225, 226, 228
Ancona, M.G., 194, 208
Andalauer, T., 171, 228
Ando, K., 8, 10, 11
Ando, T., 45, 105, 124–126, 144, 153, 201
Andreev, A., 15, 170, 208
Andrew, R., 171
Andrieu, F., 9, 170
Ang, K.W., 13
Antreasyan, A., 6
Arghavani, R., 11
Armstrong, K., 6, 12, 170
Armstrong, M., 6, 8, 10, 170
Arnold, J.C., 6
Asenov, A., 155, 157, 194, 225, 228
Ashcroft, N.W., 28
Ashley, T., 15, 170, 208
Aubry-Fortuna, V., 194
Aulnette, C., 9, 170
Aur, S., 10, 11
Auth, C., 6, 8, 10–12, 170

Babich, I., 9
Bai, P., 8, 10, 12
Baiocco, C., 8, 10
Balakrishnan, S., 10, 12
Balasubramanian, N., 13
Baldereschi, A., 79
Balestra, F., 200, 201, 203
Balseanu, M., 11
Balslev, I., 91, 92, 107, 112, 185
Bandy, K., 6
Banerjee, S.K., 15, 194
Bardeen, J., 7, 91
Barker, J.R., 155, 157, 225, 228
Barr, A., 9
Bassani, F., 58
Bastard, G., 124
Bate, R.T., 184, 188, 189
Baumgartner, O., 106, 137, 145, 157, 158, 162
Baym, G., 173
Beattie, B., 6, 11, 12, 170
Beenakker, C.W.J., 155
Bellenger, F., 208
Beltram, F., 58
Belyansky, M., 6
Benachir, M., 200, 201, 203
Benedict, J., 8
Bennett, B., 208
Bergstrom, D., 6, 11, 12, 170
Besson, P., 9
Bevan, M., 10, 11
Biegel, B., 225, 226
Bierstedt, H., 10, 11, 13
Biesemans, S., 10, 11
Bir, G.L., 7, 37, 67, 83–86, 88, 97, 98, 100, 101, 106, 107, 118, 119, 132, 137
Blase, X., 58, 102, 107
Blatchford, J., 10, 11
Bliznetsov, V., 13
Blotekjaer, K., 174
Boardman, A., 188
Bogumilowicz, Y., 9

Bohr, M., 6, 8, 10, 12, 170
Bollani, M., 9
Bonar, J., 9
Bonnoit, A., 6
Boos, J., 208
Bosi, S., 195
Bost, M., 6, 8, 10–12, 170
Boucaud, P., 68
Bourgade, J.P., 218
Bowen, C., 10, 11
Bowen, R.C., 58, 225
Boyd, D., 9
Boykin, T.B., 58, 59, 61, 69, 70, 107, 141, 142, 144, 152, 153, 226
Brain, R., 6, 10, 12, 170
Brasen, D., 8, 9
Brazier, M., 6, 11, 12, 170
Brevard, L., 9
Brini, J., 200, 201, 203
Broido, D.A., 79
Bronner, G., 6
Brown, A.R., 194, 225
Bryant, A., 10
Bu, H., 10, 11
Bude, J., 188
Buehler, M., 6, 11, 12, 170
Bufler, F.M., 194
Buin, A., 109
Buller, J., 6
Bulsara, M.T., 8
Buot, F., 218
Burbach, G., 6

Canali, C., 188
Canaperi, D., 9
Cao, M., 194
Cappellani, A., 6, 11, 12, 170
Car, R., 217
Carceller, J.E., 74
Cardona, M., 38, 40, 41, 43, 46, 50, 53, 55, 72, 73, 107
Cardone, F., 9
Carruthers, R., 9
Casida, M.E., 49
Castro, L.C., 226
Caymax, M., 208
Cea, S., 10, 95, 96, 170, 207, 208
Celler, G., 201
Chakravarthi, S., 10, 11
Chakravarti, A., 8, 10
Chan, K., 9, 14
Chan, V., 6, 8, 10, 11, 14
Chang, C.H., 6, 12, 170
Chang, P., 8, 13
Chao, C.Y.P., 74
Charvat, P., 6, 8
Chau, R., 6, 10–12, 15, 16, 170, 208
Chelikowsky, J.R., 51, 54
Chen, C., 11
Chen, C.H., 11
Chen, H., 9, 10
Chen, J., 8, 10, 162
Chen, S., 11
Chen, W.Q., 194
Chen, X., 6
Chen, X.D., 8, 10
Chen, Z., 6
Cheng, K., 11
Chidambaram, P.R., 10, 11
Chidambarrao, D., 6, 8, 13, 185
Chikarmane, V., 6, 10, 12, 170
Childs, M., 6, 12, 170
Chiu, Y., 11
Cho, M., 15
Choi, C.H., 6, 11, 12, 170
Choi, J., 8
Choi, S.M., 8
Choi, W.H., 15
Chou, A., 6
Chu, J., 9
Chu, J.O., 152–154
Chu, M., 6, 208
Chu-Kung, B., 170, 208
Chuang, S.L., 74
Chui, C.O., 15
Chui, K.J., 13
Chun, J.S., 12
Chutia, S., 153
Ciampolini, L., 68, 106, 131
Clark, W., 6
Cobb, M., 9
Cohen, M.L., 51, 54
Colman, D., 184, 188, 189
Coomber, S., 15, 170, 208
Coppersmith, S.N., 152–154
Corcoran, S., 170, 208
Coss, B., 15
Costrini, G., 10, 11
Cousins, C.S.G., 87
Cowley, A., 8
Cristoloveanu, S., 200, 201, 203
Crowder, S.W., 6
Curatola, G., 211
Curello, G., 8
Currie, M.T., 8

Dalis, A., 12
Dang, H., 10, 14
Datta, S., 15, 56, 59, 171, 211, 224–226
Daval, N., 170
Davis, A., 12
Decker, S., 193
Degond, P., 218
Degraeve, R., 10, 11
Delabie, A., 208
DeLoach, J., 10, 11
Delprat, D., 170
DeMari, A., 171
Deshpande, H., 6, 12, 170
Dev, K., 6, 12, 170
Dewey, G., 15, 170, 208
De Jaeger, B., 208
De Keersgieter, A., 10, 11
De Meyer, K., 10, 11, 208
Dhar, S., 7, 61, 70, 87, 88, 102, 106, 107, 109, 111, 112, 118, 131–137, 186, 189, 192, 208
Diaz, C., 11
Dimov, I., 220
Ding, G., 6, 11, 12, 170
Divakaruni, R., 6
Do, V.N., 220
Documaci, O., 169
Dokumaci, O., 6
Dolgopolov, V.T., 152
Dollfus, P., 194, 220
Domenicucci, A., 8, 13
Donaton, R.A., 8, 13
Donetti, L., 201
Doris, B., 169
Dow, J.D., 58
Dowsett, M., 9
Du, A., 13
Duch, E., 9
Dutton, R.W., 194

Eastwood, J.W., 187
Ebert, A., 8
Egley, J., 185
Ehrichs, E., 6
Eimori, T., 11
El-Mansy, Y., 10, 170
Elewa, T., 200, 201, 203
Eller, M., 8
El Kurdi, M., 68
Emeny, M., 15, 170, 208
Eminente, S., 170, 194, 216
Eneman, G., 208
Eneman, S., 10, 11
Engel, B., 6
Engelmann, H.J., 10, 11, 13
Eriksson, M.A., 152–154
Erlebach, A., 194
Ernst, T., 9
Esseni, D., 106, 132–135, 144, 155, 157, 159, 170, 194, 195, 198, 199, 201, 208, 216

Fan, X.F., 194
Fang, F.F., 153
Farber, D., 10, 11
Fastenau, J., 15, 170, 208
Fawcett, W., 188
Faynot, O., 9, 170
Fearn, M., 15, 170, 208
Feher, G., 7
Feraille, M., 68, 106, 131, 155
Fermi, E., 49
Ferry, D., 171, 193, 194, 218–220
Ferry, D.K., 193, 195, 219, 220
Feudel, T., 6, 10, 11, 13
Fichtner, W., 211, 226, 228
Fiegna, C., 170, 194, 201, 216
Fiori, G., 211
Fischer, A., 8
Fischer, K., 6, 11, 12, 170
Fischetti, M., 10, 11, 188, 210, 211
Fischetti, M.V., 86, 94, 102, 105, 188, 195, 216, 217
Fishman, G., 68
Fitzgerald, E., 8, 9
Fitzgerald, E.A., 8
Foisy, M.C., 194
Foreman, B.A., 153
Fossum, J.G., 10, 11, 14
Fowler, A.B., 45, 105, 124–126, 144, 153, 201
Foxon, C.T., 155
Frank, M.M., 6
Frensley, W., 211, 216–218
Fried, D., 6, 10, 11, 14
Friedel, P., 102
Friesen, M., 152–154
Frohberg, K., 6, 10, 11, 13
Fuchs, M., 49
Fujiwara, A., 152
Fukuhara, N., 15

Gámiz, F., 105, 188, 201
Galdin-Retailleau, S., 194
Gamble, L.J., 58
Gannavaram, S., 8

Gantsevich, S., 197
Gao, W., 8
Gebauer, R., 217
Gehring, A., 175, 211, 219, 220
Gerhardt, M., 6, 10, 11, 13
Gerward, L., 87
Ghani, T., 6, 8, 10–12, 170
Ghetti, A., 68, 106, 131
Ghosh, A., 211, 226, 228
Ghosh, B., 194
Ghyselen, B., 9
Gibbons, J., 8, 9, 184
Gibbons, J.F., 105
Gilbert, M., 171
Giles, M., 208
Giles, M.D., 95, 96, 207
Glass, G., 6, 8, 10, 12, 170
Glassman, T., 12
Goad, S., 6
Goldsman, N., 88
Golonzka, O., 6, 12, 170
Gonze, X., 49
Goo, J.S., 8
Goodnick, S., 193
Goolsby, B., 10
Goswami, S., 152–154
Govindan, T.R., 225, 226
Grasser, T., 173, 175, 177, 178, 186, 193
Grasshoff, G., 6
Green, M., 8, 9
Greenlaw, D., 6, 10, 11, 13
Gribelyuk, M., 169
Grider, T., 10, 11
Griffin, C., 8
Gritsch, M., 175, 177, 178
Grover, R., 6, 11, 12, 170
Grubin, H.L., 193
Guarin, F., 8
Gummel, H., 171, 173
Guo, D., 6
Gurba, A., 10, 11
Gurevich, V., 197
Gusev, E.P., 6
Gös, W., 87

Hachimine, K., 8, 10
Hackenbuchner, S., 211
Haddad, G.I., 218
Haensch, W., 6, 169
Hall, H.H., 7
Hall, L., 10, 11
Hamaguchi, C., 188
Han, W., 6, 11, 12, 170
Hanken, D., 6, 11, 12, 170
Haran, B., 6
Harifuchi, H., 6
Harper, M., 12
Hartmann, J.M., 9
Hasegawa, H., 7, 68, 106–108, 113, 114, 118, 131, 132, 135, 137
Hata, M., 15
Hattendorf, M., 6, 11, 12, 170
Hayes, D., 15, 170, 208
He, J., 6, 11, 12, 170
Hedin, L., 49
Heinz, F., 211
Heitzinger, C., 173, 175
Hellings, G., 208
Hellmich, A., 10, 11, 13
Helmholz, D., 58
Hempel, K., 10, 11, 13
Hensel, J.C., 7, 68, 106–108, 113, 114, 118, 131, 132, 135, 137
Henson, W.K., 8, 13
Hentges, P., 12
Herring, C., 7, 50, 91, 94, 178, 192
Herzog, O., 10, 11, 13
Hess, K., 188
Heussner, R., 6, 10, 12, 170
Hewson, M., 10, 11
Heyns, M., 208
Hicks, J., 6, 11, 12, 170
Higman, J., 188
Hilton, K., 15, 170, 208
Hinckley, J., 29, 93
Hiramoto, T., 162, 200, 202–206
Hirano, Y., 10, 14
Hirashita, N., 8
Hirata, T., 8, 10, 11
Hirayama, Y., 152
Hjalmarson, H.P., 58
Hockney, R., 187
Hoffmann, T., 6, 8, 10, 170
Hohage, J., 6, 10, 11, 13
Hohenberg, P., 47
Holt, B., 8
Holt, J., 8, 10, 11, 13
Hong, D., 8
Hook, T., 8, 10
Hooper, K., 8
Horiguchi, S., 202
Horiuchi, T., 8, 10, 11
Hornung, B., 10, 11
Horstmann, M., 6, 10, 11, 13
Hou, T., 11
Houghton, D.C., 73–76, 80, 94, 95
Howard, W.E., 124, 125, 129, 153, 202

Hoyt, J., 8, 9, 184
Hoyt, J.L., 105
Hsu, J., 11
Huang, C.L., 11, 14
Huang, H.J., 169
Huang, J., 15
Huang, S.F., 6, 8, 10
Hudé, R., 194
Hudait, M., 15, 16, 170, 208
Huebler, P., 6
Huessner, R., 6, 11, 12, 170
Hueting, R., 155, 157, 159
Hussein, M., 10, 12
Hwang, J., 10, 12
Hybertsen, M.S., 102
Hänsch, W., 105, 188, 193
Höntschel, J., 10, 11, 13
Hübler, P., 10, 11, 13

Iafrate, G.J., 193
Iannaccone, G., 211
Ieong, M., 9–11, 169
Ikezawa, N., 8, 10, 11
Ingerly, D., 6, 10–12, 170
Inoue, Y., 10, 11, 14
Ipposhi, T., 10, 14
Irie, H., 206, 207
Irisawa, T., 8, 208
Ito, S., 8, 10, 11
Itokawa, H., 8
Iwai, A., 14
Iwamatsu, T., 10, 14
Iwase, M., 201

Jackson, J., 217
Jacoboni, C., 67, 69, 105, 186–189, 195, 198, 220
Jacobs, S., 8
Jagannathan, R., 6
Jain, P., 6, 11, 12, 170
Jaloviar, S., 12
Jamenez, F., 201
James, D., 14
James, R., 6, 10–12, 170
Jamil, M., 15
Jamin, F., 8
Jamin, F.F., 6, 169
Jammy, R., 15
Jan, C.H., 8
Jan, C.T., 170
Jancu, J.M., 58
Jaouen, H., 58, 68, 102, 106, 107, 131, 155
Javorka, P., 10, 11, 13
Jefferies, R., 15, 170, 208
Jenkins, K., 9
Jensen, K., 218
Jeong, J., 8, 10, 12
Jin, I., 6, 12, 170
Jin, W., 8, 10
Jin, Y., 11
John, D.L., 226
Johnson, G., 8
Johnson, J., 8, 13
Johnson, K., 6, 8
Jones, D., 8
Jones, E.C., 169
Jong, L., 6, 11, 12, 170
Jopling, J., 6, 170
Joshi, S., 6, 11, 12, 170
Jovanovic, D., 225, 226
Joynt, R., 152–154
Jungel, A., 194
Jungemann, C., 175, 178, 188, 189, 193, 194, 196, 208
Jurczak, M., 10, 11

Kaczer, B., 10, 11
Kadanoff, L.P., 173
Kalna, K., 155, 157
Kalpat, S., 9
Kaltalioglu, E., 8
Kamakura, Y., 188
Kammler, T., 10, 11, 13
Kanarsky, T., 9–11, 169
Kaneshige, C., 10, 11
Kang, C.Y., 15
Karlowatz, G., 7, 61, 70, 88, 102, 106, 107, 109, 111, 112, 118, 131–137, 192, 208
Karner, M., 137, 145, 158, 193
Karunasiri, G., 11
Kasai, K., 14
Kathawala, G., 194
Katilius, R., 197
Kavalieros, J., 15, 170, 208
Kawaguchi, M., 10, 11
Kennedy, D., 171
Kenyon, C., 6, 8, 10–12, 170
Kepler, N., 6, 10, 11, 13
Khamankar, R., 10, 11
Khare, M., 6
Khrapai, V.S., 152
Kim, D., 194
Kim, J.P., 8
Kim, S., 8

Kim, Y., 10, 11
Kinoshita, A., 67, 69, 106, 131, 132, 157, 160
Kircher, R., 193
Kirkner, D., 211
Kirkpatrick, B., 10, 11
Kirsch, P., 15
Kita, K., 206, 207
Kittel, C., 27
Klais, J., 6, 10, 11, 13
Klaus, J., 6, 8, 10, 12, 170
Klee, V., 8
Klein, L.J., 152–154
Kleinman, L., 51, 87
Klimeck, G., 58, 59, 61, 69, 70, 107, 141, 142, 144, 152, 153, 225, 226, 228
Klopcic, S., 6, 8, 12, 170
Kluksdahl, N., 218
Klymko, N., 9, 10
Knoefler, R., 8
Kobayashi, M., 208
Koerner, G., 10, 11, 13
Koester, S., 9
Koga, J., 200
Koguchi, M., 8, 10
Kohn, W., 7, 42–44, 46, 47, 63, 64, 74, 76, 131, 142
Kohyama, Y., 6
Komoda, T., 14
Kong, Y., 109
Kortan, A., 8, 9
Kosemura, D., 15
Kosevich, A., 26
Kosik, R., 175, 219
Kosina, H., 7, 61, 70, 87, 88, 102, 106, 107, 109, 111, 112, 118, 131–137, 144, 145, 158, 173, 175, 177, 178, 186, 188, 189, 192, 196, 197, 208, 211, 216, 218–220, 222, 226
Koster, G.F., 57, 58
Kotlyar, R., 95, 96, 207, 208
Kouwenhoven, L.P., 155
Koyama, S., 8, 10, 11
Kriman, A., 218
Krishnamohan, T., 7, 105, 112, 113, 131, 157, 194, 207, 208
Krishnan, A., 10, 11
Krishnan, S., 6
Ku, J.H., 8
Ku, V., 8, 10
Kuhn, K., 6, 11, 12, 170
Kumar, A., 210, 211
Kunikiyo, T., 188
Kuroda, H., 6
Kuroki, S., 8, 10, 11
Kurosawa, T., 188
Kyuno, K., 206, 207

Lacey, D., 9
Lafond, D., 9
Lai, C.W., 6
Lai, K., 152
Lai, W., 6
Lake, R., 58, 225
Landau, L., 26, 43
Lauer, I., 6
Lauwers, A., 10
Laux, S., 188, 210, 211
Laux, S.E., 86, 94, 102, 105, 188, 195
Lavric, D., 12
Lea, D., 8, 10
Leake, G., 8
Lee, E., 10, 12
Lee, H.D., 15
Lee, H.K., 6
Lee, K., 6, 9, 11, 12, 170
Lee, K.W., 8
Lee, M.L., 8
Lee, S.H., 6, 10, 12, 15, 170
Lee, T., 11
Lee, W., 11
Lee, W.H., 6
Lenski, M., 6, 10, 11, 13
Lent, C., 211
Leong, L., 8
Levinshtein, M., 27, 29
Lew Yan Voon, L.C., 58
Li, J., 8, 10, 13
Li, M.F., 13
Li, X., 8, 13
Li, Y., 6
Lian, J., 8
Liang, M.S., 11
Liew, S., 8
Lifshitz, E., 26, 43
Likharev, K.K., 5, 170, 198, 209–211
Lim, E.H., 6
Lim, J., 8
Lim, J.S., 10
Lin, J., 8
Lin, M.R., 8, 14
Lin, W., 8
Lin, Y.H., 8
Lindblad, G., 217
Lindert, N., 8, 10, 12
Linton, T., 208
Lio, A., 8
Lipari, N.O., 79

Liu, H., 6, 11, 12, 170
Liu, M., 6, 10, 12, 170
Liu, W., 15, 170, 208
Liu, Y., 8, 13
Lochtefeld, A., 8
Lodha, S., 6, 12, 170
Loeb, H., 171
Loehr, J.P., 58
Loh, W.Y., 15
Loo, R., 208
López-Villanueva, J.A., 74
Louisell, W.H., 217
Love, W., 171
Löwdin, P.O., 56
Lu, J.P., 10, 11
Lu, M., 12
Lubyshev, D., 15, 170, 208
Lucci, L., 195
Lugli, P., 195, 197
Luisier, M., 226, 228
Lundstrom, M., 10, 171, 211, 228
Lundstrom, M.S., 226
Luning, S., 6
Luo, Y., 6, 170
Luttinger, J.M., 7, 42–44, 46, 63, 64, 74, 76, 131, 142
Lutze, J., 10
Lyon, S., 152

Mühlberger, M., 152
Ma, Z., 10, 12, 170
Machala, C., 10, 11
MacWilliams, K., 8
Madan, A., 8, 13
Maeda, S., 10, 14
Maegawa, S., 10, 14
Maex, K., 10
Magyari-Kope, B., 208
Mahan, G., 218, 222
Mains, R.K., 218
Maiz, J., 6, 11, 12, 170
Majewski, J.A., 211
Majhi, P., 15
Majni, G., 188
Malik, R., 6, 11
Mallikarjunan, A., 6
Manku, T., 73–76, 80, 94, 95, 185
Manley, M., 10
Mann, R., 8
Mariappan, H., 12
Marieb, T., 10, 12
Martin, R.M., 87, 102
Martin, T., 15, 170, 208
Martinez, A., 155, 157, 225, 228
Mascarenhas, A., 11
Massey, G., 8
Mastrapasqua, M., 201
Matagne, P., 95, 96, 207
Matsumoto, K., 6
Matsumoto, T., 10, 14
Matsuoka, F., 14
Mattis, B., 6, 170
Matusiewicz, G., 8
Mayer, H., 79
Mayuzumi, S., 15
Mazure, C., 170
McFadden, B., 6, 12, 170
McGregor, J.M., 73–76, 80, 94, 95
McGuire, L.M., 152–154
McIntyre, B., 6, 8, 10–12, 170
McIntyre, P.C., 15
McKnight, A., 6
McIntyre, B., 170
Mehats, F., 218
Meinerzhagen, B., 178, 188, 189, 193, 196, 208
Melchor, I., 74
Merckling, C., 208
Mermin, N.D., 28
Metz, M., 170, 208
Meuris, M., 208
Mezzapelle, J., 169
Michaillat, M., 155
Michel, J., 8, 9
Mii, Y., 8, 9
Miles, D., 10, 11
Milic, O., 14
Millard, K., 170, 208
Miller, R.J., 169
Minder, R., 188
Minondo, M., 68, 106, 131
Mistry, K., 6, 8, 10–12, 170
Mitard, J., 208
Mize, J.P., 184, 188, 189
Mizuno, H., 188
Mizushima, I., 8
Mo, R., 8
Mocuta, A., 9, 169
Moglestue, C., 187
Moon, P., 6, 11, 12, 170
Mooney, P., 9
Mooney, P.M., 152–154
Morifuji, M., 188
Morimoto, H., 11
Moroz, V., 10, 11
Morris, R., 9
Morschal, R., 171, 228

Mukherjee, N., 170, 208
Muller, R., 171
Murthy, A., 6, 8, 10, 12, 170

Nagashima, N., 14, 15
Nagata, K., 15
Nagisetty, R., 10, 12, 170
Nagumo, T., 200, 202–204, 206
Nainani, A., 208
Nakaharai, S., 8
Nakano, Y., 15
Nakayama, M., 7, 68, 106–108, 113, 114, 118, 131, 132, 135, 137, 153
Namba, H., 8, 10, 11
Narasimha, S., 6
Narayanan, V., 6
Natarajan, S., 6, 8, 10, 12, 170
Nathan, A., 73–76, 80, 94, 95, 185
Natori, K., 209, 211
Natzle, W., 169
Naujok, M., 8
Nava, F., 188
Naveh, Y., 209
Nayak, D., 8
Nayak, J., 6
Nedjalkov, M., 188, 196, 197, 216, 218–220, 222
Nee, T.W., 201
Neiberg, L., 6, 12, 170
Neinhüs, B., 193
Neirynck, J., 6, 8, 10–12, 170
Neu, A., 10, 11, 13
Neugebauer, J., 48, 49
Newbury, J., 9
Ng, H., 8, 10
Ng, H.Y., 6
Nguyen, B., 10
Nguyen, B.Y., 170
Nguyen, C.D., 193
Nguyen, P., 8, 10, 170
Nicollian, P., 10, 11
Nielsen, O.H., 87
Niimi, H., 10, 11
Niquet, Y.M., 49, 58, 102, 107, 155
Nishi, Y., 7, 105, 112, 113, 131, 157, 194, 207, 208
Nishida, T., 6–8, 15, 95, 207, 208
Nishmura, T., 10, 14
Noel, J.P., 73–76, 80, 94, 95
Noguchi, T., 14
Nonaka, Y., 8, 10
Norris, B., 12
Nouri, F., 10, 11
Nowak, E., 8, 10
Nowotny, H., 43
Numata, T., 8

Obradovic, B., 10, 95, 96, 170, 207
Oda, H., 10, 11, 14
Ogura, A., 15
Oh, J., 15
Ohkawa, F.J., 144, 153
Ohki, N., 8, 10
Ohno, K., 14
Ohno, T., 15
Ohta, H., 8, 10
Oishi, A., 14
Ok, I., 15
Okano, K., 8
Oldiges, P., 6
Olsen, J.S., 87
Ono, Y., 152
Ootsuka, F., 8, 10
Orellana, P., 152–154
Orlowski, M., 193
Osman, M., 218
Osternaud, B., 9
Ota, K., 10, 11, 14
Ott, A., 10, 12
Ott, J., 9, 10
Ottaviani, G., 188
Otterbach, R., 10, 11, 13
Ouyang, Q., 10
Ouyang, Q.C., 11
Oyafuso, F., 58, 59, 61, 69, 70, 107, 141, 142, 144, 152, 153
Ozturk, C., 8

Paasch, G., 193
Packan, P., 6, 8, 12, 170
Pae, S., 6, 11, 12, 170
Paige, E., 188
Palankovski, V., 186, 189, 220
Palestri, P., 106, 132–135, 144, 155, 157, 159, 170, 194, 195, 216
Pan, J., 8
Pan, W., 152
Panda, S., 6, 10
Park, J., 8
Park, K., 8
Parker, C., 6, 10–12, 170
Parsons, D., 6, 11, 12, 170
Parthasarathy, S., 6, 15
Patel, M., 8
Pathasarathy, S., 208

Paulsson, M., 226
Pearson, G.L., 7
Pelto, C., 6, 12, 170
Perdew, J.P., 49
Pereira, P.J.S., 226
Pesovic, N., 8
Petrarca, K., 9
Pham, A., 178, 208
Pham, A.T., 196
Phillips, J.C., 51
Pikus, G.E., 7, 37, 67, 83–86, 88, 97, 98, 100, 101, 106, 107, 118, 119, 132, 137
Pillarisetty, R., 15, 170, 208
Pipes, L., 6, 11, 12, 170
Pitaevskii, L., 26
Poli, P., 188
Polizzi, E., 211, 226, 228
Pompl, T., 8
Post, I., 8
Pötz, W., 218
Pourfath, M., 226
Prange, R.E., 201
Prasad, C., 6, 11, 12, 170
Press, P., 10, 11, 13
Price, P.J., 187, 198
Prince, M., 6, 11, 12, 170
Pulfrey, D.L., 226
Pötz, W., 218

Qteish, A., 48, 49
Quaranta, A., 188
Querlioz, D., 220

Raab, M., 6, 10, 11, 13
Rachmady, W., 15, 170, 208
Radosavljevic, M., 15, 170, 208
Raghunathan, S., 208
Rahhal-orabi, N., 12
Rahman, A., 228
Rakshit, T., 15, 170, 208
Ramanathan, S., 15
Ramanchandran, V., 8
Ramey, S., 8
Ranade, P., 6, 11, 12, 170
Rathsack, B., 10, 11
Ravaioli, U., 194, 218
Rayssac, O., 9
Reese, P., 8
Reggiani, L., 67, 69, 105, 186–189, 197, 198
Register, L.F., 194
Reichel, C., 10, 11, 13
Ren, Z., 169, 188, 226
Rengarajan, R., 6, 8, 10
Reynolds, T., 6, 11, 12, 170
Rideau, D., 58, 68, 102, 106, 107, 131, 155
Rieger, M.M., 49, 53–56, 59, 102, 105, 107, 132, 136
Rim, K., 8, 9, 11, 13, 188
Ringhofer, C., 218–220
Rinke, P., 48, 49
Risch, L., 212
Rockford, L., 8
Rodríguez, S., 74
Rodriguez, N., 201
Rooyackers, R., 10, 11
Roskowski, A., 8
Rössler, U., 79
Rota, L., 188
Roulston, D.J., 73–76, 80, 94, 95
Rovedo, N., 8, 10
Roy, R., 9
Roy, R.A., 169
Rozeau, O., 9
Rubin, M., 10
Ruelke, H., 6, 10, 11, 13
Rumyantsev, S., 27, 29
Rupp, K., 9

Saad, H., 11
Sabathil, M., 211
Sacks, G., 8
Sadaka, M., 9
Sadana, D., 6
Saint-Martin, J., 220
Saito, M., 14
Saitoh, M., 15, 67, 69, 106, 131, 132, 157, 160, 200, 202–204, 206
Saitoh, T., 8, 10, 11
Salmani-Jelodar, M., 226
Salz, H., 6, 10, 11, 13
Samoilov, A., 10, 11
Sampedro, C., 201
Samudra, G., 13
Samudra, G.S., 13
Sandford, J., 6, 8, 11, 12, 170
Sangiorgi, E., 170, 194, 201, 216
Santa Clara, C.A., 171
Sanuki, T., 14
Saraswat, K., 194, 208
Saraswat, K.C., 7, 15, 105, 112, 113, 131, 157, 207
Saraya, T., 162, 200, 202–206
Sassman, B., 15
Sato, H., 8, 10
Saunders, P., 9

Sauvage, S., 68
Sayama, H., 10, 11, 14
Schäffler, F., 152
Schaller, M., 6, 10, 11, 13
Schanovsky, F., 106
Scharfetter, D., 171
Scheer, S., 8
Scheffler, M., 48, 49
Schenk, A., 211, 226, 228
Schiml, T., 8
Schlüter, M., 102
Scholz, R., 58
Scholze, A., 211
Schreutelkamp, R., 10, 11
Schroeder, J., 171
Schwan, C., 6
Scott, G., 10
Sebastian, J., 6, 10–12, 170
Seiple, J., 6, 11, 12, 170
Selberherr, S, 70
Selberherr, S., 7, 9, 61, 87, 88, 102, 106, 107, 109, 111, 112, 118, 131–137, 144, 145, 157, 162, 169, 171, 173, 175, 177, 178, 186, 188, 189, 192, 194, 196, 197, 208, 211, 212, 216, 218–220, 222
Sell, B., 6, 12, 170
Selmi, L., 155, 157, 159, 170, 194, 195, 201, 216
Selsmark, B., 87
Serra, N., 208
Shafer, P., 8, 10
Shah, U., 170, 208
Shaheed, R., 10, 12, 170
Sham, L.J., 47, 79, 153
Sharma, A., 6, 170
Shashkin, A.A., 152
Sheldon, B.J., 87
Sheraw, C., 6, 11
Sherony, M., 8
Shi, L., 9, 14, 169
Shichijo, H., 188
Shifren, L., 6, 10–12, 95, 96, 170, 207, 208, 219, 220
Shimizu, A., 8, 10
Shimizu, K., 203, 205
Shiri, D., 109
Shluger, A., 155, 157
Shockley, W., 7, 91
Shoji, M., 202
Shull, H., 56
Shur, M. (eds.), 27, 29
Siew, Y., 8
Silberstein, M., 6, 8
Simoen, E., 208
Simon, D., 6, 11, 12, 170
Singh, D., 6
Singh, J., 29, 93, 94
Sivakumar, S., 6, 8, 10–12, 170
Slater, J.C., 57, 58
Sleight, J., 6
Slinker, K.A., 152–154
Slotboom, J., 171
Smirnov, S., 196, 197
Smith, B., 10, 11
Smith, C.S., 7, 183, 188, 189, 191
Smith, L., 10, 11, 15
Smith, P., 6, 8, 11, 12, 170
Smith, S., 15, 170, 208
Smith, S.B.A.K., 171, 228
Solomon, P.M., 188
Somervell, M., 10, 11
Song, B., 6, 12, 170
Souw, V., 12
St. Amour, A., 6, 170
Stanley, J., 88
Stathis, J., 14
Steegen, A., 8, 10
Steen, M., 9
Steigerwald, J., 10, 12
Steigerwalt, M., 6
Stephan, R., 6, 10, 11, 13
Stern, F., 45, 105, 124, 125, 129, 144, 153, 201, 202
Stettler, M., 95, 96, 207, 208
Stiles, P.J., 153
Stratton, R., 173, 174
Stucchi, M., 10
Su, L.T., 6
Subbanna, S., 6
Subramanian, K., 6
Sudijohno, J., 8
Sudijono, J., 6
Sudo, G., 6
Sugihara, K., 11
Sugiyama, M., 15
Sugiyama, N., 8
Sun, G., 6–8, 15, 95, 207, 208
Sun, M.C., 8
Sun, S.P., 6
Sun, Y., 6–8, 15, 95, 207, 208
Sung, C., 6, 10
Sushko, P., 155, 157
Suthram, S., 6, 208
Suzuki, T., 8, 10, 11
Sverdlov, V., 7, 61, 70, 106, 107, 109, 111, 112, 118, 131–137, 144, 145, 157, 158, 162, 192, 196, 208, 211, 220

Sverdlov, V.A., 5, 144, 170, 211, 212, 216
Svizhenko, A., 225, 226, 228
Swain, S., 188

Tahan, C., 152–154
Tai, L., 8, 9
Takagi, H., 15
Takagi, S., 8, 15, 200
Takagi, S.I, 8
Takagi, S.I., 8, 105, 201
Takahashi, Y., 152
Takashina, K., 152
Takei, M., 15
Takenaka, M., 15, 188
Talwar, D.N., 58
Tambwe, F., 12
Tan, W., 8
Tango, H., 201
Taniguchi, K., 188
Tao, H., 11
Tateshita, Y., 15
Tavernier, C., 58, 68, 102, 106, 107, 131, 155
Teh, Y.W., 8, 11
Tekleab, D., 9
Tessier, B., 6
Tezuka, T., 8
Thean, A., 9
Thomas, A., 8
Thomas, C., 6, 11, 12, 170
Thompson, A., 12
Thompson, S., 6, 8
Thompson, S.E., 6–8, 10, 11, 15, 95, 170, 207, 208
Tiberj, A., 9
Tiersten, H.F., 194
Tilke, A., 8
Tinkham, M., 65
To, B., 9
Toffoli, A., 9
Tone, K., 6, 12, 170
Toriumi, A., 201, 206, 207
Towner, D., 12
Toyoda, E., 8
Toyoshima, Y., 6
Tran, J., 10, 11
Tran, P., 6
Trellakis, A., 171, 228
Trentsch, M., 10, 11, 13
Triplett, B.B., 15
Troeger, T., 6, 11, 12, 170
Trui, B., 10, 11, 13
Truitt, J.L., 152–154
Tseng, H.H., 15
Tsui, D.C., 152
Tsukamoto, M., 15
Tsutsui, G., 200, 202–204, 206
Tufts, B., 170
Tung, C.H., 13
Turkot, B., 8
Tyagi, S., 10, 12, 170

Übensee, H., 193
Uchida, K., 7, 67, 69, 105, 106, 112, 113, 131, 132, 157, 160, 200, 207
Uchida, T., 10, 11, 14
Uemura, Y., 144, 153
Ukraintsev, V., 10, 11
Ungersboeck, E., 7, 53–55, 59, 61, 70, 85, 87–89, 93, 102, 106, 107, 109, 111, 112, 118, 131–137, 144, 145, 186, 189, 192, 194, 196, 208
Uppal, S., 9
Uren, M., 15, 170, 208
Usuda, K., 8
Utomo, H., 10

van Bentum, R., 6
van Dort, M.J., 193
van Houten, H., 155
van Wees, B.J., 155
Vandervoorn, P., 6, 11, 12, 170
Van de Walle, C.G., 102
van der Marel, D., 155
van der Steen, J.L., 155, 157, 159
van der Weide, D.W., 152–154
Varghese, A., 10, 11
Vasileska, D., 193, 194, 220
Vayshenker, A., 8
Venugopal, R., 225, 226
Verheyen, P., 10, 11
Vietzke, D., 8
Vitanov, P., 188
Vogelsang, T., 193
Vogl, P., 49, 53–56, 58, 59, 102, 105, 107, 132, 136, 171, 211, 228
Vogt, E., 7, 91, 94, 178, 192
Voisin, P., 58
Vook, D., 194
Voorde, P.J.V., 194
VSP1.0, 211

Wagner, M., 193
Wakabayashi, H., 15
Walker, A.J., 193

Walker, G., 9
Wallace, C., 12
Wallis, D., 15, 170, 208
Walls, T.J., 5, 170, 211
Wang, E., 95, 96, 207
Wang, G., 208
Wang, H., 14
Wang, J., 10, 211, 228
Wang, K., 8
Wang, W., 14
Wang, X., 6
Wang, Y., 8
Wann, C., 8, 10, 13
Wann, C.H., 6
Washington, L., 10, 11
Watling, J.R., 194
Weber, C., 6, 10, 12, 170, 208
Wei, A., 6, 10, 11, 13
Wei, L., 8
Weir, B., 8, 9
Welser, J., 8, 9, 184
Welser, J.J., 105
White, T., 9, 10
Wieczorek, K., 6
Wiedemer, J., 12
Wiegand, C., 12
Wigner, E., 216, 218, 219
Wilding, P., 15, 170, 208
Wildman, H., 10
Williams, S., 6, 11, 12, 170
Williamson, J.G., 155
Willoughby, A., 9
Winstead, B., 194
Wise, R., 6, 10
Witte, D., 208
Woerlee, P.H., 193
Womack, S., 169
Wong, H., 9
Wong, H.S., 208
Wong, H.S.P., 169
Wong, P., 194
Wong, R., 6
Woo, J., 8
Woolery, B., 10, 12
Workman, G.O., 11, 14
Wrschka, P., 8
Wu, K., 8

Xia, L., 11
Xiang, Q., 8, 14
Xie, Y., 8, 9

Yamada, H., 15
Yamaguchi, K., 8, 10, 11
Yamaji, M., 188
Yamakawa, S., 15
Yang, H.S., 6
Yang, I., 8
Yang, I.Y., 6
Yang, J.H., 8
Yang, J.W., 11, 14
Yang, M., 6, 10–12, 14, 170, 188
Yang, S., 8, 10, 11
Yasuda, T., 15
Yeo, Y.C., 13
Yeoh, A., 6, 10, 12, 170
Yip, J., 8
Yoder, P., 188
Yokoyama, M., 15
Yoshimura, H., 14
Young, I., 8
Yu, B., 8, 14
Yu, P., 38, 40, 41, 43, 46, 50, 53, 55, 72, 73, 107
Yu, Z., 194

Zahid, F., 226
Zandler, G., 211
Zawadzki, K., 6, 8, 11, 12, 170
Zhang, D., 10
Zhang, F., 8
Zhang, J., 8
Zhang, K., 6, 8, 10, 12, 170
Zhang, Y., 8, 9, 169
Zheng, X., 194
Zhu, H., 9
Zhuang, H., 8
Zibold, T., 171, 228
Zunger, A., 49

Subject Index

Alternative channel materials, 6, 14
Auxiliary tight-binding model, 141

Ballistic transport, 171, 209
Band structure calculation methods
    first-principle, 46, 49
    k.p methods, 63
    pseudopotential, 49, 52–55, 58, 59, 102
    tight-binding, 56, 69, 70, 153, 222
Bloch functions, 42–44, 55, 57, 67, 70, 100, 102, 132–134, 141, 142
Brillouin zone, 39, 43, 45, 46, 58, 66, 88, 92, 93, 105, 106, 118, 119, 131, 137, 184

Compliance tensor, xiii, 28
Conduction band of relaxed silicon
    degenerate valleys, 91
    effective masses, 111, 115, 117, 118

Drift-diffusion transport model, 169, 177, 178

Electron subbands
    dispersion relations, 133–135, 137, 141, 195
    double primed, 125
    at (001) interface, 126
    at (110) interface, 127
    at (111) interface, 128
    primed, and, 128, 156
    unprimed, 127, 128, 137, 147
Energy transport model, 169, 174–182

Hybrid orientation technology, 6, 14

Internal strain parameter, 86, 87, 103

Linear combination of bulk bands method, 132
Linear deformation potential theory, 91

Mobility modeling, 174, 184
Monte Carlo Methods for Transport Calculations, 187
    degeneracy effects, 192
MOSFET scaling, 1

Non-equilibrium Green's functions, 222

Piezoresistance, 7, 8, 183, 184, 188, 191, 192, 196

Quantum transport models, 208, 216

Shear strain dependence of
    electron effective masses, 111, 149–151
    non-parabolicity parameter, 115–118
    valley splitting, 92, 108, 135, 144, 146, 147
Single- and double-gate MOSFETs, 173, 200
    intersubband scattering, 192, 202
Six-moments transport model, 173, 175, 178
Stiffness tensor, 27, 28, 31
Strain tensor, 23–25, 27–33, 83, 85–88, 92, 93, 95, 106, 107, 110, 158, 185
Strain-induced symmetry reduction, 83, 102
    of Brillouin zone, 102
Stress techniques
    global, 9
    local, 10, 105
Stress tensor, 25–28, 32, 183

Stress-induced enhancement
drive current, 6, 8, 11–13
mobility, 207
Subband structure, 3, 7, 67, 105–107, 123, 126, 129, 131–133, 136, 137, 145, 159, 195, 196, 202–205, 208

TCAD Tools, 3, 132, 157, 169, 171, 173, 174, 226
Thickness dependence of the effective mass in (001) thin films
primed subbands, 155–157
unprimed subbands, 137–143, 157–162

Valence band of relaxed silicon
heavy hole band, 46, 70, 76, 77, 96–99
light hole band, 46, 70, 76, 77, 96, 99
Luttinger parameters, 76, 79, 80
spin-orbit coupling, 55, 58, 59, 76, 94, 132, 133
split-off band, 59, 76, 78, 95, 96
Valley splitting in thin films
due to shear strain, 91, 135, 144, 146, 147
in magnetic field, 152, 154
in point contact, 152, 154

Wigner equation, 218, 219, 222
Wigner function, 173, 193, 209, 216, 218–220, 227

Printing and Binding: Stürtz GmbH, Würzburg